Breed Differences in Dog Behavior: Why Tails Wag Differently

R.R.H.: I would like to dedicate this book to my husband, Jim, my everything.

T.L.B.: I would like to dedicate this book to my mother, Janis Pegg, who instilled a love and understanding of animals from a very early age, and to my daughters—may they grow to love and understand animals, too.

J.C.H.: I would like to dedicate this book to my mother Margaret Clark Ha (an avid reader!) and my brothers: Steve, Tom, Dan, and Peter Ha.

Breed Differences in Dog Behavior: Why Tails Wag Differently

by

Renee Robinette Ha, PhD, University of Washington

Tracy L. Brad, MS

James C. Ha, PhD, CAAB, University of Washington

CABI

CABI is a trading name of CAB International

CABI
Nosworthy Way
Wallingford
Oxfordshire OX10 8DE
UK

Tel: +44 (0)1491 832111
E-mail: info@cabi.org
Website: www.cabi.org

CABI
200 Portland Street
Boston
MA 02114
USA

Tel: +1 (617)682-9015
E-mail: cabi-nao@cabi.org

A catalogue record for this book is available from the British Library, London, UK.

ISBN-13: 9781800624542 (paperback)
9781800624559 (ePDF)
9781800624566 (ePub)

DOI: 10.1079/9781800624566.0000

Commissioning Editor: Alexandra Lainsbury
Editorial Assistant: Helen Elliott
Production Editor: Rosie Hayden

Typeset by Straive, Pondicherry, India
Printed and bound in the UK by CPI Group (UK) Ltd, Croydon, CR0 4YY

Contents

Prologue

The Dogs of War

Cry 'Havoc!' and let slip the dogs of war.

Julius Caesar (Act 3, Scene 1)

Dogs have been at our side during our brightest and darkest times, including accompanying us on the front lines of war. This famous utterance from Mark Antony in William Shakespeare's *Julius Caesar* referenced Antony's reaction to Caesar's assassination and his incitement to avenge Caesar's death (Shakespeare, 1996). Beyond deeper interpretations of Shakespeare's prose, dogs were a familiar warfare companion. "Letting them loose" referred to allowing them to attack the enemy. War dogs helped the Ancient Egyptians and Greeks, the Persians and the Prussians, the Britons and the Romans, the Germans and the Americans. Military officials chose specific dogs for specific jobs, including those who accompanied them into battle, those who served as sentries, those who packed gear or sent messages, and those who tracked down the enemy. Some researchers have hypothesized that dogs were key to the decline of the Neanderthals and the rise of *Homo sapiens* (Shipman, 2015), while others wonder if other hominids, too, cohabitated with these canines. There are historical records that dogs have been a significant factor in helping one civilization prevail over another. In 231 BCE, Roman Marcus Pomponius Matho used dogs to locate Sardinian natives who had hidden in caves. In 55 BCE, English Mastiffs were by Julius Caesar's side as he invaded Britain. In the 1500s, large dogs were used by the Spanish conquistadors against Native Americans (Varner and Varner, 1983). In more modern times, dogs have accompanied soldiers during every major battle, including World War I, where more than 1 million dogs died in action (Thompson, 2014), World War II, the Vietnam War, and the mission to kill Osama bin Laden.

The "Molosser" dog (also referred to as the Molossus) was perhaps the original dog of war, and one from which many similar-looking modern dogs likely descended. Molossers were large-bodied, aggressive, and revered by their people; they were the most well-known war dog of the ancient world. Philosophers, including Aristophanes, Aristotle, Horace, and Virgil, wrote about the Molossus. According to Aristotle, the Molossians bred two types of dog: a broad-muzzled hunting dog and a larger livestock guardian dog. In his *History of Animals* (Aristotle, 1887), he wrote: "In the Molossian race of dogs, those employed in hunting differ in no respect from other dogs; while those employed in following sheep are larger and more fierce in their attack on wild beasts." The former dog is the likely progenitor of modern Mastiffs. Modern war dog breeds include, but are not limited to, German Shepherds, Belgian Malinois, Mastiffs, Doberman Pinschers, and those under the "Pit Bull" umbrella, with specific breeds chosen for each type of job.

A Dog Is a Dog Is a Dog?

A multitude of different roles have shaped the direction of dogs' evolution—and our own. In recent centuries, we have shaped dogs into hundreds of different breeds, each with its own proclivities and predilections. In this book, we, PhD animal behaviorists Dr Renee Robinette Ha and Dr James C. Ha and science writer Tracy Brad, MS (Fig. P.1), will delve into the differences and similarities between these breeds. We assert that breed and pedigree are just as important as the environment, including epigenetics and early life experiences, in shaping a dog's behavior. While this might sound like a bold claim, we'll provide ample evidence of the interrelationship of those factors. And in doing so, we'll try to answer the question: is a dog … just a dog? We aren't the first people to ask this question, but we're asking it in a different way, using recent behavioral and genetic research and previously unpublished case studies.

Dr Renee Robinette Ha received her PhD in Animal Behavior from the University of Washington in 1999. She's currently a Teaching Professor in the

Fig. P.1. The authors (left to right): Tracy Brad, James Ha and Renee Robinette Ha. Photograph courtesy of Holly Cook Photography, LLC.

Department of Psychology (Animal Behavior Program) at the University of Washington where she has taught courses in animal behavior, statistics, introductory psychology, developmental psychology, laboratories on animal learning, and behavioral studies of zoo animals. She has worked with a wide range of species, including humans, monkeys, birds, cats, and dogs. Renee was the Director of the Rota Avian Behavioral Ecology Program in Micronesia from 2005 to 2022, where she worked to save the critically endangered Mariana crow. More recently, she's the co-developer (with Dr James Ha) of a Certificate Program in Applied Animal Behavior at the University of Washington, a program with a focus on companion animals. She has worked with companion animals since 2014 and has been close to her husband's work in this field since 2001. Through this, she learned how little is understood about dog breeds and behavior. Fortunately, the science is moving quickly, and we can update and educate the dog community about the role of genetics in companion animal behavior. During her research on companion animals, she saw a growing need for a book discussing the differences and similarities between dog breeds. Dr Renee Robinette Ha brings a strong background in animal behavior and learning theory, statistics, and human psychology to this project.

Dr James C. Ha earned a Bachelor's degree in Biology from Millersville University in 1980, a Master's in Biology from Wake Forest University in 1983, and his PhD in Zoology (with an emphasis on behavioral ecology) from Colorado State University in 1989. Dr Ha has professional credentialing as a Certified Applied Animal Behaviorist (CAAB; from 2005 to present), the highest level of certification in applied behavior that is currently available. During his graduate training, Dr Ha interned with Dr Philip Lehner (at Colorado State University) and was active in his early behavior consulting business, Animal Behavior Associates of Colorado. Dr Ha started his own consulting business in companion animal behavior in 1999, Animal Behavior Associates of Washington, LLC, and currently consults, lectures, and advises under the DrJimHa.com label. He has performed in-home evaluations and treatments on hundreds of behavior issues (80% of which involved instances of aggression) in dogs, cats, and parrots (averaging 45 cases per year), and advises on additional cases seen by his colleagues. He has acted as a consultant to major pet food companies, several shelter organizations, and the Center for Animal Welfare Science at Purdue University. Currently he co-directs the University of Washington's Certificate in Applied Animal Behavior Program and is active in expert legal consulting in dog bite and dog tracking cases. He also works on behalf of courthouse facility dog organizations at the federal, state and local level, serves on the Board of Directors of the Courthouse Dogs Foundation, and co-hosts the "Dogged Justice" podcast. He served as a member of the Animal Behavior Society's Executive Committee for many years, and is currently serving as co-Chair of the Animal Behavior Society's Board of Professional Certification. His most recent book, *Dog Behavior: Modern Science and Our Canine Companions*, co-written with Tracy Brad (née Campion), was the culmination of more than a decade of planning and research and was published by Elsevier's Academic Press in 2018 (Ha and Campion, 2018).

Drs Ha are responsible for the Certificate in Applied Animal Behavior Program at the University of Washington, the first solely online academic program in animal behavior from an accredited academic institution. The three-course program examines principles of animal behavior, including behavioral differences between animal species and how these reflect their evolution and habitat adaptation.

Tracy L. Brad earned her Bachelor's of Arts in Social Sciences, minoring in Anthropology, at the University of Washington in 2006. She received her Master's of Science in Primate Behavior at Central Washington University in 2012 and her Postgraduate Diploma in Journalism at the London School of Journalism in 2014. Tracy has worked with domesticated animals, including dogs, cats, and horses, and with non-human primates, including baboons in South Africa, spider monkeys in the Yucatan Peninsula of Mexico, and with chimpanzees who had learned the signs of American Sign Language. She adopted a Labrador Retriever mix named Jack in 2015, a month after he'd been hit by a car and had to have an eye and a leg removed. Tracy was recovering from a broken back and pelvis when she adopted him; the Mutual Rescue initiative created a film about their story entitled *Tracy and Jack*. Living with Jack and seeing his joy for life inspired her to become the co-publisher of *Pet Connection Magazine*, a bi-monthly Pacific Northwest-based publication that focused on people, pets, community, and the connections we share. As co-publisher of *Pet Connection Magazine*, Tracy had the opportunity to meet and interview animal behavior experts such as Dr Jane Goodall, Jackson Galaxy, Buck Brannaman, Dr Temple Grandin, Dr Patricia McConnell, Dr Jan Pol, and Dr James C. Ha. From 2019 to 2023, Tracy co-founded The Limelight Pet Project, a multimedia campaign featuring harder-to-adopt animals, airing their segments on local television.

In Chapter 1, we will answer the not-so-simple question: what is a dog? *Canis familiaris* is a member of the genus *Canis*, which is the most abundant of the terrestrial carnivores. As Darwin and so many scientists since have pointed out, dogs vary widely in their physical appearance, proclivities, and temperaments. Dogs are genetically, physically, and behaviorally different than their wild cousins, the wolves and the coyotes, but they can still produce viable offspring with them. So how are they different from their free-living cousins, and what species definition should we use for dogs?

Species is the principal taxonomic unit, ranking below genus and denoted by its Latin binomial (e.g. *Canis familiaris*). A species can be defined as a group of living organisms consisting of similar individuals that are capable of exchanging genes or interbreeding. But questions and controversies remain when it comes to domesticated dogs. "Species" is a man-made concept; nature doesn't always adhere to our neat demarcations and definitions. Again, all species within the genus *Canis* can reproduce with one another, but using this biological definition (being able to produce viable offspring within one group) doesn't always apply. Sometimes, organisms don't reproduce because of behavioral differences (such as different bird calls with the Western and Eastern meadowlarks, differences in seasonality and estrus cycles, or differences in social structure, e.g. social and cooperative versus solitary), the lack of co-occurrence due to different ecological preferences (lions and tigers can interbreed, but lions live on the grassland, while tigers live in the jungle and, as far as anyone knows, there have never been any wild tigers in Africa), an environmental barrier, such as a mountain range, a wide river, or even islands in the ocean (as in the Galapagos finches and mockingbirds), and sometimes they don't reproduce because of morphological barriers, such as differences in genitalia structure or extreme body size variation (such as a Chihuahua trying to reproduce with a Great Dane. It might be ambitious, but it's not necessarily impossible).

In Chapter 2, we will discuss how breeds of dogs came about and how they proliferated, as well as how breeds are related to one another. We also address the issue of domestication across species, and for the canine in particular.

In Chapter 3, we will discuss how genetics works, beginning with a crash course in genetic science. We'll define some common terms and provide examples of these, including single-gene effects, linked genes, multiplicative effects, genetic variability, mutations, and epistasis as they pertain to dogs. We'll use recent findings in genetics research to examine what we know about the first animal to be domesticated.

In Chapter 4, we will discuss how individuals vary using case study examples. Individuals can have very similar genetic backgrounds, such as belonging to the same breed or even the same family, but have different temperaments, behaviors, and sensitivities. In biology, "variation" indicates any differences in cells, individual organisms, or groups of organisms within a species caused by genetic differences (genotypic variation) or by the effect of environmental factors in the expression of genetic potentials (phenotypic variation). We'll examine recombination, variation within a species, and variation within a breed where different lines have been bred for different purposes, such as field and show Golden Retrievers. We'll examine epigenetics, looking at the environmental factors that can lead to changes in an individual's genes. We'll also look at the different personality types and why legislation banning certain breeds is neither scientific nor effective.

In Chapter 5, we will ask: what is the range of dog breeds? Are dog breeds almost like different species? For the evolutionary biologist, a dog is just a dog, but is it really that simple? It usually takes thousands of years—or longer—for speciation to occur. A slow accumulation of mutations causes inheritable changes to the phenotype. Most dog breeds originated during the Victorian era, falling far short of that typical timeline, although humans were the catalyst behind this accelerated process.

The vast majority of differences in dogs, even among the most dissimilar of the species, is driven by relatively few loci, or regions, in the genome. These loci have a large phenotypic effect, yielding strong variation among breeds. We'll examine where different breeds came from and the effect of inbreeding on population structure from the pedigree analysis of purebred dogs, as well as looking at genetically linked dog diseases.

In Chapter 6, we will discuss how behavior came along for the ride. While we might have been breeding Pugs for a phenotypic characteristic, a truncated muzzle, a behavioral characteristic, loud, laborious breathing, arose in the offspring, as well. For the breeds that have been studied, we'll examine which are most likely to have behavioral effects, how long the breed has been isolated, when each breed originated, and how extreme the artificial selection was. We'll use case studies where a bad environment, such as puppy mills, abuse or abandonment, and bad genetics can lead to a predisposition for aggression.

In Chapter 7, we will discuss breed differences as they pertain to temperament and reactivity, examining individual differences versus breed differences as well as variation within a breed and variation between breeds.

We will also look at case studies examining separation anxiety and client mismatches. For example, a Border Collie (a herding breed with a strong desire to have a job and be active) living with an owner in an apartment that limits their breed-specific behavior, and the ensuing behavioral issues associated with this mismatch.

In Chapter 8, we will discuss social behavior, dominance, and breed differences. How do multiple factors, such as genetic lineages and the environment, influence certain behaviors? We will discuss how "primitive" and "ancient" breeds and breed groups tend to show more dominant characteristics than recently created breeds with more derived characteristics. We'll take a look at case study examples of ancient breeds and any examples of ancient and modern breed household mismatches, as well as mismatches between dogs and their humans.

In Chapter 9, we will discuss aggression and breed differences. We will discuss the issues that complicate the interpretation of dog bite data, and thus our understanding of dog aggression by breed. We will also examine aggression examples from Dr Jim Ha's case studies, including "loaded gun" scenarios with an aggression-prone breed, a dog breed that's typically friendly, but showing aggression due to an underlying health issue, and a case study for a breed that has a prior history for specific aggression. We will examine the heritability of certain personality traits, including dominant and aggressive behavior (Pérez-Guisado *et al.*, 2006) and address the fact that there are many types of aggression, including dominant, territorial, possessive, protection-of-litter-, pain-, fear-, play-motivated, predation, redirected, intraspecific, idiopathic, physiopathological, and learned aggression (Landsberg and Horwitz, 1998).

In Chapter 10, we will discuss training and breed differences: how do different dog breeds and breed types react differently to training? How do the motivations between breeds differ? For example, which breeds are considered to be more food-driven than others, and what breeds will likely exhibit more prey drive? Why is it important to keep breed in mind? We'll examine a survey of breed differences (Eken Asp *et al.*, 2015) to provide a framework for tendencies within each breed. We'll also provide case study examples where training is comparatively "easy" and short in duration for some dogs and some breeds, and where training is longer in duration and takes more repetition for others.

In Chapter 11, we will discuss threads and inklings and tie everything together by revisiting the stories in the case studies and review the evidence that breeds are under selection and that breed genetics can affect behavior. We'll also talk about the future of this species, including the lasting repercussions of continued inbreeding and welfare implications. While we have some evidence about breed differences, we still don't have enough information to make sweeping statements about all members of a particular breed. This lack of evidence, paired with a propensity to classify dogs based upon physical appearance and assumed recent genetic heritage (breed) rather than on their behavior, is why breed-specific legislation (BSL) is still so dangerous. BSL is breed discrimination; it's canine eugenics. Breed bans and eugenics likely fuel the fire against belief in the possibility of breed-specific differences in behavior. Many breeds, such as Golden Retrievers, are bred for certain vocations (e.g. show or field), and in recent years, dogs falling under the "Pit Bull" umbrella have been bred for show or, unfortunately, for fighting. This has been used to bolster BSL arguments, but again, we lack the scientific evidence for differential temperament based upon different breeding; this remains anecdotal.

Our fear of human eugenics and racism may lead us to automatically reject breed differences because they smack of these distasteful topics, but it's important to note that because of artificial selection, lines of dogs have been bred specifically for useful behavioral traits and for desirable morphological traits, as well. While some humans practice arranged marriages, purpose-breeding is not the case with humans; these unions are for cultural or financial reasons. Natural selection is likely to select the most advantageous traits for survival and reproduction. Therefore, it's not only morally distasteful, but scientifically unlikely, that one "race" is superior to another. It's reasonable to think, however, that lines of related dogs bred for herding may excel in behaviors related to herding, and that some other traits may have been inadvertently selected (or come along for the ride) with the herding gene. We will examine the gaps in knowledge, as well as the timing of adolescence and breed differences. We will also re-examine Scott and Fuller's landmark longitudinal dog behavior study (Scott and Fuller, 1965). Even decades later, this study remains the most comprehensive reference of its kind.

Many, if not all, of the topics we discuss in this book come back to one important point: we can't define a species, much less a breed. And depending upon their geographic location, breeds continue to be in flux—German Shepherds in the Pacific Northwest aren't the same as those originating from Germany. It's the "same" breed, but an examination of the genes of lineages from these disparate locations shows that they're almost as different from one another as two breeds are from one another. While we've attempted to quantify breeds as accurately as possible, different organizations recognize different numbers of breeds and breed groups. A breed might be recognized by one organization, but not by another; it's a dynamic process. The processes of artificial selection we're using on dog breeds is like natural selection on steroids. We already have dogs that are so dissimilar in behavior, appearance, and recent selection that they appear to be different species. If you follow that logic through, and some breeds become isolated from other breeds or intermix with wolves or coyotes, it's highly likely that some breeds will diverge enough from *Canis familiaris* to become new species. So, if we could time travel, which of the domesticated dog breeds might become a new species?

Throughout these chapters, we'll walk you through what is and is not currently known about breed differences in dogs and how this pertains to our relationship with them. We'll also discuss how we can continue to improve their overall welfare today and well into the future.

In addition to updating you on the latest science, we'll introduce you to case studies from Dr Jim Ha's interesting in-home client cases. We're going to introduce you to two of those in-home case studies now. One is a case of inappropriate elimination in a Pomeranian, and the other is a case of severe separation anxiety in a Chow Chow.

The Mysterious Case of the Peeing Pomeranian

When assessing a behavior issue with a dog, you have to think about complexities. You need to think about the factors that contribute to the problems you might have with an animal family member. You can ask a client, "What kind of problem are you having?" And use their initial description, paired with the dog's history, as a point of departure to investigate the dog's specific behavioral issue. One set of clients, a couple in Redmond, Washington, just outside of Seattle—we'll call them Jeff and Julia Oleson—had an ongoing problem with their dogs. They said, "Our Pomeranian pees on our feet." Now, this isn't a behavioral issue you'll hear about often, much less with a Pomeranian. Prior to the in-home consultation, I had no idea what might be causing this unusual issue; this is atypical behavior for a Pomeranian.

For those who aren't familiar with Pomeranians, they're a toy breed dog of the Spitz type, originating in the Pomerania region of Northwest Poland and Northeast Germany, flanking the Baltic Sea. While stone-age dog fossils throughout Central Europe share the same physical characteristics that Spitz dogs do, Pomeranians and their closest relatives are likely a far more modern lineage whose forebears originated from larger working dogs in the Arctic regions. Pomeranian-type dogs were first referenced in the 16th century. King George III of England and his wife, Queen Charlotte, brought two Pomeranians named Mercury and Phoebe to England. In the 1800s, their granddaughter Victoria also took a liking to this breed. Victoria's ownership of Pomeranians made them *en vogue*, and they became well known for their amiable temperaments. A dog with a predilection for peeing on its people isn't going to remain a favorite of anyone, much less the Royal family, whose animals have to behave in a manner fitting their esteemed place. Thus, Pomeranians aren't associated with a dog that would pee on its owner. But the inheritance of temperament can be less predictable than the inheritance of physical characteristics; behavior essentially "came along for the ride" as humans bred dogs for a particular appearance.

Jeff and Julia both worked long hours at Microsoft and lived in a townhouse close to Microsoft's main campus. The living space was up two flights of stairs from the foyer. At the top was a living room and a kitchen. There was another staircase that went up to the bedrooms with a slider to the backyard. In this living room, there was a couch, a television, and a couple of chairs. Everything about the setup of this house was pertinent to solving the mystery of the peeing Pomeranian.

During the initial consultation, Jeff and Julia sat together on the couch. The owners discussed their dogs' history, which was very "normal"—the dogs came from a respected breeder who had genetically and temperament-tested them. The dogs entered their home at the "right" time, having received sufficient early socialization; they hadn't been confined; they hadn't been deprived of food, abused, or neglected, and they were taken out of the home for ongoing socialization opportunities. The male dog, Banjo, was 1.5 years old, while the female, Pepper, was several years older. Nothing unusual here; this was too perfect. You don't get a lot of Pomeranian behavior cases; they were bred to be companions, aren't inherently aggressive, and aren't particularly active.

Once the obvious potential issues are addressed, you are left to delve into other potential contributing variables. They weren't being fed anything out of the ordinary. But then, after observing the dogs and their owners during the interview, we began to get to the heart of the issue. Jeff said, "We'll be sitting on the couch in the evening, watching a movie or TV show, and halfway through the evening, our feet get warm. One of the dogs comes over and pees on our feet." This is apparently even after walking them prior to sitting down, and not having fed them recently. This was puzzling.

Behavioral observations are the most important part of an intake session—even more so than the owner's assessment of the situation. In fact, it's better to watch the dog than to make a lot of eye contact with the owners.

While Jeff and Julia were sitting on the couch the male dog, Banjo, came over and put his paw up on Jeff's left knee. Jeff was just talking away and slipped over on the couch seat, making space for Banjo, and Banjo jumped up. Then he jumped down and went running around the house while the interview continued. Then Banjo came over and put his paw on Jeff's right knee. Jeff was talking away, not paying any attention. Banjo jumped up. The female dog, Pepper, came over. She put her paw up on Jeff's left knee and he squeezed himself over into the corner of the couch. Pepper then jumped up and curled up on the couch, and then she jumped down again. Jeff was talking, oblivious to the canine choreography taking place on his couch. This was while they were answering questions to ascertain the dog's history: who was their breeder, what kind of food did they eat and had it changed recently, what was the dogs' medical history, and had anything recently changed in the household?

As the intake went on, the dogs continued to dance around the conversation. Pepper then came back, put her paw up on Jeff's right knee ... and then both dogs jumped up on Julia. Jeff and Julia were moving all over because the dogs were moving them all over the couch. They finished with the initial intake information and said, "We don't know why they pee on our feet."

Of course, to the behaviorist, it was clear by then. "I bet that you get involved in watching TV, get all settled into the couch and you're watching TV and you don't pay attention to the tap on the knee." They looked at each other, and said incredulously, "What tap on the knee?" "The dogs have been tapping your knee. Between the two of you, you have moved six times to accommodate the dogs." There was sufficient space on the couch for Jeff, Julia, Banjo, and Pepper to all sit together, but the dogs wanted to be in between their owners, and next to them, and on them, and all over the place. "You've moved six times, the two of you, in response to a tap on the knee."

The owners hadn't been aware of the behaviors that led up to their dogs eliminating on their feet. To elicit the response from the dogs, and convince the owners, it was necessary to engage the owners in conversation. And right on cue, Banjo came over to us. Banjo was getting up and down, he came back over to the couch, and then he tapped Jeff on the knee. "Did you feel that? You are going to move over and the dog is going to move up into the spot where you'd been sitting. So, you're sitting here in the evening, you're really tired, the two of you are curled up together on the couch, you're deeply involved in your TV show and you don't pay attention to the tap." They looked at one another. "We never noticed it," they said in unison. "We never noticed that we moved. Did you notice that we were moving?" they asked. They hadn't realized that they'd moved at all; they thought that they had sat down on the couch and not moved.

"What you have here is a little social structure issue." What does this mean, exactly? Well, it was particularly Jeff who was peed on. This was an interesting challenge. Pepper never peed inappropriately; it was always Banjo that peed on Jeff. The solution was first, to alter the greeting at the door. When asked who Jeff greeted first when he came through the door he said, "Well, I come through the door, I come upstairs and usually Julia is making dinner or something, and I give her a kiss and I greet her and then I take the dogs outside." When questioned about the two sets of stairs from the landing and who he greeted from the landing, he modified his answer. He thought for a moment. "Oh, I greet the dogs," he said. When asked which dog, he said, "Oh, Banjo is always first." So the recommendation was made that next time you come home, you should come in the door and completely ignore the dogs, without crushing them or stepping on them, and then come up the stairs and greet your wife first and then greet the dogs afterward, and take them out to the backyard, or whatever you want.

The next step was to address what to do when the dogs peed on their feet. The clients indicated that when the dogs had peed on their feet in the past, Jeff and Julia immediately jumped up to get paper towels and to clean. When asked where the dogs were during that time they said, "They're always sitting on the couch, watching us clean it up."

The advice to modify this behavior was to come through the door, come up the stairs and give Julia this big greeting. Then you're going to completely ignore the dogs until you're ready to take them out into the backyard. And in the evening, you aren't going to move. Don't move. You can put pillows up or something so you don't move, just don't move. You have to focus on this for a while and not move.

Jeff and Julia followed this advice, and Banjo peed on Jeff's feet twice more, once each of the following nights. Jeff refused to move when this happened. He said, "I didn't even get up to clean it." And that was the last time that the Pomeranian peed on his peoples' feet. Two nights and that was the last time. These slight changes just completely altered the whole social dynamic. There was an issue with Banjo trying to assert social dominance over his people—and Jeff in particular—but more than anything, it was mostly a learned contingency. They had learned how to train the owners to move out of the way.

The Curious Incident of the Dog in the Daytime

Sometimes, a case can be resolved rather quickly, like Jeff and Julia's, but for other clients, the behavioral issue is far harder to address. That was what Ken and Pamela Mandrake experienced with their two dogs. As Ken revealed, one night when he came home from work, he could hear an ear-piercing buzzer long before he reached his front door. He had left that morning before sunrise, and it was dark once again, but he could see that his backyard was illuminated by the motion detector light. He hurriedly turned the key in his front door lock; as the door opened, the throb of the buzzer became almost unbearable. Closing the door quickly, he passed the threshold into the entryway. A familiar metallic smell wafted toward his nostrils. Ken hurried toward his living room and rounded the corner to his dining room. The buzzer grew ever louder as he went.

"Charlie? Annie?" he called out, uncertainly. A gorgeous Chow Chow, head held low, tail tucked between her legs, approached him with tentative steps. "Hey, Annie girl," he said, looking at her with concern. Annie panted, saliva coating her neck and the sides of her mouth. Her eyes were dilated and her face and neck were smeared with red. Alarmed, he studied her for a moment, then with a shaking hand, he gently turned her head from side to side, brushing it with his hand and examining his palm. It wasn't hers. He scanned the house for his other Chow. "Charlie!" he called out, passing through his kitchen. The buzzer grew ever louder as he approached his stove. He turned it off brusquely, but even after it was silenced, his ears still rang. "Charlie!" he called out again, dashing down the hallway toward the garage. He stopped mid-stride, trying to interpret what he was seeing.

Where there had once been a door there hung a shredded, splintered frame, splashed with blood and edged in fur. Ken gasped. "Oh, Charlie ..." On the opposite side of the garage, facing toward their yard, was a second demolished door, this one smeared with even more blood. "Charlie?" he called out softly. There was no response. Annie padded cautiously behind him, sniffing nervously at each door frame and along the ground. She whined, her tail still tucked.

He stared out into the semi-darkness, looking where the arc of brightness from the motion detector porch light illuminated his lawn. There, 10 yards away, Ken saw a still, golden lump in the yard. It was Charlie. Ken saw only a darkened tail; most of Charlie's body was somehow wedged beneath the fence. Dirt had been flung out from either side of the enclosure and his motionless body was surrounded by a semicircle of bloody mud. Ken stood transfixed, his arms dangling numbly by his sides. Annie padded up silently and whined quietly behind him, touching his hands with her damp nose.

"Oh my God," a voice gasped from behind him. Ken turned and saw his wife's colorless face. "Charlie, Charlie," they called out to him together, grabbing ahold of him and pulling him back into the yard.

Pam grabbed Charlie's hind end while Ken gingerly supported his head, and they lifted him from the ground. Together, they carried him toward their SUV, hoisting him in as gently as they could. Pam's hands shook as she tried to comfort her bloodied dog, her eyes blurred with tears as the doors slammed shut and they sped toward the vet's office. She ran her hands across Charlie's body, putting pressure on his wounded neck. "Hi," Ken's voice said, louder than he intended. Pam and Charlie both flinched. "Yes, yes, this is Ken Mandrake ... we're bringing in Charlie. There's been an accident ... We're about 5 minutes out ..." Charlie began to whine and pant. Pam re-focused on her injured dog.

As they entered the parking lot, Ken tapped the horn; they were promptly met by a veterinary assistant who helped carry Charlie in. Then, their arms and work clothes stained red, they sat in the waiting room, huddled together, waiting for word on their dog. The veterinarian finally met them in the lobby.

"I had to put 70 stitches into Charlie," he'd said quietly to Ken. "He's going to be okay, but you need to call a Certified Animal Behaviorist."

Canine 911

When there's a behavioral issue that can't be solved by the pet owner, they're often referred to a Certified Applied Animal Behaviorist. While Charlie was a particularly urgent case, cases range from self-injurious behaviors to inappropriate elimination, and every issue in between. Often, the dog's breed can help troubleshoot the underlying cause of his or her behavioral issues. Let's talk about what's important to know in order to diagnose a case like this.

While a dog's breed can provide insight into their predilections and behaviors, individual differences exist as well. Labrador Retrievers are "water dogs," but you will find individuals who are hydrophobic. German Shepherds are renowned for their work as guard and police dogs, but you'll find individuals who are meek and retiring. Border Collies are adept at herding, but every once in a while, you'll find one who shows no propensity to round up anything. Animal behaviorists have to consider the dog's breed and a suite of other variables, including early life experiences and socialization, behavioral contexts, any past traumatic events or recent lifestyle changes, and personality differences among individual dogs.

A Series of Unfortunate Events

The Chow Chow, unlike the Pomeranian, is a rather ancient breed that originated in Northern China. The Chow Chow is a large dog with a robust frame covered in a dense double coat, and distinctively thick hair in their neck region. They have broad skulls, deep-set eyes, trademark purplish-black tongues and lips, erect triangular ears, and curly tails. The breed likely dates back 2000–3000 years, and they were initially bred as "war" and sled-pulling dogs. In northern China, they're referred to as Tang Quan, which translates to "Dog of the Empire," or Songshi Quan, which translates to "Puffy Lion Dog." Chows have often been referred to as "cat-like" in personality, frequently exhibiting reticence toward unfamiliar people, and they can become protective of their people and property, often experiencing intense bouts of separation anxiety. Interestingly, Queen Victoria also had a Chow and, reportedly, stuffed bears were modeled after her Chow Chow puppy. While the Pomeranian is known for its gregariousness, Chows are comparatively aloof in the presence of unfamiliar people. Separation anxiety—such as the kind that Charlie demonstrated—is a form of distress that occurs when a dog is separated from their owner. Separation anxiety can take many forms, including, but not limited to, excessive vocalization, destruction of property, self-injurious behaviors, inappropriate elimination, and escaping or attempting to escape from the area where they're being held. Because they are genetically prone to separation anxiety, you can get very severe cases with them.

But *why* are they genetically prone to it? Chow Chows aren't one of the more common dog breeds, so the data on their rates of behavioral issues, including separation anxiety, is still lacking, but definite correlations between breed and separation anxiety have been found. A study conducted at a behavior clinic in Norway found evidence to support a genetic link with behavioral issues (Storengen *et al.*, 2014). The researchers found that mixed breed dogs, Cocker Spaniels, Gordon Setters, Schnauzers, long- and short-haired Dachshunds, Jack Russell Terriers, German Shepherds, Tibetan Spaniels, and Rhodesian Ridgebacks had the highest rates of separation anxiety. Among all the dogs at the behavior clinic, mixed breed, German Shepherds, and Cocker Spaniels were the three most common breeds seen, while among dogs registered in the Norwegian Kennel Club, German Shepherds, Norwegian Elkhounds, and Golden Retrievers were the most common breeds. While Cocker Spaniels had the highest rate of separation anxiety among purebred dogs in the study, they only comprised 1.8% of the dogs that were registered in the Norwegian Kennel Club, and were not one of the ten most popular dog breeds between 2005 and 2010. Similarly, Schnauzers represented only 0.3% of those dogs registered in the Norwegian Kennel Club, but 2.8% of all dogs seen at the behavior clinic. Animal behaviorists have found a similar correlation between breeds and specific behavioral issues.

It should come as no surprise that separation anxiety is being diagnosed with dogs at increasing rates. The place of dogs in our homes has evolved over time, from farm dogs with specific occupations primarily living their lives outdoors with freedom to roam, to family members who spend larger chunks of time indoors. Not only have they been moved indoors, but they're also being left alone for longer periods of time. For dogs living in homes where their owners are often gone for 8, 10, or 12 hours per day, separation anxiety is an increasingly prevalent issue. In 1970, only 31% of two-parent households had two parents that worked outside of the home full-time. By 2018, that percentage had doubled: 65% of households had both parents working outside the home full-time (US Bureau of Labor Statistics, 2023). Among couples, dual-income households have increasingly become a financial necessity over the past 50 years, and whether households comprise single persons, couples without children, or parents, the trend has been to spend increasing amounts of time outside of the residence. Moving our canines indoors and increasing the time they've spent alone has created a perfect storm for increased rates of separation anxiety. A 2001 study found that dogs who lived in apartments had higher rates of separation anxiety than dogs who lived in homes with yards (Takeuchi *et al.*, 2001). The authors hypothesized that this was perhaps due to decreased opportunities for social interaction, limited space, and greater rates of attachment due to smaller rooms in conjunction with greater tendencies for owner absence.

While dogs can't tell time, per se, a growing body of literature provides evidence that dogs who have been left alone for increasing amounts of time show higher rates of behaviors indicative of anxiety than those who aren't. Rehn and Keeling (2011) studied the effect of time left alone on dog welfare by examining behavior and cardiac activity for durations of 0.5, 2, and 4 hours with 12 dogs who had previously exhibited no indications of separation anxiety. After longer periods of separation, the dogs exhibited more anxiety-related behaviors, including body shaking and lip licking, and had higher rates of interaction and tail wagging when reunited with their owners than the dogs who had been separated for half an hour. Those dogs who had been left alone for longer periods of time demonstrated more intense greeting behaviors than their shorter-interval peers, even between the 0.5- and 2-hour separation sets. The study provided evidence that dogs are affected by the duration of time that they are left home alone; for dogs who have a predisposition for separation anxiety, it's easy to see how an 8-, 10-, or 12-hour separation could have a dramatic impact on their welfare and eventually lead to psychological problems.

Why would "time spent alone" be an important area to research, particularly when it pertains to our dogs? Questions like these—the effect of time spent alone and its relation to welfare—have been framed by the concept of the "five freedoms," which provide a framework for gauging the quality of animals' lives. Animal welfare researchers first conceptualized the "five freedoms" that are under human control in 1965. In the following decades, these freedoms, which include freedom from hunger and thirst, freedom from discomfort, freedom from pain, injury, or disease, freedom from fear or distress, and freedom to express normal behavior, have helped us measure and subsequently improve the welfare of animals. When an animal has these freedoms, they are said to have good welfare. Dogs that are unable to socialize and exercise because they're kept in small apartments for one-third of a day or longer don't have the freedom to express their normal, species-specific behaviors. Dogs that don't have humans to reassure them when they begin to feel anxious because they're alone don't have freedom from fear or distress. So, while these dogs are likely receiving food, shelter, relative comfort, and care for any physical issues, their emotional needs aren't being met. Animals whose emotional needs aren't met often present with various issues, including separation anxiety and behaviors that can harm themselves and others.

Why do some dogs exhibit separation anxiety? Multiple factors come into play, including breed, changes in familiarity and predictability (such as schedule changes, moving residence, additions or deletions from the family, or changing families altogether), individual temperament, and prior life experiences. Some dogs begin to exhibit anxiety prior to their guardian's departure; they use "pre-departure cues," such as picking up keys, putting on shoes, grabbing a purse, or putting on a jacket to deduce that their person is going to be leaving. These pre-departure cues can lead to anticipatory anxiety, which can continue to build. Behaviorists recommend teaching your dog that these actions aren't always necessarily followed by departure, thus breaking the association and alleviating their stress levels. They also generally recommend pairing pre-departure signals with a highly valued reward. This helps change the emotional association from a negative one to a positive one. For example, giving a dog a frozen (and long-lasting) peanut butter-filled Kong toy as you leave to help create a more positive emotional association. This is classical conditioning, also commonly referred to as Pavlovian conditioning. This can be followed by leaving the home for short durations of time, gradually increasing the length of the absence until the dog's separation anxiety diminishes.

All dogs can experience separation anxiety, but why is the Chow Chow particularly susceptible to it? Somewhere along the line, it's in their genetic heritage. As an ancient breed, they have a storied history with humans; they were originally bred to be guard dogs, herd dogs, and hunting assistants; Chinese Han Dynasty-era artwork often depicts them running alongside game. Chow Chows are described as "loyal," and they bond quickly with their owners; issues with separation anxiety appeared to have come along for the ride.

While separation anxiety is common in Chow Chows, unlike Banjo's unusual pedestrian peeing, the catalyst behind Charlie's issues would prove to be particularly complex. During the interview with Ken and Pamela about what led to Charlie's issues, they said, "Well, you're going to have to come over to really see this. We have a male and a female Chow Chow, and our male Chow Chow, Charlie, just had 70 stitches and it's because of separation anxiety." Again, the layout of this home was pertinent to the behavioral issues that they were experiencing. When you go in the door, it's one level, and then you head straight into a sort of U-shaped living room. Next to it, there's a wall sticking out and kind of a U-shaped dining room with windows looking out into the backyard. There's a garage door, an open kitchen with an island, and then a door on the other side over there going to the bedrooms. The interview took place in the living room, and around the wall was the dining room. The garage door was right there, and it was in shreds. It looked like a forensic scene. There was blood everywhere and it was shredded and splintered. It's a hollow framed door, but it looked like somebody just blew through there, like a gunshot just splintered it all up.

They said, "Well, you have to see Charlie." Charlie was back up on his feet, tottering around and he had all of these stitches. He had 70 stitches on his face and on his body; his shoulders were all ripped up, and huge patches of fur were gone. They said, "Yeah, he went right out through the door into the garage." In addition, the dog had destroyed, blown out, and finally busted the lock on the back door on the garage. When they got home, following the trail of blood, that's when they found Charlie halfway under the back fence. They saw just a tail in the air, under the back fence, heading out, leaving their property, as fast as he could. They took him to the veterinarian, who got him all cleaned up and then called an expert.

While going over Charlie's history with them, the beautiful female Chow Chow, Annie, came over and went to sleep. Charlie, in contrast, was pacing around, and then he disappeared into the bedrooms. Later, Charlie came out of the bedroom and slid down the wall of the dining room, under the window, along the wall and slipped as tight as he could around the couch, and into our area. Then he was petted by us and then he paced back and went around and around and around ... and he just wouldn't go near the kitchen. This odd behavior prompted questions about what was happening prior to this incident? What series of events led to these odd behaviors? Pam and Ken, like the owners of the Pomeranians, also worked at Microsoft, typically for 12 hours a day. There were no problems with their dogs for 2 years in the house since they'd both been gone working. No problems whatsoever. And then they decided to stay home and remodel the house over Christmas, so they were home for a week. They were at home all the time, and took the dogs in the car with them everywhere they went. Then they both got the flu and they were really at home, and they didn't go anywhere for another week.

The day that Charlie went out through the hollow frame door and got the stitches was Ken and Pam's first day back at work. Now this was dramatic separation anxiety! Separation anxiety is common among Chow Chows, but that's pretty dramatic separation anxiety. The additional issue was why Charlie wouldn't go through the kitchen or go past the kitchen, but always went through the dining room all the way around to get to the living room. Charlie's owners had to think about this.

One thing that this case had in common with many of my other cases was the dynamic between the husband and the wife. Often, the husband would be explaining the factors associated with the behavioral issues, and the wife would interject, nudge him, and say, "Go ahead, tell him the rest of the story." That's what happened in this instance. Ken said, "Well, we have this smoke alarm in the kitchen. You're not supposed to have a smoke alarm in the kitchen, as it goes off too easily. It goes off any time we make toast, but the dog now acts extremely anxious when we open the plastic bag that the bread is in. We keep the bread in the refrigerator. The dog now panics and has anxiety attacks when we open the door to the refrigerator to get the plastic bag, open the bag, and put in the bread, to have the toast burn, to have the smoke alarm go off."

There's an association between the sound of the plastic for the bread, the toaster, and the alarm—the dogs can't discern causation from correlation, but the alarm has gone off often enough after the bread has been removed from its wrapper that the dogs have become anxious about this. This is Pavlovian conditioning at its best—or worst, depending upon how you see it. Russian physiologist Ivan Pavlov's pioneering work unveiled Pavlovian conditioning. During the 1890s, Pavlov was researching dogs' salivatory responses to food. He predicted that the dogs would begin to salivate when food was placed in front of them, but they actually began to salivate far before they saw the food. Anticipating the food reward, the dogs would salivate when they heard Pavlov's assistant approaching. The dogs had learned to associate the sound of this person's footfalls with the receipt of food. Once Pavlov realized what was occurring, he built upon this experiment, pairing a metronome and later a bell with the receipt of food. Soon, the presence of the metronome or bell alone elicited drooling. Similarly, Charlie had learned that the sound of the plastic wrapper sometimes precedes the sound of the alarm; thus, he began to exhibit anxious behaviors with the mere crinkle of the cellophane, much like Pavlov's dogs salivated from the sound of a bell. But how could we undo this response?

It was pretty clear that this could be contributing to Charlie's anxiety around the house, and particularly around the kitchen. But, then Pam said, "Well, tell him the rest of it." Ken said, "Really?" And Pam said, "I think he needs to know."

So, it turns out that they have a very old oven with a very old, mechanical timer. It's very loud, so that you don't burn your pies when you're away from the kitchen. It goes off randomly, and it doesn't shut off. And the dog is at home. It's a Chow Chow, and he's at home without his owners. And periodically, every week or 2 or 3 weeks, this thing goes off, and it's going off when they get home. They don't know for how long it's been going off. The dog is usually slobbering, and they don't know what that is. Well, that's anxiety. A loud stimulus is occurring at random points in the day when the dog is all alone, and he can't properly anticipate it, stop it or escape from it. Essentially, this is torturous—the dogs can't accurately predict the alarm, nor can they make it stop once it starts.

Multiple factors play into the development of fear responses and separation anxiety, including a dog's early social experiences and environment, individual personality characteristics, including breed differences, and exposure to certain loud noises. Cases of separation anxiety can develop or worsen when a dog experiences a loud noise—such as a loud, unpredictable, intermittent alarm—while separated from their owners. Separation can become a discriminative stimulus for exposure to noises, especially when a dog's primary means of coping is seeking reassurance from their owners (Blackwell *et al.*, 2013). In the absence of their owners, a dog's ability to cope with loud noises is diminished, and feelings of anxiety can intensify rapidly.

For the Mandrakes, this was truly a series of unfortunate events that led to these aberrant behaviors and the incident that resulted in 70 stitches for Charlie. Separating out each contributing factor was critical in determining the catalyst and a possible solution to these issues. It turns out that the day the dog went out through the door was the first day back at Microsoft, first day back to work after 2 weeks, and …? They said, "Yeah, the timer went off." So they left, they were gone all day after being home permanently with their dogs, day in, day out for 2 weeks, and the timer went off, and that was it. The dog went out through two doors and was digging his way to China.

Healing from Anxiety

The Mandrakes worked to help Charlie overcome his separation anxiety, beginning with pre-departure cues associated with positive and long-lasting treats left behind, using high-value rewards, and leaving the home in increasingly longer increments of time, beginning with very brief departures. It took about 1.5–2 years of treatment, where they were staying away from home for increasingly long periods of time, and they were finally able to be at work for 12 hours at a time and leave them alone again.

Charlie's case had multiple facets: his owners had to help him become comfortable being alone for long timespans again and they had to remove any aversive stimuli. Fortunately, Pam and Ken did get rid of the timer, but Charlie had applied his anxiety to other objects, as well. He got anxious when the phone rang in the kitchen, so they moved the phone out of the kitchen, too. And they moved the smoke alarm and ensured that this series of events wouldn't reoccur.

We Can Work It Out

Banjo the Pomeranian and Charlie the Chow Chow were two very different dog breeds with two very different problems—but they had one thing in common: learning how to negotiate living with humans during modern times. In both cases, only one of the two dogs had the behavioral issue, thus demonstrating the importance of individual differences. Annie and Charlie were both subjected to the same long periods of separation from their people, accompanied by the loud timer that went off at unpredictable intervals, but only Charlie had severe separation anxiety. Pepper and Banjo both maneuvered their humans in a show of dominance, but only Banjo urinated on his owner. You cannot link breed and behavior issues 100% together. Breed does tell us predispositions to certain behaviors for particular breeds. It doesn't mean a particular behavior problem is destiny for an individual dog of a certain breed, but it can help point you in the right direction in terms of what the problem might be, and it can help inform you to just how challenging the behavior may be to correct. It took about 2 years of work to cure Charlie of his separation anxiety because of the breed and the severity of the response. The only way it was possible was that Ken and Pam could take him to work and not leave him home alone before he was specifically involved in lengthening departures.

The Chow Chow is an ancient breed compared to the Pomeranian, but the Pomeranian has many genes in common with ancient breeds, as well (Wayne and Ostrander, 2007; vonHoldt *et al.*, 2010). Ancient breeds tend to have more issues with dominance, so while Pomeranians aren't "ancient," per se, they may share in common many of the genes that are related to dominant behavior. This helps explain why the male Pomeranian, Banjo, may have had a propensity to start the peeing behavior. Basically, Pomeranians have a few more ancient breed characteristics as a whole, than say, Victorian-era Poodles do. Dominance behavior is one characteristic that's seen in some ancient breeds. Jeff and Julia weren't aware that their dogs were acting dominant, or that they were being maneuvered around the couch, but Banjo interpreted their behavior as submissive—that they were yielding to him. Still, the Pomeranian's behavior was atypical for the breed in general, but not completely unheard of, and the behavior was inadvertently reinforced by the humans moving around on the couch, and even giving the dogs the run of the couch after they peed.

What can we learn from Charlie, Annie, Pepper, and Banjo? All dogs belong to the species *Canis lupus familiaris*, but given their widely varying genetic histories, it wouldn't be accurate to say, "a dog is a dog," even within the same breed. Even within one breed, such as Golden Retrievers, some breeders specifically select dogs for their hunting abilities, while others select dogs for show or for temperament as family pets. Field- or hunting-bred Golden Retrievers tend to be leaner, smaller, darker- and shorter-coated, less gentle, and less sociable than their show-bred counterparts. Goldens from field-bred lines, however, tend to be the more intelligent of the two. Show- or conformation-bred Goldens tend to be more robust bodied, larger, lighter- and longer-coated, and possessing full "feathering," or furry plumage, on their tails. While Goldens are considered to be "friendly" dogs, show-bred Goldens tend to be the more gregarious of the two lines. All Goldens are the same breed, but these different lines will have a higher likelihood of exhibiting the traits they have been selected for.

When a dog's recent breeding is known, the breed and the specific lineage can provide insight, but individual differences will still exist. Charlie and Annie were both Chow Chows, and Pepper and Banjo were both Pomeranians, but only Charlie suffered from severe separation anxiety, and only Banjo eliminated on Jeff's feet as a sign of dominance. The relationship between genetics, the environment, and other potential factors is complex; the breed can provide a baseline to inform the difficulty and the length of treatment, but each treatment plan has to be individualized. Thus, as we discuss breed differences and similarities, remember that there will always be both trends and exceptions to the rule.

Conclusion

Think of the dog's breed as their blueprint—a design plan for an individual. With a house, a blueprint shows the general layout, the potential, the square footage, but not the more specific details, like paint colors, roofing material, countertops, flooring, appliances, and landscaping, that would differentiate this structure from others that have the same initial blueprint. With dogs, their breed would provide the general blueprint (e.g. with Golden Retrievers, a propensity to retrieve and to be social, or with Border Collies, the drive to herd other animals) but it wouldn't take into account individual factors, such as personality traits, temperament, early learning, and socialization. The blueprint for the house is the starting point, or template, but as construction progresses, you get a home with additional influences, such as last-minute changes and decisions about characteristics that weren't on the original blueprint. This is what we're witnessing with dogs; more than any other species, their construction has deviated from the original blueprint. Humans, as the masterminds behind these departures from the original plan, need to understand the differences between breeds and breed groups as we help them navigate our world.

References

Aristotle (1887) *History of Animals*, IX. I. 2. Translated by Richard Cresswell. George Bell & Sons, London.

Blackwell, E.J., Bradshaw, J.W.S. and Casey, R.A. (2013) Fear responses to noises in domestic dogs: prevalence, risk factors and co-occurrence with other fear related behaviour. *Applied Animal Behaviour Science* 145(1–2), 15–25.

Eken Asp, H., Fikse, W.F., Nilsson, K. and Strandberg, E. (2015) Breed differences in everyday behaviour of dogs. *Applied Animal Behaviour Science* 169, 69–77.

Ha, J.C. and Campion, T.L. (2018) *Dog Behavior: Modern Science and Our Canine Companions*. Elsevier/Academic Press, London.

Landsberg, G.M. and Horwitz, D.F. (1998) *Aggression: Diagnosing and Treating Aggressive Behavior in Dogs and Cats. Custom Care on Disk, Behavior Version*. Lifelearn Inc./University of Guelph, Guelph, Canada.

Pérez-Guisado, J., Lopez-Rodríguez, R. and Muñoz-Serrano, A. (2006) Heritability of dominant–aggressive behaviour in English Cocker Spaniels. *Applied Animal Behaviour Science* 100(3–4), 219–227.

Rehn, T. and Keeling, L.J. (2011) The effect of time left alone at home on dog welfare. *Applied Animal Behaviour Science* 129(2–4), 129–135.

Scott, J.P. and Fuller, J.L. (1965) *Genetics and the Social Behavior of the Dog*. The University of Chicago Press, Chicago, IL, p. 4.

Shakespeare, W. (1996) *The Complete Works of William Shakespeare. Julius Caesar*, Act 3, Scene 1, line 273. Wordsworth Editions, Stansted, UK, p. 597.

Shipman, P. (2015) *The Invaders: How Humans and Their Dogs Drove Neanderthals to Extinction*. Harvard University Press, Cambridge, MA.

Storengen, L.M., Boge, S.C.K., Strøm, S.J., Løberg, G. and Lingaas, F. (2014) A descriptive study of 215 dogs diagnosed with separation anxiety. *Applied Animal Behaviour Science* 159, 82–89.

Takeuchi, Y., Ogata, N., Houpt, K.A. and Scarlett, J.M. (2001) Differences in background and outcome of three behavior problems of dogs. *Applied Animal Behaviour Science* 70(4), 297–308.

Thompson, M. (2014) 9 million unsung heroes WWI: dogs, horses and carrier pigeons made victory possible. *Daily Mirror*. Available at: www.mirror.co.uk/news/real-life-stories/9-million-unsung-heroes-ww1-3939895 (accessed 21 May 2018).

US Bureau of Labor Statistics (2023) Employment characteristics of families. Available at: www.bls.gov/news.release/pdf/famee.pdf (accessed 25 October 2023).

Varner, J.G. and Varner, J.J. (1983) *Dogs of the Conquest*. University of Oklahoma Press, Norman, OK.

vonHoldt, B.M., Pollinger, J.P., Lohmueller, K.E., Han, E., Parker, H.G. *et al.* (2010) Genome-wide SNP and haplotype analyses reveal a rich history underlying dog domestication. *Nature* 464(7290), 898–902.

Wayne, R.K. and Ostrander, E.A. (2007) Lessons learned from the dog genome. *Trends in Genetics* 23(11), 557–567.

Acknowledgments

There are so many friends and family that helped make this book possible.

We'd like to thank our friends and photographers, Holly C. Cook (Holly Cook Photography, LLC) and Marika Moffitt (SoulDog Creative) for numerous images used in this book. Jenni Pfafman and Carol Harris also contributed images of behaving animals. Anne-Lise Nilsen Knight kindly assisted us in obtaining permission for the use of the University of Washington's photographs of their mascot, Dubs.

Arianne Taylor generously donated her time and artistic talent to numerous illustrations in the text and on the book cover.

Wikipedia Commons was a wonderful resource for free images, and benefited this academic effort.

Melissa Cole, Abbie DeLeers, and Margaret C. Ha provided crucial editorial comments and suggestions. Any mistakes that remain are our responsibility.

Numerous professional colleagues agreed to be interviewed for their take on dogs, breed differences, or wolves. We thank them for their time and for the knowledge they shared for this book (order of appearance in the text): Angela Perri, Nick Jans, Monique Udell, Wendy Spencer, Charlotte Lindqvist, Janice Koler-Matznick, Vladimir Beregovoy, Mike Powers, Wendy Dahl, James Serpell, Maria Muradas, Adam Winston, Kevin Yeo, Lori Theis, Celeste Walsen, Marina Hall Phillips, Carly Loyer, Cheryl Franz, Julie Forbes, Sarah Weideman, Gary Pegg, Jessica Meaghan (Meg) Mas, and Jennifer Hartman.

Throughout the book, some names have been changed, where appropriate, to protect privacy.

Introduction

Across the Universe

Johann Georg Palitzsch wiped his forehead, pushing an errant curl behind his ear as he placed the eyepiece against his face. Palitzsch, a self-taught naturalist who would later be referred to as the "peasant astronomer," had been raised to be a farmer, but abandoned the fields for the tools of science—in particular, the telescope. This night, he aligned the scope meticulously, watching as distant points of light oscillated in and out of focus, until the stars popped out at him like bubbles breaking the surface of a calm sea. He paused, taking the eyepiece away and using the back of his hand to rub the frost-etched glass, then he scanned the night sky in a slow, sweeping arc. Holiday carols lilted up from below. Celestial bodies clicked into view, their names and stories familiar to him, and he slowly noted each one, jotting them down with a quill pen as he had the night before. He pursed his lips as he scanned, his right hand, stained with ink, slowly guiding the scope. And then, right on the edge of his peripheral vision, he saw it: still far distant, but unmistakable: a blue-white, peanut-shaped interstellar visitor whose return Edmond Halley had prognosticated decades before. The much-anticipated re-appearance of earth's most famous planetary interloper finally occurred on Christmas Day, 1758. That same year Swedish biologist Carl Linnaeus, working from his office at Uppsala University in Sweden, introduced binomial nomenclature to his system of Linnaean taxonomy. It was there that *Canis familiaris* made its classification debut in the tenth edition of his unabridged guide to all known living things, the *Systema Naturae* (Linnaeus, 1758). Of all Linnaeus' entries in the *Systema Naturae*, *Canis familiaris* was perhaps the most diverse species, molded by humans' affinity for them.

Linnaeus gained his fame through his scientific works, but it was love (well, and the pursuit of money) that inspired him to excel at the highest levels of science. Ruddy cheeked, dark-eyed, and dark-haired before he began to don the customary white wig he wore for his portraits, Linnaeus wasn't unattractive. A dark-eyed beauty named Sara Elizabeth Moraea caught his eye in the mid-1730s; perhaps just as attractive was her rich bourgeoisie heritage. In February 1735, Linnaeus, then 27 years old, proposed to Sara Elizabeth, who was the daughter of physician Johan Moraeus, a member of the bourgeoisie. While Moraeus wasn't inherently opposed to the union, the elder statesman physician didn't want his daughter to marry a pauper, so he placed a condition on this engagement: Linnaeus had to be financially stable first. To do so, Linnaeus would have to pass his MD. Eager to wed, Linnaeus tried taking the path of least resistance. The University of Harderwijk in the not-so-nearby Netherlands had a reputation for issuing MD examinations rapidly. After the father of his friend, Claes Sohlberg, hired him to tutor his younger sons, Linnaeus was able to fund his journey. He arrived at the University on May 5 and received his MD only 4 days later. While Linnaeus would forever after be known as a taxonomist, he wrote his Doctor of Medicine dissertation on malaria, arguing (incorrectly, it would be found not long afterward) that the disease arose from drinking water that contained clay particles that irritated the body and resulted in this deadly "swamp fever." (Malaria, which is now both preventable and curable, is an infectious disease that comes from mosquitoes.) Nonetheless, it has claimed billions of lives over the course of human history, making it a particularly apt area of research. Disease, along with war, has been a major shaper of human evolution and history. Almost 300 years after Linnaeus studied malaria, the world's leading scientists again faced with an infectious foe, trying to determine the cause and abatement of the worldwide health threat of a novel coronavirus, COVID-19.

After his brief foray into epidemiology, Linnaeus could have returned to Sweden where his sweetheart

 Breed Differences in Dog Behavior: Why Tails Wag Differently. (R.L. Ha, T.L. Brad and J.C. Ha)
DOI: 10.1079/9781800624566.intr

was waiting, but instead, he was on the hunt for money again. The pursuit of funding was a driving force in Linnaeus' life, and others took note: he earned a reputation for being rather stingy after declaring, "All the beasts are splendid, but money more splendid." He spent the next 3 years traveling, researching, and writing—and earning as much money as he could. During this time, the senator of Leiden, Dr Johan Friedrich Gronovius, paid the fee for the first edition of the *Systema Naturae* to be published (Fig. I.1).

After leaving Leiden, Linnaeus continued to travel and write, visiting Amsterdam, where he

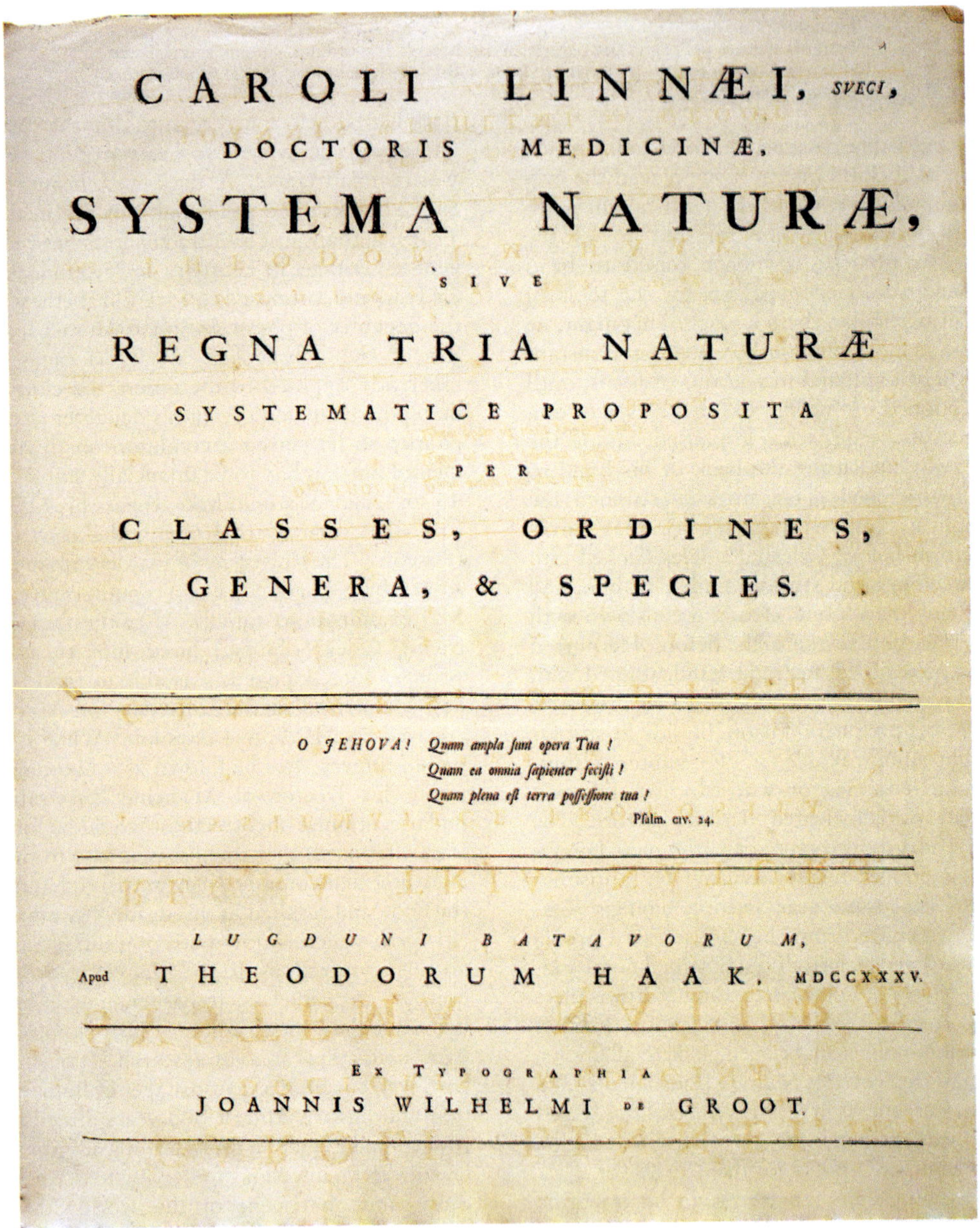

CAROLI LINNÆI, *SVECI*,
DOCTORIS MEDICINÆ,
SYSTEMA NATURÆ,
SIVE
REGNA TRIA NATURÆ
SYSTEMATICE PROPOSITA
PER
CLASSES, ORDINES,
GENERA, & SPECIES.

O JEHOVA! Quam ampla sunt opera Tua!
Quam ea omnia sapienter fecisti!
Quam plena est terra possessione tua!
Psalm. CIV. 24.

LUGDUNI BATAVORUM,
Apud THEODORUM HAAK, MDCCXXXV.

Ex Typographia
JOANNIS WILHELMI DE GROOT.

Fig. I.1. *Systema Naturae*, first edition (1735). Image is in the public domain and was provided by Wikipedia Commons (CC-PD-Mark). Available at: https://commons.wikimedia.org/wiki/File:Systema_naturae.jpg.

published *Bibliotheca Botanica* and *Fundamenta Botanica* and, later, *Genera Plantarum* and *Flora Lapponica*. The last work resulted from his 4600-mile trek from Lapland in Northern Sweden to the Arctic Ocean and then back again. Funded by Georg Clifford, then-director of the Dutch East India Company, Linnaeus traveled on to England, where he collected fauna samples and wrote *Hortus Cliffortianus* and *Critica Botanica*. In 1737, Linnaeus traveled to Paris via Leiden, and survived a bout of cholera. In 1738, he finally headed for home, where he became formally engaged to Sara and moved to Stockholm to secure employment as a physician. While Linnaeus would one day be known, along with Charles Darwin, as one of the two most well-known categorizers of species, securing that position wouldn't come easy. His initial reception as a doctor in Stockholm during 1738 was rather frosty: as a relatively unknown in the field, people were loath to entrust him with the medical care of their dogs, much less with human patients. The tide would soon turn for Linnaeus, though. He was one of the founders of the Swedish Academy of Sciences (today they are responsible for selecting the Nobel Prize laureates in Chemistry and Physics). The fledgling organization held its first meeting in May 1739. One month later, he finally wed the eternally patient Sara, with whom he eventually had seven children, five of whom survived past infancy. But he would be remembered most for his scientific progeny.

Animal, Vegetable, or Mineral

The *Systema Naturae*, whose full, translated title was *System of Nature Through the Three Kingdoms of Nature, According to Classes, Orders, Genera and Species, With Characters, Differences, Synonyms, Places*, was Linnaeus' magnum opus, a listophile's guide to the living world (Linnaeus, 1758). Linnaeus included the Animal, Plant, and Mineral Kingdoms in his work; today, this has been modified to the Plant, Animal, Fungi, Archaebacteria, Eubacteria, and Protist kingdoms (we'll revisit taxonomy in more detail in Chapter 1). What began as an 11-page treatise on the hierarchical classification of organisms continued to grow with each new version. Published in Latin, as all scientific books of that time were, even then, it was a *magnus paciscor* (that's Latin for "big deal"). The tenth edition of this voluminous tome (it would reach 2400 pages by its 12th edition, the last produced by Linnaeus himself) was the point of departure for zoological nomenclature, featuring thousands of species of plants and animals, including canids. The *Systema Naturae* was under continuous revision. While the tenth edition rectified prior errors (he changed the placement of manatees and whales to the mammal class, rather than the fish class, where they were first categorized), there were still a few missteps. Bats were recognized as mammals, and not birds, but they were classified under primates (imagine if the flying monkeys of "Oz" were science and not fiction), hyenas were classified as *Canis hyaena* and included under the *Canis* genus with dogs (hyenas are more closely genetically related to felines, but behaviorally, more similar to canines, due to convergent evolution, so this mistake is understandable), and sloths were paradoxically placed in the same family as elephants.

Despite the errors in this tenth edition, Linnaeus got many things right; lacking knowledge about genetics and evolution, he used the physical traits of plants and animals, including canines, to classify them. But his work wasn't without contemporary criticism. French naturalist Georges-Louis Leclerc, the Comte de Buffon (1707–1788), questioned Linnaeus's approach to natural history. Buffon wondered whether the vast variety seen in canines could be attributed to the environment. He believed that there were 17 "races" of domesticated dogs, and that their differences were due to climatic differences throughout their geographic distribution. While Buffon wasn't necessarily on the right track, he hypothesized that all dogs could be traced back to a shepherding dog progenitor, and that dogs—whether it was the parent dog or the offspring, he wasn't sure—would transform into new dog morphs when they were introduced to different climates. In his popular treatise, the *Histoire Naturale*, Buffon (1830) wrote:

> We may, therefore, suppose, with some degree of probability, that the shepherd's dog approaches nearer to the primitive race than any of the other kinds; for in every country inhabited by savage or by half civilized men, the native dogs resemble this race more than any other … If it be farther considered, that this dog, notwithstanding his ugliness, and his wild and melancholy aspect, is superior in instinct to all others; that he has a decided character, independent of education; that he alone is born fully trained; that, guided solely by natural powers, he applies himself spontaneously to the keeping of flocks, which he executes with amazing fidelity, vigilance, and assiduity; that

he conducts them with an admirable and uncommunicated intelligence; that his talents, at the same time astonish and give repose to his master, while other dogs require the most laborious instruction to train them to the purposes for which they are destined; we will be confirmed in the opinion, that the shepherd's dog is the true dog of Nature; that he has been preferably bestowed on us for the extent of his utility; that he has a superior relation to the general order of animated beings, who mutually depend on each other; and, lastly, that he ought to be regarded as the origin and model of the whole species.

It was Buffon who first wondered (on the record, at least), where dogs fit into the picture of human evolution, as well. He wrote: "To conceive the importance of this species in the order of Nature, let us suppose that it never existed. Without the assistance of the dog, how could man have conquered [and] tamed ... the other animals ... The training of the dog seems to have been the first art invented by man; and the result of this art was the conquest and peaceable possession of the earth." (Buffon, 1830)

An influential naturalist in his own right, the Comte de Buffon would inspire later scientists, including natural selection co-discoverers Charles Darwin and Alfred Russel Wallace and French naturalist Jean-Baptiste Chevalier de Lamarck. The Comte didn't win over Linnaeus, however; the two had a rather contentious relationship, and the latter dismissed the concept of the environment potentially shaping species. In his own version of diplomacy, the ever-humble organizer said that the Comte was "Always eloquent, often incorrect."

In the tenth iteration of the *Systema Naturae*, Linnaeus coined the term *Canis familiaris*, meaning "the intimate, familiar dog." It isn't clear whether dogs were familiar to Linnaeus himself (or if he was even fond of them), but he made it clear that he believed that all species, including *Canis familiaris*, had been created by God in an absolute, permanent shape. Thus, for Linnaeus, all *Canis* species, and all breeds within *Canis familiaris*, were precisely as they were supposed to be, always had been, and always would be. While he supported the Linnaean system of classification, the Comte de Buffon argued that species *weren't* immutable: they changed over time, or died out, and their places were taken in the ecosystem by other animals, as evidenced by discoveries such as long-extinct woolly mammoths in the Siberian permafrost. Mammoths weren't the only animals preserved in Ice Age snow, though; scientists also found now-extinct cave lion cubs and woolly rhinos. The Comte wasn't without his followers, however. Lamarck (1744–1829) was one of the Comte's protégés, and a famous scientist in his own right. He was a French biologist, naturalist, soldier, and academic who is perhaps best known for some of his scientific missteps. Lamarck promoted the idea that parents could pass on traits that they had *developed* over their lifetimes. We know now, of course, that that's not true, but in Lamarck's day, it made perfect sense, and was the accepted idea of the time. He wrote about this in his 1809 book on zoology (Lamarck, 1809). For example, Lamarck believed that species such as giraffes had such long necks due to continued stretching that showed up in subsequent generations. But this isn't how evolution actually works; if you get a tattoo, or if your wisdom teeth are pulled, that doesn't mean that subsequent generations will have that same tattoo or won't have wisdom teeth. And if a dog's ears or tail are cropped, that doesn't mean that their offspring, or generations after that, will also have shortened tails or ears and be spared that procedure. Lacking knowledge of the *actual* mechanism for these changes, Lamarck's revolutionary concept of evolution fell short. It was a gap in knowledge that the Comte and Lamarck shared in common.

Linnaeus, the Comte, and Lamarck were all equally right ... and equally wrong. Had Linnaeus realized that dogs (and all other species) weren't immutable, but had changed (often dramatically so) over time, he may have been more open to the idea of evolution. Had the Comte realized that the environment had a complex working relationship with one's genes, he may have had a clearer idea of evolutionary processes. And had Lamarck known that dogs had descended from a common ancestor of the gray wolf, *Canis lupus*, and had changed over time due to these same factors, and not something akin to what Rudyard Kipling would later write in his "Just So" stories, he may have understood the mechanisms of evolution.

Rudyard Kipling was the king of esoteric explanation. When you're the father of young children, as Kipling was at the time he created his 1895 masterpiece, *The Jungle Book*, and his 1902 collection of "Just So" stories, novelty and creativity make good bedmates and even better bedtime stories. Kipling's "Just So" collection, which originated as bedtime stories for his daughter, Josephine (Effie to those who knew her), offered morality-fable explanations for evolution; "literary Lamarckianism," if you will. A modern take on Kipling's popular pieces might be "Pokémon-like evolution." The "Just So" stories earned their name because Effie asked her father to tell them "just so," without deviating by even a single word. It was a tall order, but Kipling responded by putting this prose into print. Of the 13 original "Just So" stories, nine explained how animals were modified from their original forms into their modern ones. These stories included "How the Whale got his Throat" (the whale only eats tiny prey because he swallowed a sailor whose raft prevented the whale from swallowing other sailors), "How the Leopard got his Spots" (an Ethiopian man painted the spots), "How the Camel got his Hump" (the camel was given humps as punishment for refusing to work), and "How the Elephant got his Trunk" (Fig. I.2), a story that entailed a juvenile elephant who, as all elephants apparently did during his time, had "only a blackish, bulgy nose, as big as a boot, that he could wriggle about from side to side; but he couldn't pick up things with it" (Kipling, 1942) but an insatiable sense of curiosity that eventually resulted in one very long nose. The juvenile elephant's burning question was: what does the crocodile have for dinner? The answer: anything the crocodile could reach. And then the crocodile lured the juvenile elephant to the side of the river and took ahold of the elephant's nose. Kipling (1942) wrote:

> ... the Elephant's Child sat back on his little haunches, and pulled, and pulled, and pulled, and his nose began to stretch. And the Crocodile floundered into the water, making it all creamy with great sweeps of his tail, and he pulled, and pulled, and pulled.
>
> And the Elephant's Child's nose kept on stretching; and the Elephant's Child spread all his little four legs and pulled, and pulled, and pulled, and his nose kept on stretching; and the Crocodile threshed his tail like an oar, and he pulled, and pulled, and pulled, and at each pull the Elephant's Child's nose grew longer and longer ...

While the Elephant's Child survived the encounter, his nose remained long. He waited for his nose to return to normal, but it never did, and all elephants onward sported the new, long nose. Sadly, Effie would succumb to pneumonia in 1899 at the age of 6 years, but the stories that were penned in her honor continue to endure for generations afterward.

So at this intersection of literature and science, one might ask: did Kipling *believe* Lamarckian ideology? Fortunately for us, Kipling discussed his scientific leanings: he despised Darwin's work with a capital D. Of the father of evolution's work, Kipling wrote: "I've been trying once more to plough through *The Descent of Man* and every fiber ... of my body revolted against it" (Kipling *et al.*, 1996). Kipling's opinion of Lamarck, however, was more sweet than salty. And with works like *Organic Evolution as the Result of the Inheritance of Acquired Characters According to the Laws of Organic Growth* (Eimer, 1890), Kipling's world was one where Lamarck *might just be* evolution's father.

English professor Allen MacDuffie (2014) discussed this in his piece, "*The Jungle Books*: Rudyard Kipling's Lamarckian Fantasy," where he points to "Red Dog", one of the stories in *The Second Jungle Book* as the prime example of Kipling's Lamarckian leanings. In "Red Dog", Mowgli, the protagonist, has an encounter with a group of wild dogs. Mowgli cuts off the tail of the pack leader. Afterward, Mowgli taunts the dog, saying, "Nay, but consider, wise rat of the Dekkan. There will now be many litters of little tailless red dogs, yea, raw red stumps that sting when the sand is hot. Go home, Red Dog, and cry that an ape has done this." (Kipling and Kipling, 1906).

The Origin of "Species"

The *Systema Naturae* was the scientific debut of the *Canis* genus, which includes the dog-like carnivore species. Now the scientific terms "genus" and "species" are commonplace enough to have entered the vernacular, but what do we mean when we say them? We've been using the words "genus" and "species" without operationally defining them, which is particularly important for the latter term. The word *genus*, which is the Latin plural of *genera*, originated in the mid-16th century to refer to a class or kind of thing; in the 17th century, genus picked up a biological connotation, denoting kind, race, stock, birth, descent, family, and origin.

The word *species* originated in the late 14th century to denote a particular type or kind of organism that had a specific physical appearance; species, too, gained a biological connotation during the 17th century. Thus, we had categories to classify organisms, in smaller and more specific groupings, but how *clear* were these classifications? The

Fig. I.2. "How the Elephant got his Trunk": *Just So Stories* by Rudyard Kipling. This image is in the public domain and was provided by Wikipedia Commons. Available at: https://commons.wikimedia.org/wiki/File:Illustration_at_Cover_of_Just_So_Stories_(c1912).jpg.

definition of species, in particular, is an issue. There remains today a "species problem" when scientists try to define this word. As a result, there are at least as many "species concepts" as there are letters in the English alphabet (Frankham *et al.*, 2012). These species concepts primarily refer to issues pertaining to morphology (that which pertains to an organism's structure, form, and specific features), biology (whether organisms actually or potentially interbreed), and phylogeny (evolutionary history of a taxonomic group or species). These species concepts differ in the number of recognized species, which can have numerous ramifications pertaining to genetics, biology, conservation, and even legal and financial issues, when trying to determine whether an animal is an endangered species or "merely" a subspecies of a prolific organism. The situation is further obscured by species with thousands (or even millions) of years of evolutionary separation still being able to produce viable (and sometimes fertile) offspring. Take, for example, all species of the *Canis* and *Panthera* (so-called "big cats") genera, which can interbreed within their respective genus and produce viable and fertile offspring, despite the millions of years since some species last shared a common ancestor. Coyotes and domesticated dogs have been separated for 5 million years, and lions and tigers last shared a common ancestor 3.7 mya (million years ago), yet they can interbreed and produce viable offspring.

Visual signs are the first indication that most early observers had to decide who fit into which grouping. Morphology was commonly used by early scientists, including Darwin and Linnaeus, to demarcate one species from another. But appearances can be deceiving; two individuals may share many morphological traits in common, but not belong to the same species. Recall Linnaeus's mistakes: classifying bats as primates and hyenas as canines. This happens across a wide range of animals, from mammals to birds to reptiles. The jaguar (*Panthera onca*) and the leopard (*Panthera pardus*), for example, share many morphological traits in common, including large body size and a "rosette" coat pattern. They're often mistaken for the same species, even though they do have physical differences (the jaguar's tail tends to be shorter than the leopard's, and jaguars are larger bodied and more muscular than leopards). Even though the Atlantic Ocean separates them, jaguars and leopards, like all members of the *Panthera* genus, can interbreed, sometimes producing fertile offspring. The Eastern meadowlark (*Sturnella magna*) and the Western meadowlark (*Sturnella neglecta*) share many morphological features, including general body shape, beak shape and length, feather coloration and distinct markings, but their songs (which are all-important for attracting the opposite sex) differ greatly, so even if they live in the same geographic location, they rarely interbreed. Conversely, a species might have morphological differences depending upon gender (sexual dimorphism, where males and females have different body sizes), or dimorphism within one gender, which is exhibited by a wide range of species, including orangutans and ants. With orangutans (*Pongo pygmaeus*, *Pongo abelii*, and *Pongo tapanuliensis*), males can be either fully "flanged," with distinct, protruding, fatty cheek pads, or these features can be absent. (Females tend to prefer to mate with the fully flanged males). With ants (such as the species *Pheidole barbata*), two sisters can vary threefold or more in size, depending upon the role that they have in their colony. Most *Pheidole* species are dimorphic, comprising "major" workers with oversized heads and mandibles and "minor" workers without exceptional physical characteristics. So, to the casual observer, morphologically similar species can look like they belong to the same species, while those with dramatic physical differences can look like they're from separate ones. That's why morphological categorizations can get us close, but they're often just approximations; most scientists adhere to biological and phylogenetic species concepts.

Who Let the Dogs Out?

While there's still debate regarding what constitutes a species and a subspecies in any genus, much less the *Canis* genus, we are going with the modern classification that comprises eight extant (not extinct) species, including *Canis adustus*, the side-striped jackal; *Canis aureus*, the golden jackal; *Canis mesomelas*, the black-backed jackal, *Canis latrans*, the coyote, *Canis simensis*, the Ethiopian wolf, *Canis dingo*, the Dingo, *Canis lupus*, the gray wolf (whose ancestors were the progenitors of modern domesticated dogs), and *Canis familiaris*, the domesticated dog. Of these eight species, six of them are closely phylogenetically related and interfertile; the other two species, *Canis mesomelas* and *Canis adustus*, are not as closely related, and probably could not interbreed with the other species in the genus (Lord *et al.*, 2013). While dogs have shared their lives with humans for millennia, the tenth iteration of the *Systema Naturae* was the first time that they were scientifically classified and recognized as a *species*, distinct from their free-living cousins.

Let's revisit that debate regarding species versus subspecies. Domesticated dogs descended from the ancestors of gray wolves, tens of thousands of years ago. Depending upon how you choose to classify domesticated dogs, they are listed as *Canis lupus familiaris* (a subspecies of the gray wolf) almost as often as they are listed as *Canis familiaris*. Now why does this matter? Well, given their genetic similarities, it could be argued that domesticated dogs are, in fact, "just" a subspecies of gray wolves, but given their differences in development, behavior, nutritional needs, and unique genetic mutations distinct from their wolf cousins, the argument that they're different species is even stronger. We follow the latter viewpoint and will present a strong body of evidence supporting the fact that domesticated dogs are a different species than gray wolves.

A century after the *Systema Naturae*'s debut, Charles Darwin published *On the Origin of Species by Means of Natural Selection, or, The Preservation of Favoured Races in the Struggle for Life* (Darwin, 1859). While Linnaeus struggled financially for years, Darwin was born into a financially secure family (Fig. I.3). This fiscal security enabled his 5-year voyage on the ship HMS *Beagle*, starting in 1831. It also allowed him to observe, research, and write some of science's greatest works, many of which discussed humanity's best friend. Darwin, who had more than a dozen dogs of his own over his lifetime, was fascinated by the diversity within the species. He wrote (Darwin, 1859):

> ... domestic [dogs] throughout the world ... descended from several wild species, [and] it cannot be doubted that there has been an immense amount of inherited variation; for who will believe that animals resembling the Italian Greyhound, the blood hound, the bull-dog, pug-dog, or Blenheim spaniel—so unlike all wild *Canidae*—ever existed in a state of nature? ... all of our dogs have been produced by the crossing of a few aboriginal species ... [and] the possibility of making distinct races by crossing has been greatly exaggerated ...

Darwin wasn't only interested in the differences in dogs' skulls, body sizes, and coats; he was also aware of their emotional lives and the unique relationship humans had with them. In his 1871 work, *The Descent of Man, and Selection in Relation to Sex*, Darwin (1871) wrote:

> In the agony of death a dog has been known to caress his master, and everyone has heard of the dog suffering under vivisection who licked the hand of the operator; this man, unless the operation was fully justified by an increase of our knowledge, or unless he had a heart of stone, must have felt remorse to the last hour of his life.

Dogs weren't just another animal to be classified; Darwin, and many of his contemporaries, already recognized the unique relationship that man had with the first domesticated species. And while it's a running joke that many of us have questioned whether we wanted to invest our time in a long-term relationship or just spend our time with our pets, Darwin, that prolific creator of journal entries and lists, actually enumerated this quandary in his notebooks. His thoroughly thought out (and often belly-laugh-inducing) list of reasons for marrying or not marrying (and perhaps spending more time with dogs) is shown in Table I.1. Let's be glad that Darwin's union resulted not in "banishment and degradation", but in scientific and literary fertility.

Part of Darwin's prolific success happened the year following the *Descent of Man*'s debut, with his third major work, *The Expression of the Emotions in Man and Animals* (Darwin, 1872). This work was filled with illustrations of dogs demonstrating various emotional states as well as the social dynamic between humans and their dogs. Dogs aren't just fascinating scientific models of behavior and morphology, though—their genetics have proven to be equally intriguing. In their landmark longitudinal dog study, John Paul Scott and John L. Fuller called the dog a "veritable genetic goldmine," writing, "besides the enormous differences *between* breeds, all sorts of *individual* differences appear at the stroke of the geneticist's pickaxe, in this case the mating of two closely related animals. Anyone who wishes to understand a human behavior or hereditary disease can usually find the corresponding condition in dogs with very little effort" (emphasis ours) (Scott and Fuller, 1965).

Linnaeus categorized Canidae first, but for Darwin, they were a pivotal point of departure for the study of evolution. He wrote extensively about inherited traits after parents were artificially selected, and he naturally worried about his own children's vigor, as he and his wife, Emma Goodwood, were first cousins. Despite this close

Fig. I.3. Charles Darwin. This image is in the public domain and was provided by Wikipedia Commons. This image was taken from Flickr's The Commons. Available at: https://commons.wikimedia.org/wiki/File:The_descent_of_man,_and_selection_in_relation_to_sex_(1874)_(20679361230).jpg.

relation (which was commonplace for the era), they married and had ten children, seven of which survived to adulthood. Today, consanguineous (closely related) marriages aren't as uncommon as one would think; in the US, at least 0.2% of marriages are between those who are second cousins or closer (Chalabi, 2015). To examine this, let's look at something called the coefficient of relatedness (designated as *r*) between two partners and the coefficient of inbreeding (designated as *F*), a measure of consanguinity for an individual. The *r* of two individuals is the rate (designated with a fraction) of alleles (bits of DNA coding at a particular position, called a locus, on a chromosome) that they will share at all loci. The more closely related two individuals are, the higher that likelihood will be, and thus, the higher the *r* will be. When two individuals are not closely related to one another, their *r* approaches 0 (which is a good thing, genetically speaking, as it decreases the likelihood of carrying on a potentially harmful condition). For example, the *r* of identical twins is 100%, designated as "1." For parents and their offspring and for full siblings, *r* is 0.5. For grandparents

Table I.1. Charles Darwin's list of reasons to marry/not marry.

Reasons to marry	Reasons to not marry
Children (if it please God)	Freedom to go where one liked
Constant companionship (and friend in old age) who will feel interested in one	Choice of society and little of it
Object to be loved and played with	Conversation with clever men at clubs
Home and someone to take care of house	Not forced to visit relatives and bend to every trifle
Charms of music and female chit-chat	[not having] the expense and anxiety of children
These things are good for one's health, but a terrible loss of time	Quarreling
Better than a dog, anyhow	Loss of time
	Cannot read in the evenings
	Fatness and idleness
	Anxiety and responsibility
	Less money for books
	If many children forced to gain one's bread—but then is very bad for one's health to work too much
	Perhaps my wife won't like London; then the sentence is banishment and degradation into indolent idle fool

Compiled from *The Correspondence of Charles Darwin, Volume 2: 1837–1843* (Darwin, 1987).

and their grandchildren, r is 0.25, for first cousins, it's 0.125, and for second cousins, it's 0.0625. You can then calculate the inbreeding coefficient based upon the number of connecting links between two parents with a common ancestor. Approximately 0.2% of the population is in a marriage with a second cousins or closer. The F for second cousins is 0.015625, so we have at least 250,000 people in relationships with a relatedness coefficient of 0.0625 (r = 0.0625), which means their offspring would have an inbreeding coefficient of more than 0.015 (F = 0.015625). Because 25 states ban consanguineous marriages and seven have restrictions against it, that number is likely even higher due to underreporting. Darwin was always concerned about his children's health. When his daughter, Anne Elizabeth (Annie) caught scarlet fever, Darwin doted on her at her bedside. And when she died at age 10 years, he was anguished; it's said that he never recovered from the loss. This loss was compounded by Darwin's concern that her death may have been due to his consanguinity with his wife. Two of his other daughters caught scarlet fever at the same time, but they survived; it was believed that Annie also had tuberculosis (TB). There *is* a genetic component for the susceptibility of developing TB; those who have one or more high-risk copies of the genetic region *NRAMP1* are more likely to develop the disease than those who don't have high-risk copies (Stagas *et al.*, 2011). It's hard to say whether the Darwins both had this gene, and thus passed it on to Annie, although it would be likelier for the child of two closely related individuals to have these copies than it would be with the child of a non-consanguineous union. Genetic concerns likely kept Darwin up at night; sadly, for centuries, unscrupulous dog breeders have not shared this same concern, but recent research has begun to look into the harmful side effects of all of this inbreeding.

The Creation of Breeds

In 1878, 6 years after Darwin published his third major work, nine dogs were being ushered in as the first to be recognized by the American Kennel Club (AKC). This was an important scientific contribution as it would help us begin to classify the many iterations of dogs by their breed, a category defined by the AKC as "a domestic race of dogs (selected and maintained by man) with a common gene pool and characterized appearance and function." (AKC, 2023). That nonet of breeds included the Cocker, Clumber, Irish Water, and Sussex Spaniels, the English, Irish, and Gordon Setters, the Chesapeake Bay Retriever, and the Pointer. The AKC, which provided a registry of recognized "purebred" dogs in the US, joined the Kennel Club of the UK (now the Royal Kennel Club), founded in 1873, and later, the Canadian Kennel Club, founded in 1888, the United Kennel Club (UKC), founded in 1898, and the Fédération Cynologique Internationale (FCI), founded in 1911. Kennel clubs were a clear point of departure for many breeds, which would soon change in response to newly designated breed standards. The AKC would later categorize dogs into eight groups: Hound, Terrier, Working, Herding, Sporting, Non-sporting, Toy, and Miscellaneous, describing physical standards for each breed within each group. The Kennel Club of the UK recognizes seven breed groups: Hound, Working, Terrier, Gundog, Pastoral, Utility, and Toy. The Canadian Kennel Club recognizes seven groups, including Sporting Dogs, Hounds, Working Dogs, Terriers, Toy Dogs, Non-sporting Dogs, and Herding Dogs. The UKC recognizes eight breed groups: Guardian Dogs, Scenthounds, Sighthounds and Pariah Dogs, Gun Dogs, Northern Breeds, Herding Dogs, Terriers, and Companion Dogs. The FCI, however, recognizes the most distinct breed groups, with ten: Sheepdogs and Cattledogs (excepting Swiss Cattledogs); Pinscher and Schnauzer-Molossoid and Swiss Mountain and Cattledogs; Terriers; Dachshunds; Spitz and Primitive Types; Scenthounds and related breeds; Pointing Dogs; Retrievers, Flushing Dogs, and Water Dogs; Companion and Toy Dogs; and Sighthounds. While there's some overlap among the groups for each kennel club, there are large differences, as well. And if it feels a little bit disjointed, that's because it is, and that's not even taking into consideration the Australian National Kennel Council, the Kennel Club of India, or the Kennel Union of Southern Africa.

Not all of the organizations recognize one another's registries, either. While the FCI officially represents more than 80 countries and has reciprocal agreements with its affiliate members, national kennel clubs don't always see eye to eye. The AKC is the largest purebred dog registry organization in the world, and does not recognize UKC-registered dogs, while the UKC, which is the second largest organization, does recognize AKC-registered dogs. Regardless of which kennel club or breed group, to meet the standards that were set by each respective organization, breeders looked to those dogs with the desired physical traits, and bred dogs who also had those physical traits (and who were often closely related to them), without an awareness of the genetics (genotypes) behind these physical traits (phenotypes).

Categorizing dogs based upon these standards might make one think that they're all standardized, but that's not the case. With so many countries having their own kennel clubs in addition to the aforementioned organizations, discerning differences between breeds, whether it be genotypic, phenotypic, or behavioral, can be problematic. And while dog breeds are typically thought to be uniform groups, there isn't always genetic uniformity within one breed. A dog of a specific breed that's been genetically isolated in the US will have a different genetic lineage than one of the same breed that's been bred for many generations in Europe. Björnerfeldt *et al.* (2008) examined patterns of genetic diversity in 164 Poodles, comparing them with 133 dogs representing eight other breeds, including the Siberian Husky, Miniature Schnauzer, Giant Schnauzer, German Shepherd, Smooth and Wire Fox Terriers, Labrador Retriever, and Bull Terrier. The study found that assortative mating imposed by breed standards and preferences created five novel subgroups within Poodles—subgroups that aren't recognized by any kennel clubs, but which stand out because of their color and size combinations. These subgroups provide support for breed fragmentation, using the same process that creates new breeds.

The confusion isn't just within breed subgroups, either. To further complicate matters, different kennel clubs may have defined breeds dissimilarly in the past, or may do so presently, as they do with breed group classifications, and the breed may have a different genetic history as a result of this, as well, even though it's classified as the "same" breed across multiple kennel clubs (Turcsán *et al.*, 2011). Then there are the designations outside of recognized kennel clubs: non-kennel club groups and organizations that recognize breeds, even within the US, including the American Rare Breed Association (ARBA).

So how do we get around all this confusion? If we're interested in examining behavioral differences, it may be better to look at the function of working groups that were selected on the basis of behavioral traits, such as "herding" or "guarding" or "retrieving," rather than looking at "breed" (Turcsán *et al.*, 2011). This is because behavior, and not other characteristics, was the trait under selection with working breeds. Some "working" breeds were historically under selection for behavioral traits that were vital for jobs that they performed for us; the ability to herd animals was more important than the slope of the back or the length of the muzzle. More recently, though, this selection may have been relaxed as we breed more dogs to be companions than we do for having specific jobs. Of course, some are still bred for work, or for show performance, but fewer than before. But with continued urbanization and the decreased need for working dogs, this has declined, compared to the past (Turcsán *et al.*, 2011). Even breeds in groups such as those recognized by the AKC have relatively poor historical records, in terms of longitudinal genetic accounts. According to Brynn White, archivist for the AKC Library and Archives, the AKC has been keeping detailed pedigree records since the 1800s. In the AKC's first constitution, dated October 22, 1884, Article II states, in part (AKC, 1884a):

> The object of this Association shall be to secure uniformity in rules governing Bench Shows and Field Trials, the revision of standards, the decision of such appeals as may be taken from the decisions of the managers of Bench Shows and Field Trials held by members of this Association ... [and] the advancement of fellowship and a higher standard of action among breeders, Exhibitors and Sportsmen ...

The AKC's corresponding dog show rules from that same date still allowed dogs with unknown parentage; it states, in part (AKC, 1884b):

> 2. If the names of sire and dam are not known, it may be entered "Pedigree unknown."
> 3. If a dog shall be entered without being identified, as directed in Rule 2, it shall be disqualified from competition.

From their earliest days, kennel clubs like the AKC based their recognition on physical traits and morphology (Turcsán *et al.*, 2011), which, as aforementioned, we've found from a wide range of species can be a highly problematic way to classify animals. We currently can't test the genetics of a "mutt" and tell you what breeds compose it; we can only find approximations, based upon prior reporting of known breeds. We can't go "backwards" and determine breed, but there are genetic breed differences in "purebreds." And there can be large variation in behavior, even within breeds. So where does this leave us? Well, in a word: hazy. While the scientific study of canine behavior and genetics is a relatively new (and hazy) discipline, dogs have intrigued us for millennia. Dogs are a part of the human cultural experience and evidence of our longstanding relationship with them is found in their physical variation, fossils, and ancient artwork.

Tell Me What You See

Denis Peyrony gingerly ran his hand along the wall, his fingers tracing the ancient depictions of animals, both foreign and familiar. One of them was a quadrupedal animal with a long snout, erect ears, and a long tail. Peyrony gazed at the images, not knowing that they were some of the earliest surviving artistic depictions of canids. These 17,000-year-old cave paintings of wolves in Font-de-Gaume, France, and 8000-year-old rock carvings of dogs from the Arabian Peninsula revealed that dogs have been an important part of our lives for thousands of years. The artwork of Font-de-Gaume, located in Les Eyzies in southwestern France, was first discovered by Peyrony, a modest, mustached schoolmaster, in September 1901, and includes a fantastic bestiary of bison, mammoths, reindeer, horses, a now-extinct woolly rhinoceros, and a lone wolf (Fig. I.4). Peyrony was also the director of the excavation of the 1909 unearthing of a skeleton that would be dubbed La Ferrassie 1 (CLF1). The skeleton, estimated to be 50,000–70,000 years old, was the most complete Neanderthal skeleton discovered. Examining its placement in the earth, Peyrony felt that it had been buried in a grave, indicating that Neanderthals shared cultural traits in common with *Homo sapiens*. A total of eight skeletons have been discovered at La Ferrassie Cave, France, including children.

The rock carvings discovered in the Arabian Peninsula depict a far different scene: a hunter with bow and arrows accompanied by 13 dogs, some of them on leashes—and all of them likely domesticated. The carvings, found in rock faces in Jubbah and Shuwaymis in modern-day Saudi Arabia, closely resemble the Canaan breed, also known as the Bedouin Sheepdog or Palestinian Pariah dog, an ancient Middle Eastern dog. Like the Canaan, the dogs depicted in the carving have a medium build, relatively short muzzle, and upwardly curled tails. Similarly aged cave paintings in Tennessee's Cumberland Plateau appear to depict numerous canid species, including foxes, jackals, and wolves. The Sulawesi artwork also depicts charcoal drawings of dogs, but they are likely far more recent, as dogs are thought to have been introduced to the island only a few thousand years ago. Magura Cave, located in Northwestern Bulgaria, also has paintings that are up to 8000 years old; the artists used bat excrement to create various images, including geometric shapes, humans, and a wide range of animals, including birds, bovines, and dogs.

The Walk

Footfalls echoed through the cave as a small girl walked slowly along, her feet pressing gently into the soft clay. Beside her walked a long, lean canine,

Fig. I.4. Polychrome painting of wolf in the Font-de-Gaume cavern. This image is in the public domain and was provided by Wikipedia Commons. Available at: https://commons.wikimedia.org/wiki/File:Men_of_the_old_stone_age_(1915)_Wolf.png.

its paws also pressing into the malleable earth. The girl paused, glancing down at the tracks that had been left before: large, hand-like prints, with claws that extended into the ground. Nearby, she saw circular depressions where bears had once lain. As she exited the cave, she gave no more thought to any of the prints, and neither did anyone else, for almost 26,000 years. Not long after she left, a landslide closed off the entrance, leaving the prints preserved until someone found another access point to the cavity.

Etched in the substrate of this almost-forgotten cave is this intriguing evidence of the human–canine relationship. The famous footprints of Chauvet-Pont d'Arc Cave in the Ardeche region of Southern France accompany burial sites including both humans and canids. The 26,000-year-old Chauvet prints are thought to be some of the oldest physical evidence of the partnership between humans and canines and show a child of approximately 10 years of age walking alongside a canid. But this wasn't any canid. While the length of the stride has made some state that the prints were made by a wolf, the track line left behind by this ancient animal shows paws with a shortened middle digit, which is a trait shared in common with domesticated dogs, but not with any wild canid species, past or present.

In addition to evidence carved in stone and imprinted in the substrate, we have also found fossil remains of ancient man and dog where they co-occurred in both the Old and New Worlds. In 1848, Captain Edmund Flint of the British Royal Navy discovered a hominid skull in the Gorham's Cave Complex of Gibraltar. While the find was unassuming at the time, this skull from a middle-aged hominid was one of only two specimens of its kind that would be discovered before Charles Darwin's *On the Origin of Species* was published. Hidden in a cave not far from where her skull was discovered were the far more recent remains of her descendants: a man, his wife, and the still unfound remains of an ancient canine. The three died not far apart, tens of thousands of years ago. The husband had succumbed to cancer; the dog to injuries sustained while protecting her remaining family; and the wife from "old age," as it was measured in prehistoric terms.

Canine fossils reveal the varied dynamic between humans and dogs. The 14,000-year-old remains of a young dog were found in the grave site of what is present-day Bonn-Oberkassel, Germany. The Bonn-Oberkassel site is the oldest known evidence of burying a canine, and accompanying the dog were two humans, indicating a close relationship between the three. The remains of three dogs were found in what is modern-day west-central Illinois and show a familial relationship, as the dogs have no signs of being killed by people and were each buried in their own graves. At 10,000 years old, they are the oldest known domesticated canines found in the Americas to date (Bower, 2018). The dogs were originally found at two different sites in 1960 (the Stilwell II site) and the 1970s (the Koster site), around 20 miles apart. Researchers recently radiocarbon dated the remains, finding that they were even older than the remains of dogs at Hinds Cave in southwestern Texas. Those 9300-year-old bone fragments, discovered in 2009, hint at a far less benevolent relationship: they were discovered inside human feces, indicating that these dogs had been killed for human consumption. While appalling to the dog lover, it's not unheard of; historical accounts include stories of people eating their dogs during times of famine, and dogs are still consumed in numerous countries, including China, Korea, Indonesia, Vietnam, the Philippines, Taiwan, Polynesia, Cameroon, Liberia, Ghana, the Arctic, and in some parts of Switzerland. Thus, while much of the oldest evidence of the human–canine relationship points to dogs being partners and beloved companions, there's also evidence that dogs have shared a wide range of roles with humans, including sometime food source. We will explore the complexities of this relationship, from its earliest roots to its current varied forms throughout the world, examining the impact that breed has on dog behavior.

References

AKC (1884a) Constitution and by-laws of the American Kennel Club. Available at: https://images.akc.org/pdf/about/depts/archive/Constitution1884.pdf (accessed 25 October 2023).

AKC (1884b) Dog show rules and regulations. Available at: https://images.akc.org/pdf/about/depts/archive/show_rules/1884.pdf (accessed 25 October 2023).

AKC (2023) Glossary. Available at: www.akc.org/about/glossary (accessed 25 October 2023).

Björnerfeldt, S., Hailer, F., Nord, M. and Vilà, C. (2008) Assortative mating and fragmentation within dog breeds. *BMC Evolutionary Biology* 8, 28.

Bower, B. (2018) Dogs lived and died with humans 10,000 years ago in the Americas. *Science News* 193(8), 22.

Buffon, C. de (1830) *The Natural History of Quadrupeds*. Thomas Nelson and Peter Brown, Edinburgh, UK, pp. 191–192.

Chalabi, M. (2015) How many Americans are married to their cousins? *FiveThirtyEight*. Available at: https://fivethirtyeight.com/features/how-many-americans-are-married-to-their-cousins (accessed 25 October 2023).

Darwin, C. (1859) *On the Origin of Species by Means of Natural Selection, or, The Preservation of Favoured Races in the Struggle for Life*. J. Murray, London, UK.

Darwin, C. (1871) *The Descent of Man, and Selection in Relation to Sex*. J. Murray, London.

Darwin, C. (1872) *The Expression of the Emotions in Man and Animals*. J. Murray, London.

Darwin, C. (1987) *The Correspondence of Charles Darwin, Volume 2: 1837–1843*. Edited by Burkhardt, F. and Smith, S. Illustrated Edition. Cambridge University Press, Cambridge.

Eimer, G.H.T. (1890) *Organic Evolution as the Result of the Inheritance of Acquired Characters According to the Laws of Organic Growth*. Macmillan, London.

Frankham, R., Ballou, J.D., Dudash, M.R., Eldridge, M.D.B., Fenster, C.B. *et al.* (2012) Implications of different species concepts for conserving biodiversity. *Biological Conservation* 153, 25–31.

Kipling, R. (1942) *The Elephant's Child and other Just So Stories*. The Junior Literary Guild and Garden City Publishing, Garden City, NY.

Kipling, R. and Kipling, J.L. (1906) *The Second Jungle Book*. Macmillan, London.

Kipling, R., Kemp, S. and Lewis, L. (eds). (1996) *Writings on Writing*. Cambridge University Press, Cambridge.

Lamarck, J.-B. (1809) *Philosophie zoologique, ou exposition des considérations relatives à l'histoire naturelle des animaux*. Musée d'Histoire Naturelle, Paris.

Linnaeus, C. (1758) *Systema Naturæ per Regna Tria Naturæ, Secundum Classes, Ordines, Genera, Species, Cum Characteribus, Differentiis, Synonymis, Locis*, 10th edn, Vol. 1. Holmiæ Laurentius Salvius, Stockholm.

Lord, K., Feinstein, M., Smith, B. and Coppinger, R. (2013) Variation in reproductive traits of members of the genus Canis with special attention to the domestic dog (*Canis familiaris*). *Behavioural Processes* 92, 131–142.

MacDuffie, A. (2014) *The Jungle Books*: Rudyard Kipling's Lamarckian fantasy. *PMLA/Publications of the Modern Language Association of America* 129(1), 18–34.

Scott, J.P. and Fuller, J.L. (1965) *Genetics and the Social Behavior of the Dog*. The University of Chicago Press, Chicago, IL, p. 4.

Stagas, M.K., Papaetis, G.S., Orphanidou, D., Kostopoulos, C., Syriou, S. *et al.* (2011) Polymorphisms of the NRAMP1 gene: distribution and susceptibility to the development of pulmonary tuberculosis in the Greek population. *Medical Science Monitor* 17(1), 1–6.

Turcsán, B., Kubinyi, E. and Miklósi, A. (2011) Trainability and boldness traits differ between dog breed clusters based on conventional breed categories and genetic relatedness. *Applied Animal Behaviour Science* 132(1–2), 61–70.

1 What Is a Dog?

Man selects only for his own good: Nature only for that of the being which she tends.

On the Origin of Species (Charles Darwin, 1859)

Abstract

Chapter 1 reviews how we define a species and details the evolutionary history of canids. *Canis familiaris* is a member of the genus *Canis*, which is the most abundant of the terrestrial carnivores. As Darwin and so many scientists since have pointed out, dogs vary widely in their physical appearance, proclivities, and temperaments. A dog is genetically, physically, and behaviorally different than their wild cousins, the wolves and the coyotes, but they can still produce viable offspring with them. This synopsis covers the archaeological evidence for the origins of the modern domesticated dog and its relationship with early humans.

The Peacock, the Pachyderm, and the Pomeranian

Ask 100 different amateur sketch artists to draw a peacock or an elephant and you'll likely get roughly the same result: a bird with an ornate, fanned out tail, and a large, gray quadruped with sturdy legs and a very long nose. For both the peacock and the pachyderm, the general blueprint remains the same. But if you were to ask those same 100 artists to each draw a dog, you'd likely get 100 slightly different results. Yes, they would have the same basic "formula" of a four-legged mammal, but they would come in almost every possible variety: very small and very large, brachycephalic and long-snouted, very furry and very short-haired, with droopy or erect ears, long-tailed or truncated, light or dark, speckled or solid, and every variation in between. And every one of them would very much be a "dog"—the same species, *Canis familiaris*.

While the peacock comprises three different species (*Pavo cristatus*, the Indian peafowl, *Pavo muticus*, the green or Java peafowl, and *Afropavo congensis*, the Congo peafowl) that have slight differences in their plumage and morphology, the males of the species all share that distinguishing tail plumage. Elephants, too, comprise three different species: *Loxodonta africana*, the African bush elephant, *Elephus maximus*, the Asian elephant, and the lesser known *Loxodonta cylotis*, the forest elephant. *Loxodonta africana* and *Loxodonta cylotis* shared so many physical similarities that they weren't recognized as different species until relatively recently, but genetic analyses revealed that the two species diverged between 2 and 7 mya. Asian and African elephants last shared a common ancestor 6 mya (Krause *et al.*, 2006)—the same approximate time frame that separates humans and chimpanzees from their last common ancestor. Chimpanzees and humans look very different from one another, but Asian and African elephants are still morphologically very similar. The clearest physical difference between them is the size of their ears; both species of African elephants have large, fanning ears that reach over their necks, while Asian elephants have much smaller ears in comparison. Even with their vast geographic separation, there is less interspecies variation between the three elephant species than there is intraspecies variation within the domesticated dog.

What was the driver behind their differences and similarities? The peacock, the elephant, and the domesticated dog are exemplars of sexual, natural, and artificial selection, respectively. These are the scientifically testable mechanisms of evolution that

DOI: 10.1079/9781800624566.0001

Charles Darwin presented in his 1859 page-turner, *On the Origin of Species by Means of Natural Selection, or, The Preservation of Favoured Races in the Struggle for Life* (Darwin, 1859). Evolution is the process by which different life forms developed and diversified, yielding more complex and well-adapted organisms from earlier life forms. Darwin wasn't the first to posit the theory, and even shared the stage with Alfred Russell Wallace, who "scooped" his great idea, but had the grace to allow the elder scientist to take the lead. While the theory gained traction among scientists, the *mechanism* of natural selection was still poorly understood. Even Darwin thought that other mechanisms, such as Lamarck's inheritance of acquired characteristics, might also be involved in evolutionary change. Imagine, then, how mysterious it was for people who were breeding dogs, to see changes in the offspring of certain individuals, such as shorter and shorter legs or increasingly shortened faces, but not understand *why* these changes were happening. Then, as now, people who didn't understand evolution would ask: if dogs came from wolves, then why do we still have wolves? And if humans came from monkeys, why do we still have monkeys? We'll revisit that momentarily.

For now, let's examine the *mechanisms* of evolution more closely. Natural selection is the overarching umbrella of selection, with sexual and artificial selection nested beneath it. For natural selection to occur, it requires both necessary and sufficient conditions. A sufficient condition is a circumstance or set of circumstances that must be present for an event to be able to occur. A necessary condition must also be present, but on its own it cannot provide the sufficient cause for an event to occur. The four conditions that are required for natural selection to occur are reproduction, heredity (is the trait heritable?), variation in individual characteristics within members of the population, and variation in the fitness of members in the population conferred by the trait. Let's define what each of these conditions mean. Reproduction refers to a parent producing offspring who themselves reproduce. The key is continuing to have one's genes in the gene pool. Heredity refers to a parent transmitting specific traits to their offspring, who in turn transmit them to their own offspring. Variation in individual characteristics within members of the population refers to the biodiversity of a group: are there many different "types," or genetic variants, of this animal, to increase the likelihood of surviving in different environmental conditions? And variation in individual fitness refers to the average number of offspring with a particular genotype that reproduce, relative to the average number of offspring with a different genotype that reproduce.

What does this look like when these conditions happen, though? For the gray peppered moth (*Biston betularia*), this looks like a dramatic change in wing color in response to environmental changes. This insect exhibited a case of industrialized melanism (changes in their wing color, or melanin, brought on by rapid changes due to increased industry) that continues to be the benchmark for natural selection. At the inception of the Industrial Revolution, it was relatively uncommon to see a darker-colored gray peppered moth in the UK. The phenotypic standard of that time was a lighter-gray variety. But as the decades wore on, the environmental effects of the revolution became clear: soot began to stain every surface, including tree trunks and the lichens that grew on them. Around 1760, lighter-colored moths had a selective advantage over their darker counterparts, as the trees and lichens were also lighter-colored and they blended in well. But in a very short time, this selective advantage disappeared: the contrast of light moth against sooty lichen or tree bark made them easy targets for predators. Lighter moths began to die sooner and reproduce less, while their darker conspecifics suddenly had the winning hand and experienced exponential reproductive success; correspondingly, lighter moths became very rare. Over the course of 130 years, the rate of darker-winged moths reached 98% in Manchester (van't Hof *et al.*, 2011); but just as quickly, the pendulum swung again. With the advent of more environmentally friendly technologies in the 1960s, soot accumulations have since decreased, and the melanism of the moth has followed suit; lighter-colored moths are common once again (Fig. 1.1). The phenotypic changes in the moths were clear, but this selection pressure had a surprising genetic component, as well: the gene involved with these changes, called "*cortex*," affects cell division and development. A mutation in that gene caused the growth of scales on the moths, with the darker-winged moths having a mutation that caused darker scales. The peppered moth isn't the only species with unexpected components to their genetic makeup: genetic analyses of canines have revealed surprising results about their physical appearances, as well. We'll examine this more later.

Fig. 1.1. White-bodied peppered moth (*Biston betularia*), 1931. Photography by Ben Sale. This file is licensed under the Creative Commons Attribution 2.0 Generic license and provided by Wikipedia Commons. Available at: https://commons.wikimedia.org/wiki/File:(1931)_Peppered_Moth_(Biston_betularia)_(5857613284).jpg.

Over succeeding generations, strong selective pressures, such as pollution, can result in observable changes. Individuals' genotypes (the "genotype" is the genetic makeup of an organism carried by the DNA that determines or contributes to the organism's characteristics) interact with the environment to produce morphological changes. The peppered moth went from mostly white and speckled to mostly black over just a few generations to blend into its soot-covered surroundings. With strong selective pressures, these expressed phenotypic (physical expression of traits or characteristics) changes diverged far from the founding population.

Selective pressure can act in several different ways to influence a population. We'll delve more deeply into the genetics of these selective processes in Chapter 2. In the case of the moth, though, selection was "directional", such that *one* extreme phenotype (dark color) was favored over the other phenotypes, resulting in an increase in the frequency of the extreme phenotype. To relate this to canids, let's say you started with one population of wolves, with a mix of coat colors (black, white, and gray). If there was a shift in climate and the snow no longer fell at the higher altitudes, the white-barked trees began to die out, and the seeds of the darker-barked trees began to take hold, this could result in a disadvantage for the white-coated wolves. Eventually, the wolves would all have the black coat phenotype.

Disruptive selection, also sometimes referred to as "diversifying selection," occurs when the extreme values for a trait are favored over moderate ones. When this occurs, there's an increase in the variance of this trait. We can again use the example of a population of wolves with varying coat colors; let's assume that the population split off into two different habitats, one at a lower elevation, and one higher in the mountains. The lower elevation had darker-barked trees that thrived in warmer climates, and dense foliage, providing cover and camouflage for sneaking up on their prey. This provided a selective advantage for wolves with black coats. The higher elevation had white-barked trees that thrived in colder climates, white rocks, and larger, more frequent snowfall. This provided a selective advantage for wolves with white coats. Thus, the wolves with black fur and the wolves with white fur would each have a selective advantage in their given environments, but a wolf born with an intermediate coat color—like gray—would not have an advantage at either altitude. Rather than being able to stalk its prey, it would stand out, which would likely result in lower predation success rates, and thus lower rates of reproductive success. The intermediate phenotype—the gray coat color—would be selected against and slowly disappear from each population.

Darwin's finches and mockingbirds, which we'll discuss shortly, exhibited disruptive variation in the size and shapes of their beaks, with larger and smaller beaks being selected for over medium-sized ones, which weren't suitable for breaking big seeds or retrieving smaller ones. So to recap: disruptive selection results in individuals with phenotypes at both ends of the extreme (black coats and white coats, or large beaks and small beaks) while individuals with intermediate phenotypes (gray coats or medium-sized beaks) would have a selective disadvantage.

And "stabilizing" selection occurs when a population becomes stable with an intermediate trait value. With our wolves, if both the black and white coats stood out too much in their environments, then an intermediate phenotype, such as gray coats, would have a selective advantage. The Siberian Husky, which was originally bred to pull sleds through the snow, is also an excellent example of stabilizing selection. They need to develop enough leg muscle to effectively pull sleds, but also need to be lean and light enough to stay atop the snow. Thus, Huskies with medium-sized muscles would be selected for, while those that had too much leg muscle would be too heavy for the snow, and those who didn't develop enough leg muscle would be unable to pull the sled.

Now what about the specific selective pressure that shaped domesticated dogs? Artificial selection occurs when plants and animals are bred to produce desirable traits. This process of genetic modification leads to more rapid evolutionary responses than would be seen under natural conditions—sudden changes will occur in a relatively short amount of time. If a gene pool is a bowl of white paint, then natural selection would be like adding a few drops of another color here and there, slowly changing the color of the paint until it noticeably diverges from the shade of the founding population (paint).

We've been skirting around the topic of DNA and its role in evolution, so let's explore it a little bit further. Most people are familiar with this three-letter designation, but what is it, really? DNA (deoxyribonucleic acid) is the code that provides the directions for an organism: it is the genetic blueprint for life. All cellular life, including archaea (single-celled microorganisms, distinct from bacteria and eukaryotes), bacteria, and eukaryotes (which includes plants, animals, fungi, and animals and has complex cells or a complex single-cell structure), has DNA. Visibly, DNA is a threadlike double-helix of nucleotides (organic molecules) that carries the information about an organism's form and function (Fig. 1.2). Pieces of information, called genes, reside on different areas of the DNA. Each nucleotide has a nitrogenous base, a phosphate group, and a five-carbon sugar (deoxyribose). DNA

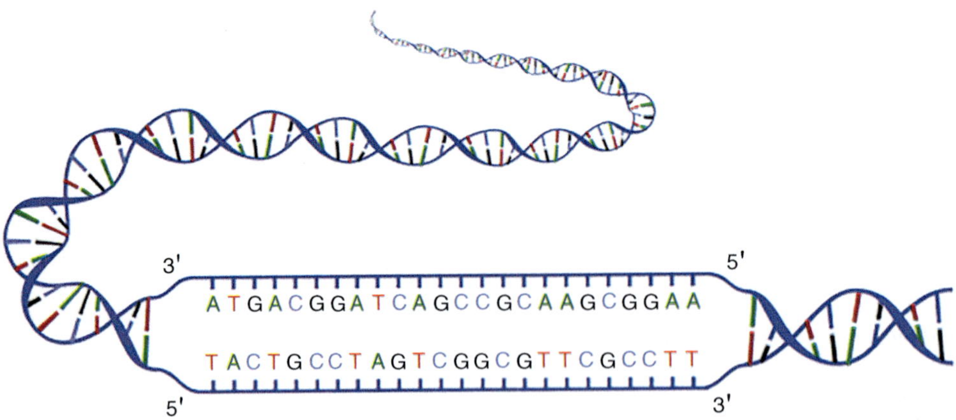

Fig. 1.2. DNA strand. Public domain image. This work is in the public domain in the US because it is a work prepared by an officer or employee of the US Government as part of that person's official duties under the terms of Title 17, Chapter 1, Section 105 of the US Code. Provided by Wikipedia Commons under CC-PD-Mark. Available at: https://commons.wikimedia.org/wiki/File:DNA_strands.png.

contains four base nucleotides, or compounds, that contain either a pyrimidine (a colorless crystalline compound with the properties of a base) or a purine (a colorless crystalline compound with the properties of a base, forming uric acid on oxidation) base. (The properties of a base include, but are not limited to, having a pH value range of 8–14, being conductive with electricity, having a bitter taste and a "slippery" feel, and changing the color of litmus from red to blue.) Deoxyribonucleotides are nucleotides where the sugar is deoxyribose. DNA's pyrimidine bases are thymine and cytosine (represented as T and C), while its purine bases are adenine and guanine (represented as A and G). While DNA doesn't control the behaviors of an individual, per se, it can include the *directions* for certain behaviors and the propensity to behave a certain way, as well as the instructions for an individual's growth, development, reproduction, and physical appearance. Sometimes, a small alteration to one's DNA (known as a mutation) can result in large behavioral and prototypical changes in the subsequent generations. Over the course of generations, these traits can provide a selective advantage, becoming increasingly more common within a population. This process is known as descent with modification: the chance arrival of new forms through DNA mutation.

When Darwin wrote about descent with modification, he was specifically thinking about the finches and mockingbirds of the Galapagos, which had beaks that were perfectly suited for their differing environments. The Galapagos birds were an example of adaptive radiation (organisms diversifying into forms that filled disparate ecological niches). One of the most well-known instances of a mutation happened with our 18th century insect friend, the gray peppered moth. But what about less familiar examples of mutation? For the peppered moth, the selective pressure was pollution, and the subsequent change in the color of the tree trunks. For some other insects, the selective pressure is pesticide. Some insects have a mutation wherein they develop resistance to the pesticides. The pesticide-resistant insects survive and pass this resistance on to their descendants. This can happen within only one generation (Hawkins *et al.*, 2019). Among mammals, that change can take a little bit longer. Take, for example, the deer mouse of Nebraska's Sand Hills; their selective pressures were the environment and predation risk. Thanks to one mutation in one gene, these deer mice experienced a change in coat color. Within approximately 8000 years, the typically dark-coated, brown-colored mouse evolved to be a sandy color to better blend in against the light terrain of the Sand Hills (and not be spotted by predators) (Linnen *et al.*, 2009).

With the "bowl of paint" gene pool, the end result would be bowls of pastel paint (or differently colored moths or mice) that are initially similar enough to be traced back to the founding color, but different enough to provide a selective advantage, given each territorial niche. Over time, with more drops of color, the changes would become more profound: the bowl of paint, like the moth's wings or the mouse's coat, would have a dramatically different hue. As separate environmental factors continued to influence organisms' reproductive success, certain traits would be favored over others, thus skewing the appearance of the entire population. Darwin's finches and mockingbirds exemplify this: birds on different islands that were separated from a founding population of similar individuals gradually developed specialized beaks for the food sources in each area, whether it was insects, seeds, buds, or cactus. Now consider artificial selection. This would be more like having multiple bowls of white paint with a big splash of another color; changes would occur much more quickly in comparison to natural selection, with the end result being bowls of bright blue and electric orange. Now imagine that even the bowls would change in shape and size. While the founders of the species would bear some resemblance to subsequent generations, the differences would have the potential to be far greater than those illustrated in natural selection. That bright blue is the flat face of brachycephalic breeds, such as the Pomeranian; the electric orange is the lithe build of Sighthounds, such as the Greyhound. Pomeranians and Greyhounds are the same species, but they represent contrasting phenotypes.

Whether it's a sudden environmental change like pollution or human intervention like selective breeding, drastic change in a species can be affected in a relatively short amount of time, altering subsequent populations from the founding population. In artificial selection, these results occur so quickly because humans are manipulating the breeding, and perhaps even the survival, of the target species so that only the extreme individuals with the desired characteristics produce the next generation. Many of these desired characteristics, such as dogs' propensity to make "puppy dog eyes," have a profound effect on us. Take, for example, the evolution of dogs' facial muscle anatomy. Kaminski *et al.* (2019)

found that all breeds of dogs have a specialized muscle for raising their inner eyebrow. While this trait is uniformly present among dogs, it is absent in wolves; meaning, this derived trait arose after dogs and wolves branched away from one another. Dogs also produce higher-intensity eyebrow movements than wolves do, and it is likely that these emotion-eliciting expressions were selected for by humans.

The first dog who had a particularly appealing trait, like the first glimmer of puppy dog eyes, would have a selective advantage over those that didn't. Humans would have chosen this dog to breed with other dogs with similarly "human" facial muscle movements. In each subsequent generation, the most extreme individuals are again the only ones selected to produce the next generation. Those who don't have these exaggerated traits will disappear from the gene pool rapidly, as they are denied opportunities to reproduce. Over the years, numerous dog species have disappeared from the gene pool when humans determined that their appearance or behavior were no longer *en vogue*. With natural selection, though, it is far more likely that a few individuals without the chosen characteristic would continue to survive and breed for a few generations. Even with strong environmental pressure, the peppered moth still had the genotypes for lighter-colored wings, so when the tree trunks were no longer darkened with pollution, the entire species didn't die out; there was enough genetic variation still that they were able to adapt. Artificial selection, conversely, removes certain individuals from the gene pool altogether. But don't let the name "artificial selection" fool you—it's still, first and foremost, selection.

Sexual selection is part of natural selection, but it's driven by intrasexual selection or mate choice. It can also lead to dramatic physical manifestations, as in the peacock's exaggerated tail plumage. With sexual selection, those with the preferred traits have more opportunities to mate than those who don't have them, thus skewing succeeding populations to have increasing members with the preferred traits (and often more exaggerated versions of these traits, too). The traits that evolve during sexual selection are referred to as "secondary sexual characteristics," as they correspond to mating opportunities. When Darwin first documented these traits, he categorized them as ornamental, like the peacock's tail, or weaponry, like an ungulate's antlers. The former attracts members of the opposite gender, while the latter provides weapons for males to fight for the opportunity to mate. These traits can be costly to grow and maintain; they're calorically expensive, impede movement, or attract predators. But over multiple generations, these traits have been chosen to be "worth it": the benefits outweighed the costs. While it's harder to be inconspicuous and flee from a predator with ornate plumage, peahens decided that they prefer peacocks to be ostentatious—and those peacocks who aren't have fewer opportunities to pass on their comparatively plainer traits to offspring.

Wild birds' adaptations exhibit the effect of limited or changing habitat, competition, and predation. Birds also provide a considerable body of evidence for the effects of sexual selection. The peacock, with its "costly" tail, demonstrates that individual success can also depend "not on a struggle for existence, but on a struggle between the males for possession of the females; the result is not death to the unsuccessful competitor, but few or no offspring." Females will only choose partners with these ornamental tails, which marks them as healthy, worthy candidates to father their offspring; thus, individuals "differed in structure, color, or ornament … mainly [from] sexual selection." (Darwin, 1859).

Building a Species

So what makes a species, and how did we get the extant (currently living) species of today? Species is the principal taxonomic unit, ranking below genus, and denoted by its Latin binomial (e.g. *Homo sapiens* for humans and *Canis familiaris* for dogs). When Linnaeus created his *Systema Naturae*, he wasn't starting from scratch; he was building upon the work of brothers Gaspard and Johann Bauhin, botanists who used the same classification system, but to a lesser extent than Linnaeus eventually would. During his time, Linnaeus recognized three kingdoms in the living world, but today there are six recognized kingdoms. The kingdom Animalia to which you and I, and dogs, and all other mammals, and birds, reptiles, and insects, and more than 1 million other species belong, is the largest of the six kingdoms (BiologyOnline, 2022). Those in the animal kingdom are multicellular, with their cells having a nucleus but no cell wall or chloroplasts. The kingdom Plantae which has more than 250,000 known plant species, including ferns, mosses, trees, flowering plants, fruits, and vegetables, is the second largest kingdom. Plants are multicellular and autotrophic (they produce their own food, unlike animals, which have to obtain food from other sources). The kingdom Archaebacteria (Archaebacterium) has single-celled members (Archaea) whose genetic material is not contained within a nucleus. They are believed to be among the first forms of life on Earth and can survive in extreme environments. The kingdom Eubacteria, whose members include the bacteria *Streptococcus*, which causes strep throat, have single cells, lack a nuclear membrane, and reproduce rapidly. Some individuals in Eubacteria can produce food, while others have to obtain it from outside sources. The kingdom Fungi, whose members include mushrooms, yeast, and some molds, consists of both multicellular and single-celled organisms that reproduce by spores. Fungi are unable to produce their own food, feeding instead on other organisms (both alive and dead; they are the great decomposers). The kingdom Protista is the "we don't really fit anywhere else kingdom"—think of it as the "junk or miscellaneous drawer" of the living world—and includes algae, amoebae, and paramecia. Protists contain a nucleus and are primarily single-celled organisms in nature but can also be found as colonies of cells.

Linnaeus, and all scientists after him, used ever more specific categories ("taxonomic ranks") to classify the living world. After kingdom, the taxonomic pattern includes the phylum (or division), class, order, family, tribe, genus, and finally, species and sometimes subspecies, as well. A species can be defined as a population of organisms that could reproduce with one another, and that are reproductively isolated (Mayr, 1942). Taxonomy is the science of classifying living things. Linnaeus is known as the "Father of Taxonomy," and Darwin's theories underpin the concepts of species and speciation (the formation of new and distinct species). Over the centuries, taxonomic classification has become more precise; Darwin and Linnaeus were the trailblazers of evolution and taxonomy, respectively, and their work still underpins the concepts of species and speciation. Where Darwin hesitated, delaying the publication of *On the Origin of Species* for two decades after he began working on it, Linnaeus hastened headlong into his newfound role as a leading scientist. When it comes to dogs, though, there's still some disagreement over their classification. For hundreds of years, scientists have attempted to properly identify and classify all life. With "species" being a man-made concept, though, nature doesn't always get the memo to adhere to our neat demarcations and definitions. But there's more than one definition of species, and the definition of being able to produce viable offspring only with others of the same species doesn't always apply. Linnaeus used the motto, "God created, Linnaeus classified," but we're going to add "*and nature defied*" to his dictum.

For many organisms, the definition of being able to produce viable offspring within one group alone doesn't always apply; sometimes, organisms don't reproduce because of behavioral differences (such as different bird calls with the Western and Eastern meadowlarks), an environmental barrier, such as a mountain range, a wide river, or even islands in the ocean (as in the Galapagos finches and mockingbirds) and sometimes they don't reproduce because of morphological barriers, such as a Chihuahua trying to reproduce with a Great Dane. It's ambitious, but not necessarily impossible: a male Chihuahua and a female Great Dane *might* be able to produce offspring, but a female Chihuahua would not be able to carry to term the offspring of a Great Dane. Interestingly, when *Homo sapiens* and *Homo neanderthalensis* coincided, genetic evidence shows that offspring were viable when a male human procreated with a female Neanderthal, but not when a male Neanderthal procreated with a female human. Thus, Neanderthals are in the lineage of many modern humans, but only through their female lines. So while Chihuahuas and Great Danes are of the same species, they represent the extremes of the size continuum, exemplifying how speciation might occur, given the time and the right circumstances.

Whether they're diminutive Chihuahuas or "Great" Danes, all dogs are of the genus *Canis*, tribe Canini, subfamily Caninae, family Canidae, order Carnivora, class Mammalia, phylum Chordata, and kingdom Animalia. Let's take a closer look at this taxonomy, from kingdom to species.

- **Kingdom Animalia:** As of this writing, more than 2 million living species have been identified as being a part of the animal kingdom (International Union for Conservation of Nature, 2023), but there are likely millions more as-yet-undiscovered members of this kingdom. Animals are multicellular eukaryotic (their cells have a nucleus that is enclosed with membranes, while prokaryotes are unicellular, lacking mitochondria and a membrane-bound nucleus) organisms that breathe oxygen, have some form of locomotion, consume organic material, reproduce sexually, and develop from embryos (small masses of unspecialized cells). Most of the species in the animal kingdom belong to the Bilateria clade, which have a bilaterally (left/right) symmetric body plan. The first of the animals arose in the Late Precambrian. All living animals share 6331 genes in common, and may have evolved from a single ancestor that lived approximately 650 mya.

- **Phylum Chordata:** There are more than 65,000 living chordate species. Chordates are characterized by their bilateral symmetry, metameric segmentation (the repetition of organs and tissues at intervals), a circulatory system, and a coelom (the body cavity located between the body wall and the intestinal canal). At some period in their development (either pre- or post-natal), chordates will have a dorsal nerve cord, pharyngeal slits (filter-feeding organs), a notochord (a cartilaginous, body-supporting skeletal rod), an endostyle (an organ that helps with filter feeding), and a post-anal tail. The phylum Chordata has three subphylums: Vertebrata, which includes fish, reptiles, amphibians, birds, and mammals; Cephalochordata, which includes lancelets; and Tunicata, which includes sea squirts and salps. The first of the chordates emerged 541 mya, during the Early Cambrian explosion.
- **Class Mammalia:** There are approximately 6400 mammal species. The mammals are characterized by the presence of milk-producing mammary glands to nurse their young, fur or hair covering their skin, a neocortex, and three bones in the middle ear. Mammals give birth to live young, with the exception of the egg-laying monotremes (e.g. the platypus). The mammals diverged from birds and reptiles during the Late Triassic, approximately 201–227 mya.
- **Order Carnivora:** There are approximately 280 species in the order Carnivora. The carnivorans are the most diverse of the mammalian order and have claws and teeth for catching live prey. Carnivorans can be obligate carnivores such as the feliforms and the pinnipeds, omnivores such as bears, or primarily herbivorous, such as the giant panda. The order Carnivora has two suborders: the cat-like Feliformia and the dog-like Caniformia. The Carnivora first appeared during the Paleocene, approximately 60 mya.
- **Family Canidae:** The dog family, Canidae, consists of 35 closely related but phenotypically diverse extant species, including 16 fox species, three wolf species, three jackal species, the dhole (*Cuon alpinus*), the culpeo (*Lycalopex culpaeus*), the chilla (*Lycalopex griseus*), the coyote (*Canis latrans*), the raccoon dog (*Nyctereutes procyonoides*), the short-eared dog (*Atelocynus microtis*), the bush dog (*Speothos venaticus*), the African wild dog (*Lycaon pictus*), the domesticated dog, the black bear (*Ursus americanus*), the giant panda (*Ailuropoda melanoleuca*), the Northern seal elephant (*Mirounga angustirostris*), and the walrus (*Odobenus rosmarus*). (Yes, the Northern seal elephant and the walrus! Canidae provides a whole new outlook on biodiversity and co-evolution. One might have thought that these species were more closely related to manatees than dogs, but like we said, it's a diverse family!) The Canidae first emerged during the Late Eocene, approximately 40 mya.
- **Subfamily Caninae:** Most of us have never had to think about extinct subfamilies of dogs, and why would we? But within the family Canidae there is one extant (still living) subfamily, the Caninae, and two extinct ones, Hesperocyoninae and Borophaginae. The subfamily Hesperocyoninae was one of the most primitive of the Canidae and are the likely ancestors of the Borophaginae and Caninae. Hesperocyoninae lived in North America during the Uintan (approximately 46–42 mya) through the Bridgerian (approximately 50–46 mya) ages. They existed for approximately 11.5 million years. The Borophaginae are called the "bone-crushing dogs" and were a highly diverse group of canines. Unlike modern Caninae, which have four-toed feet, they had five toes on their rear feet. This family lived in North America during the Oligocene (approximately 34–23 mya) through the Pliocene (approximately 5.3–2.5 mya) periods. These families existed for about 33.5 million years.
- **Tribe Canini:** Canini is a monophyletic (all organisms descending from a common ancestor or ancestral group) tribe. Members of the Canini have an enlarged frontal sinus, an enlarged mastoid process (the temporal bone behind the ear), and no lateral flare of the orbital border of the zygomatic bone (cheek bone), a part of the skull formed by parts of the temporal bone and zygomatic bone. The Canini first emerged 9 mya.
- **Genus *Canis*:** The genus *Canis* is the most abundant of the terrestrial carnivores and comprises species and subspecies including the Dingo, coyotes, jackals, wolves, and dogs. The first *Canis* individuals emerged in the Palearctic approximately 5 mya.

Before we explore the evolution of the dog family further, let's revisit the species versus subspecies debate regarding dogs and gray wolves. While it's true that domesticated dogs descended from the ancestors of gray wolves, modern gray wolves are also genetically distinct from their ancestors. Modern domesticated dogs did not descend from what we know as modern wolves; they define a monophyletic group (a group of organisms consisting of all descendants of a common ancestor) that is sister to Old World wolves (Freedman *et al.*, 2016). Thus, the two modern species are sister taxa and share ancient ancestors in common.

Despite this, domesticated dogs are still listed as *Canis lupus familiaris* (a subspecies of the gray wolf) almost as often as they are listed as *Canis familiaris*. Now why does this matter? Well, given the preponderance of their genetic similarities, it could be argued that domesticated dogs are, in fact, "just" a subspecies of gray wolves, a "tamed wolf," if you will, but given their differences in development, neurology, behavior, nutritional needs, and unique genetic mutations distinct from their wolf cousins, the argument that they're different species is even stronger. We follow the latter viewpoint that modern dogs and wolves are sister taxa, and that today's wolves have changed over time and are not the same as those that were domesticated, tens of thousands of years ago (Larson and Bradley, 2014). We'll present a strong body of evidence supporting the fact that domesticated dogs aren't just "tamed wolves," but a distinct domesticated species.

Now let's revisit the arguments that "If dogs came from wolves, then how are there still wolves," and "If people came from monkeys, how are there still monkeys?" The latter argument, of course, is forged from ignorance: humans didn't "come from monkeys;" humans and the other primates last shared a common ancestor many millions of years ago. All extant primates are "equally evolved" and adapted to their unique ecological niche, just as all contemporaneous organisms are. Humans are not "more evolved" than any of the other primates, despite how much some might like to think we are. But what about wolves and dogs? Well, to put it simply, saying, "If dogs came from wolves, then how are there still wolves," is as illogical as saying, "If I came from great-grandma, how is there still a great-grandma?" Granted, that's on a much smaller scale, but if you were to extend that argument, for tens of thousands of great-grandmothers, the logic still holds. Yes, you are a descendant of your great-grandmother, and your great-great-great-grandmother, hundreds and thousands of times over, but over millennia, there have been changes across time. Similarly, today's wolves are genetically different from their wolf ancestors, as are today's dogs.

Speciation among similar-looking animals, such as dogs and wolves, is seen throughout the animal kingdom. Sometimes these species interbreed (or attempt to do so), and sometimes they don't. *Loxodonta africana*, the African bush elephant, and *Loxodonta cylotis*, the forest elephant, the two species of elephants that live on the African continent have opportunities to breed with one another, but choose not to (The Local SE, 2018). That hasn't always been the case, however: genetic evidence suggests that these two species were interbreeding within the last 500,000 years. Something happened between them, and now one or both of these species fell out of favor with the other ... but what was it? The bush elephant weighs approximately 7 tons (6500 kg)—double that of the forest elephant, but size doesn't appear to be the delimiting factor in interspecies breeding attempts. Analysis of ancient genomes revealed evidence of gene flow among ancient elephants, mammoths, and mastodons, resulting in populations with high levels of genetic variation that could adapt to environmental changes. This is similar to the interbreeding between *Homo sapiens* and *Homo neanderthalensis* and between *Homo neanderthalensis* and the Denisovans, who, pending their taxonomic status, will either be *Homo denisova*, *Homo altaiensis*, *Homo sapiens denisova*, or *Homo sapiens Altai*. Genetic testing has revealed that some modern humans have as much as 4% Neanderthal DNA while modern Melanesians and Aboriginal Australians have between 3% and 5% Denisovan DNA (Callaway, 2011). Similarly, dogs, coyotes, and wolves, which are all in the *Canis* genus, sometimes breed with one another where they coincide geographically, yielding fertile hybrid offspring.

But for Asian and African elephants, who are so geographically distant, the only well-documented instance of interspecies breeding resulted in an infant named Motty who lived for less than 2 weeks. While offspring between the two species should hypothetically be viable, other reports of hybrid progeny were similarly unsuccessful. According to recent research, the African elephant species appear to be as genetically dissimilar from one another as woolly mammoths are to modern Asian elephants

(Palkopoulou *et al.*, 2018). The two species split off from a common ancestor between 2.5 and 5 mya, a temporal distance similar to when chimpanzees and humans last shared a common ancestor. Dogs and wolves, however, last shared a common ancestor as recently as 40,000 years ago, and have continued to interbreed from time to time where they co-occur. Hybridization between dogs and wolves in both North America and Eurasia is not a recent or rare phenomenon (Pilot *et al.*, 2018). Thus far, this hybridization hasn't diminished the distinct wolf gene pool, although conservationists are concerned that increased interspecies mating could become disadvantageous to wolf populations.

How to Get You Into My Life

Now let's take a closer look at the evolutionary history of the dog-like carnivores. Other canid species are well-defined, but what, exactly, is the domesticated dog, how did it develop, and why did they become humanity's best—and first—friend? Journalists, researchers, and investigators often tackle their topics with the Five Ws and one H: who, what, when, where, why, and how, to get the full picture of their story. The "who" is easy: we're talking about modern domesticated dogs. To answer the rest of these questions, though, we'll have to circumnavigate the world, examine the evidence from fossils and genetic analyses, and go back tens of millions of years.

To put this time frame into context, we'll use epochs, geologic time scales that fall in between an "age" and a "period." We're currently living in the Holocene epoch, but for reference, the eras, periods, and epochs that we'll be referring to for canid and human evolution are as follows:

- **Cenozoic era:** 66 mya to the present day. Cenozoic translates to "new life" and this time is also known as the "Age of Mammals". The Cretaceous–Paleogene extinction event (also known as the K-Pg extinction event) approximately 66 mya at the end of the Cretaceous saw the extinction of all of the dinosaurs excepting the neornithine birds—those feathered dinosaurs that evolved into modern birds. Those animals that did survive the K-Pg event were relatively small; this provided the animals, which included the mammals, with the opportunity to disperse into every ecological niche, where they diversified. Some species, given an abundance of resources, grew into hypermorphs—exaggeratedly large versions of the species—because resources were so plentiful.
- *Quaternary period*: 2.5 mya to the present.
  Holocene epoch: Approximately 11,700 years ago.
  Pleistocene epoch: About 2.5 mya to 11,700 years ago. The end of the Pleistocene corresponds with the end of the Paleolithic period, an archaeological term for the timespan from the earliest known tool use by hominins (early humans) around 3.3 mya until the beginning of the Holocene. This also corresponds with the end of the last glacial period.
- Neogene period: 23–2.5 mya.
  Pliocene epoch: 5.3–2.5 mya.
  Miocene epoch: 23–5.3 mya.
- Paleogene period: 66–23 mya.
  Oligocene epoch: 34–23 mya.
  Eocene epoch: 56–34 mya.
  Paleocene epoch: 66–56 mya.

Dogs are members of the order Carnivora, so let's go back to 56 mya to the start of the Eocene epoch, when the carnivorans evolved from a group of animals called the "miacoids." Miacoidea was a primitive superfamily of carnivores that thrived during the Paleocene and Eocene epochs, approximately 66–34 mya, and consisted of two families: the Miacidae and the Viverravidae. Around 43 mya, the feliforms (the cat-like mammals) and the caniforms (the dog-like mammals) emerged from the Carnivoramorpha clade (a "clade" being a group of organisms that are believed to have evolved from a common ancestor) (Wesley-Hunt and Flynn, 2005) (Fig. 1.3). The newly formed suborder Caniformia would include the current extant families: Canidae (foxes, wolves, and dogs), Ursidae (bears), Mustelidae (weasels, badgers, and otters), Ailuridae (the red panda), Procyonidae (raccoons), Odobenidae (the walrus), Otariidae (eared seals), and Phocidae (earless seals). It would also include the extinct family Amphicyonidae (bear-dogs). Caniforms are primarily omnivorous and are characterized by non-retractable claws and auditory bullae (the bony, hollow structures that enclose part of the inner and middle ear) that are partially chambered or non-chambered.

Dawn of the Dog

A small muzzle appeared from the top of a hillock, its upright ears oscillating left and right as she surveyed

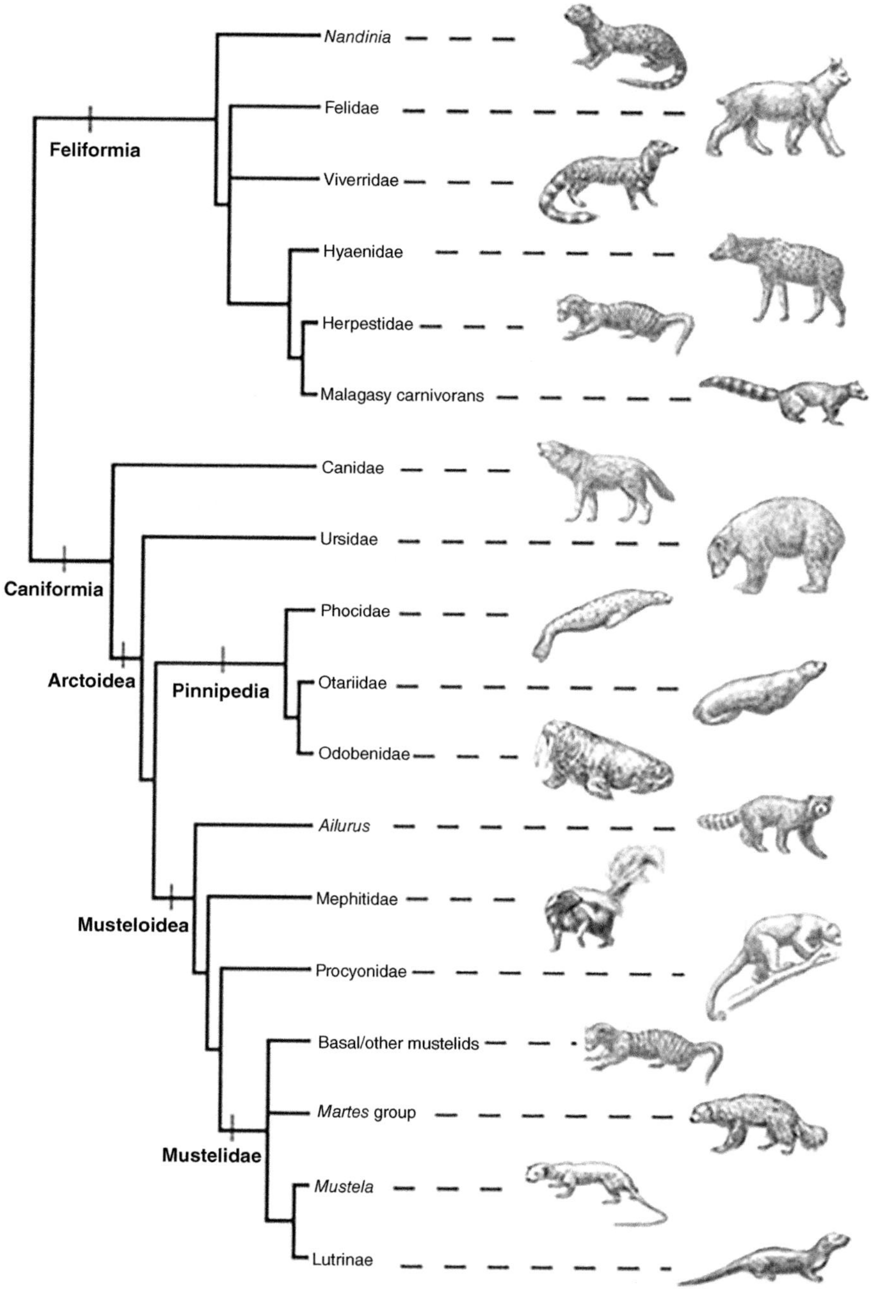

Fig. 1.3. A schematic cladogram representing the major evolutionary relationships recovered in the analysis of Carnivora by Flynn *et al.* (2005). Illustrations of representative taxa for major lineages include (from top): *Nandinia binotata*; Felidae (*Lynx rufus*); Viverridae (*Viverra zibetha*); Hyaenidae (*Crocuta crocuta*); Herpestidae (*Mungos mungo*); Malagasy carnivorans (*Eupleres goudotii*); Canidae (*Canis lupus*); Ursidae (*Ursus americanus*); Phocidae (*Phoca vitulina*); Otariidae (*Zalophus californianus*); Odobenidae (*Odobenus rosmarus*); *Ailurus fulgens*; Mephitidae (*Mephitis mephitis*); Procyonidae (*Potos flavus*); Mustelidae, basal/other mustelids (generalized schematic representing diverse taxa: African polecat and striped marten, badger, etc.); Mustelidae, *Martes* group (*Gulo gulo*); Mustelidae, *Mustela* (*Mustel afrenata*); Mustelidae, Lutrinae (*Lontra canadensis*). Figure from Flynn *et al.* (2005), by permission of Oxford University Press.

her territory. The sound of footfalls was accompanied by muffled barks, and she glanced back as her four pups bounded up the slope after her, biting at her thick, bushy tail and pushing their faces against her large, clawed paws. Far below them in a lush valley, a herd of Mesohippus grazed, their long tails swishing at insects. Sensing something, one Mesohippus raised her head, glancing up at the top of the hillock. What she saw was a short-legged creature, something that looked more like a raccoon dog or a civet than what it actually was: the first identified member of the dog family. True, this wasn't a "dog" as you or I would recognize it, but it was, nonetheless, the first member of this family.

This was *Prohespercyon wilsoni*, and 40 mya, it lived on the North American continent, including the site where its remains were discovered at an airstrip in Presidio County, Texas. You or I might not recognize this fossil as a "dog" because it was a transitional form between the miacids and the canids, exhibiting traits from both families. The fossil, which was dated at approximately 36.5–36.6 mya, included Canidae features such as an enlarged, bony bulla and no upper third molar. While *Prohespercyon wilsoni* was the first true member of the dog family, she wasn't alone. During the Eocene epoch (56–34 mya), the canid family diverged into three subfamilies: the now-extinct Hesperocyoninae (40–15 mya) and Borophaginae (34–2 mya) and the extant Caninae (34 mya to present day) to which all modern canines belong. These three subfamilies left a robust fossil record across the North American continent. During the Oligocene epoch (34–23 mya), the Hesperocyonines yielded the genera *Archaeocyon* and *Leptocyon* (which is Greek for "slender dog"), the latter an omnivorous, 4-pound fox-like animal. The genus *Leptocyon*, which lived from 31 to 10 mya, comprised the most primitive known Caninae members, and later branched into the Canini (canines) and Vulpini (foxes). During the Miocene, around 9–10 mya, the genera *Vulpes*, *Urocyon*, and *Canis* expanded from southwestern North America. One million years later, members of the genus *Eucyon* crossed the Beringian land bridge and entered Asia and Europe. One million years later, during the Pliocene, *Canis lepophagus* emerged in North America. Both morphological (Nowak, 1995) and genetic (Wayne *et al.*, 1997) evidence point to *Canis lepophagus* as the ancestor of coyotes and wolves (Tedford *et al.*, 2009). *Canis latrans* (the coyote), *Canis priscolatrans*, and *Canis etruscus* descended from *Canis lepophagus*, with *Canis armbrusteri* descending from *Canis priscolatrans*, the lineage that eventually led to several species that have since gone extinct, including the dire wolf. The descendants of *Canis etruscus* included *Canis variabilis*, the great plains wolf, and *Canis mosbachensis*, the extinct Mosbach wolf.

When the Isthmus of Panama joined North and South America 3 mya, canids radiated out from the North American continent. The ancestors of the North American wolf arrived during the Pleistocene, 800,000 years ago, entering into South America during the latter part of the Pleistocene. Approximately 300,000 years ago, *Canis mosbachensis* gave rise to *Canis rufus*, the red wolf, and several subspecies of *Canis lupus*, the gray wolf. *Canis lupus* began dispersing across northern Asia and Europe, crossing the Bering land bridge and spreading out onto the North American continent.

With multiple radiations during a relatively short amount of time, resolving the genetic relatedness of the dog-like carnivore species has been problematic. Genetic tests with nuclear genes that had "relaxed" evolutionary constraints (Bardeleben *et al.*, 2005) scarcely disambiguated the question of the canine family tree. "Evolutionary constraint" refers to the restrictions on the outcome of adaptive evolution, so tests with strict constraints would yield more precise results than those with relaxed constraints. With this in mind, researchers Robert K. Wayne and Elaine A. Ostrander examined a new approach to examine the complete dog genome sequence (Wayne and Ostrander, 2007). Using applied comparative genomics with the complete dog genome, a team of researchers identified rapidly evolving nuclear genes that revealed the relationships among phenotypically divergent taxa, including the bush dog and maned wolf, kit fox and Arctic fox, and the domesticated dog and the gray wolf (Lindblad-Toh *et al.*, 2005). Rapidly evolving nuclear genes provide a clearer picture of the evolutionary tree of the extant species of Canidae, but prior researchers hadn't used this particular approach with the canine genome before because isolating these genes can be expensive and difficult. To find these rapidly evolving genes, researchers compared 14,000 canine transcripts to their homologs (biological equivalents) among mice, rats, and humans to identify where divergent canine exons appeared to be evolving. This methodology helped identify the topology of Canidae's three main branches: the fox-like canids, the wolf-like canids, and the South American canids. Molecular dating has found a common ancestor for modern canids approximately 10 mya, which is a very recent radiation from a family originating approximately 50 mya (Wayne *et al.*, 1989). While the wolf-like canids first made their appearance in the family Canidae 6 mya, the group comprising the modern dog's closest relatives (the golden jackal, the coyote, and the gray wolf) last shared a common ancestor as recently as 3–4 mya.

A Game of Bones

The late summer sun steeped the riverbank in orange as it sunk further across the skyline. The heat radiated in waves. Undaunted, Francis A Linck licked his parched lips, glancing up at the sky, his hands searching the banks of the Ohio River. Not far from the mouth of Pigeon Creek, just outside of Evansville, Indiana, Linck had been scouring the sandy banks for fossils, brushing away the top layer of sand and gravel, but he'd found nothing yet. An egret waded downriver, illuminated in the sun, its head cocked sideways as it examined the riverbed. The rippling water glowed like liquid gold as the setting sun glinted down. And then, in the Pleistocene-age alluvial deposits, he found something more rare and valuable: a sharp protuberance in the soft bank. He swept a layer of silt off the clay with his thumb and forefinger until he'd partially unearthed a curved, white crescent. As he brushed off thousands of years of dirt, he could see the slightly serrated edge of a cheek tooth, weather-worn but unmistakable. It belonged to a canid.

Linck hadn't found just any canid, though; he'd found the jawbone of what was once thought to be the largest known wolf, to date. When it was alive, this animal would have weighed 150 pounds or more. While it was clearly canid, this jawbone was just dissimilar enough to arouse the curiosity of the local scientific community. Bald, bespectacled, and brilliant, Dr Joseph Granville Norwood was the US' first state geologist—and he was fascinated with the fossil find. Realizing that the find could have great scientific importance, Norwood asked Linck to send it to paleontologist Dr Joseph Leidy in Pennsylvania for further analysis; Linck, however, refused. Three years after Linck's find, Leidy discovered the vertebrae of a canid. After Linck's death in 1858, his family honored Dr Norwood's request and sent the jaw to Dr Leidy. Leidy determined that the jaw was a new species, officially dubbing it—and the vertebrae—as *Canis dirus* in 1858. To this day, Linck's dire wolf jaw fossil remains at the Academy of Natural Sciences in Philadelphia. For 163 years, it held onto a "wolf" classification, but in 2021, intrepid zooarchaeologist Angela Perri of England's Durham University published a paper that would give the dire wolf a delightful new identity. In a Pleistocene-era edition of a paternity surprise show, Perri discovered that the dire wolf wasn't a wolf at all, but a distantly related canine cousin—one that paradoxically hadn't interbred with other canines, as many canid species are apt to do. Because of this, the dire "wolf" was an evolutionary dead end.

"We don't know if they weren't as adaptable as other canines, or if they experienced a bottleneck, but it's clear that there was no evidence of interbreeding among other canids," Dr Perri said. "Most canines will, when given the opportunity, interbreed with other canid species; hybridization is common. But this didn't occur with the dire wolf."

Perri performed this work by collecting enough bones of dire wolves across the country to paint a clearer picture of their parentage—one that revealed the dire wolf diverging from the ancestors of the gray wolf 5.7 mya (Perri *et al.*, 2021).

"The quest for dire wolves took many years," Dr Perri said. "We took a 'sampling road trip' in 2016, where I went to Texas, Idaho, and California. Most of the prior attempts to extract dire wolf DNA came from La Brea, but those samples were too deteriorated. Finally, there were three independent researchers who had found a total of five samples. Rather than publish competing papers, we decided to work together."

Thanks to Perri and her colleagues, we now know that gray wolves (genus *Canis*) and dire "wolves" are far from being closely related. The gray wolf is more closely related to the dhole (genus *Cuon*), African wild dog (genus *Lycaon*), and Ethiopian wolf (genus *Canis*) than it is to the African jackal and the dire wolf, which are both classified as *Canis*, as well (Perri *et al.*, 2021). Perri notes that rather than classifying the dire wolf within *Canis* that there is an alternative classification, the monotypic genus *Aenocyon* (meaning "dreadful" or "terrible" wolf) (Merriam, 1918).

Like its modern wolf cousins, *Canis dirus*, which translates to "fearsome dog" (or alternatively, *Aenocyon dirus*, if one provided a new genus for this species), was probably a pack hunter. Along with the gray wolf, the dire wolf shared the honor of being the largest of the *Canis* genus. The dire wolf is also considered to be the most evolutionarily evolved *Canis* species of the Americas during its reign, exhibiting a unique dental pattern which included larger maxillary 1 (M1) teeth and a large upper carnassial (P4) with larger tooth blades than its canid cousins (Anyonge and Baker, 2006). (Basically, the dire wolf had a "toothier" grin and a more shearing bite than the canids we're familiar with.) The dire wolf's skull was also larger than its present-day descendants, with a broader frontal region, zygomatic arches, and palate in comparison to present-day wolves.

The dire wolf was a contemporary of (and competitor with) the saber-toothed cat (*Smilodon fatalis*) and the North American lion (*Panthera leo atrox*), and a slew of canine cousins, including coyotes, dholes, gray foxes, and wolves (Perri *et al.*, 2021); and perhaps, too, early humans. But it was the trio of megafauna (dire wolf, saber-toothed cat, and North American lion) that took center stage for thousands of years. During the Late Pleistocene and Early Holocene epochs, these flagship species reigned over the North American continent, preying on the continent's plentiful megaherbivores. It has been hypothesized that larger morphs of a species, called hypermorphs, could adapt in an environment during times of abundant food, but these larger species would also be more vulnerable to extinction during times of scarcity. Both the dire wolf and the saber-toothed cat died out during the Quaternary Extinction Event, around 10,000–12,000 years ago. During this time, many of earth's larger species went extinct, with North America's megafauna seeing more than a 70% loss, including mammoths, giant ground sloths, and mastodons—much of the flagship species' preferred diet. South America fared even worse, seeing extinction rates of 82%, with Australasia at 71%, Europe at 59%, Asia at 52% and Subsaharan Africa at 16%.

The Bounty of Bonn-Oberkassel

Ernst sadly patted the little pup, smoothing out her black and silver coat as her ragged breathing finally subsided. She had fought so hard. "There, Mina," he said, his voice barely a whisper, as he gently shut her eyes. Mina had been through several bouts of sickness, which her family had successfully nursed her through, but this final attack was more than her 7-month-old body could endure. Ernst lowered her lovingly into a grave with her owners and her mother, who had passed away shortly before. Working together, Ernst and his family hoisted basalt blocks gently onto the deceased and then gathered the earth around them, pushing it over the blocks until they were completely concealed. "Goodbye, Brecht, Heske, Sophey, Mina," Ernst said as he walked away.

On the eve of the World War I, workers in Oberkassel, Bonn, in Germany who were quarrying for basalt inadvertently dug up the little family's grave. They noted the discovery of human remains, assorted animal bones and antlers, and a wolf mandible. They sent the remains off for storage, and the find was largely forgotten as war swept across the continent. More than 60 years later, the Bonn-Oberkassel remains were re-examined, and researchers discovered that the mitochondrial DNA (mtDNA) for the mandible of the small pup belonged not to *Canis lupus*, but to *Canis familiaris*.

The young dog—"Mina"—who had both the genetic and morphological characteristics of a modern dog, died during the Upper Pleistocene era approximately 14,000 years ago. The Paleolithic puppy was in the oldest known grave containing both canine and human remains, providing further evidence of the longevity of the human–canine relationship and the quality of it, as well. The condition of her teeth indicated that she was a highly valued member of the family. A thorough examination of her remains (Janssens *et al.*, 2018) revealed that her teeth had evidence of suffering from canine distemper between the ages of 19 and 23 weeks, with multiple episodes of illness. Canine distemper has high mortality rates, and the puppy's initial survival was likely due to human intervention. Symptoms of distemper include lethargy, vomiting, diarrhea, dehydration, and fever. Paper co-author Liane Giemsch stated that the puppy "probably could only have survived thanks to intensive and long-lasting human care and nursing." The remains provide strong evidence that the human–canine relationship was already well-established during the Late Pleistocene, and that dogs were valued beyond their utilitarian worth: a puppy of this age would have still been untrained and highly dependent upon its family. Caring for an animal as sick as she had been would have required a great amount of time and dedication.

The 10,000-year-old dog remains that were found on the American continent at the Stilwell II and Koster sites in present-day Illinois showed a similar level of familial investment. The two sites, which are around 20 miles apart, are the oldest known fossil evidence of domesticated canines in the Americas (Bower, 2018), and each of the dogs had been placed in their own gravesite. According to Angela Perri, the teeth and lower jaws of the Stilwell II dog and a Koster dog shared similarities with modern-day wolves, while another jawbone from a Koster dog shared similarities with modern-day coyotes (Perri *et al.*, 2019), indicating possible interbreeding between the species.

"The Koster dog was gracile and small, while the Stilwell dog was much more robust," Perri recalled. Unlike the dire wolf, which appeared to avoid interbreeding with other canids, the Koster dog appeared to have recent coyote ancestry. While the Koster and Stilwell sites were around 20 miles apart, it's possible that these very different domesticated dogs were already exhibiting evidence of differentiated breeding—perhaps the robust Stilwell dog had been bred for guarding or carrying supplies, and perhaps the Koster canine had been bred to herd or be a companion. This, of course, is speculation, but it's worth a second thought. "Dogs are the Swiss army knife of animals," Perri said. "When confronted with new environments without megafauna, humans increasingly used dogs for a variety of purposes."

A Song of Ice and Mire

A small pack of dire wolves gathered together on a bluff, surveying their prey below. Their bodies were tall and lean and their heads were massive; their long canines were exposed as they panted, slack-jawed. This should be easy, their glances at one another seemed to say. There, in a darkened area of the earth below them, was their highly sought-after quarry. Tall, tusked, and with a wildly thrashing trunk, their familiar food choice was paradoxically standing in one spot, but flailing its head and trumpeting loudly. Soundlessly, two wolves broke off to the left and right, running along the slope, while three others descended toward their prey dead-on. Teeth bared and ears flattened back, their kill looked all but certain. Within moments, though, the quintet of dire wolves realized why the mammoth wasn't moving from where he stood; they, too, were caught in the thick of the La Brea Tar Pits. (While these remains were a treasure trove, the consistency and heat of the tar pits would break down the DNA of all of the animals who succumbed to it, making Perri's job tracking the dire wolf genome much more difficult.)

Dire wolf remains have been discovered in areas as diverse as the South American savannah to the fields and forests of North America, including Tarija, Bolivia, Talara, Peru, Muaco, Venezuela, and Alberta, Canada (Dundas, 1999). The species flourished in North America for more than 100,000 years. The last known dire wolf died in 9440 BCE, several thousand years after the end of the most recent Ice Age, but its mythology lives on in stories such as George R.R. Martin's *A Game of Thrones* (Fig. 1.4). In 1984, Finnish paleontologist Bjorn Olof Lennartson Kurten determined that there were two subspecies of *Canis dirus*: *Canis dirus guildayi* and *Canis dirus dirus*, with the former weighing approximately 132 pounds and the latter 150 pounds (Kurten, 1984). While its skull and jaw were similar to extant wolves, the dire wolf had larger teeth and the strongest bite force of any identified *Canis* species to date. The dire wolf likely evolved from Armbruster's wolf, *Canis armbrusteri*, which had similar features to the dire wolf, with the exception that Armbruster's wolf had a slightly narrower skull and likely a smaller bite force.

To date, the bodies of more than 4000 dire wolves have been exhumed from Los Angeles' La Brea Tar Pits. The La Brea Tar Pits and Museum has more than 3.5 million Ice Age fossils from 650 species—and counting. It's the largest collection of fossils ever found, and the museum has many of these finds on exhibit, from a Columbian Mammoth tusk to the tooth of a baby mouse. Because of the high number of fossils that have been discovered here, including five species of canids (*Aenocyon (Canis) dirus*, *Canis familiaris*, *Canis lupus*, the Pleistocene coyote (*Canis latrans orcutti*), and the gray fox (*Urocyon cinereoargenteus*)), excavations of the tar pits continue to this day, 7 days a week.

Isotope analysis of bone collagen from the La Brea dire wolves revealed that their prey included the western horse (*Equus occidentalis*), Pleistocene bison (*Bison antiquus*), grazing ground sloth, also known as Harlan's ground sloth (*Paramylodon harlani*), yesterday's camel (*Camelops hesternus*), and the dwarf pronghorn (*Capromeryx minor*), with American mastodons (*Mammut americanum*) and Columbian mammoths (*Mammuthus columbi*) being rarer (but even more enticing) finds.

Horses first evolved in North America approximately 50 mya and lived on the continent until about 11,000 years ago. The western horse was one

Fig. 1.4. Dire wolves. Illustration donated by Arianne Taylor, used with permission.

of the last native North American horse species. The Pleistocene bison, however, weren't native to the continent, and entered North America from Asia approximately 200,000 years ago. The ground sloth was endemic to North America and weighed up to 1500 pounds. Like the horse and ground sloth, yesterday's camel also evolved in North America, beginning around 45 mya, and later migrating to Africa, South America, and Asia.

Dire wolves, saber-toothed cats, and the North American lion all competed for the same prey, and were all found in the tar pits, likely trying to eat animals that had been immobilized there. To date, in addition to the 4000 dire wolves that became mired in the infamous asphalt, 2000 individual saber-toothed cats and 80 North American lions have been recovered from the La Brea pits, and there they all remained, for tens of thousands of years.

Long before the first written record of the tar pits in 1769, Native Americans used the sticky substance to seal their canoes and later, their roofs. In 1875, when the Hancock Family owned the land comprising the tar pits, Englishman William Denton visited the site, receiving a saber-toothed cat canine from the Hancocks. Denton was the first person to describe the pit's fossils, and in the following decades, even more bones would be found. In 1908, paleontologist John Campbell Merriam discovered wolf bone fragments in the tar pits, and within 4 years, he had discovered a skeleton that was complete enough to identify the bones as those of *Canis dirus*.

Dig It

Tens of thousands of years after the arrival of the dire wolf, hominids and canines began to coexist (sometimes peacefully, sometimes not). Dogs were the first species domesticated by humans, so here we'll tackle the *when*. DNA sequencing and fossil evidence continue to point to the domestication of dogs and, not much later, livestock, during the Upper Paleolithic period (between 50,000 and 10,000 years ago) (MacHugh *et al.*, 2016) with estimates ranging as widely as 15,000–100,000 years ago. Some of the most robust evidence, including numerous Paleolithic fossil finds, puts the inception of this relationship at 40,000 years or earlier. One of the sites indicating an earlier relationship is Grotte du Lazaret, France. Excavators at this site discovered wolf skulls placed at the front of numerous Paleolithic shelters. Follow-up studies using maternal DNA (mtDNA) and paternal DNA placed the wolf–dog divergence between 100,000 and 135,000 years ago (Vila *et al.*, 1997) and 68,000 and 151,000 years ago (Oetjens *et al.*, 2018), respectively. Interestingly, this predates the arrival of *Homo sapiens* into the European continent, indicating that this wasn't the only hominid species that had a relationship with canids. Many hypotheses posit that humans' partnership with dogs was a contributing factor to the "extinction" of Neanderthals, but perhaps this idea needs to be re-examined.

Numerous sites across Siberia and Europe have yielded fossil finds ranging from 40,000–17,000 years ago (Irving-Pease *et al.*, 2018). A maxillary (jaw) fragment found in Hohle Fels, Schelklingen, Germany, had the morphological characteristics of a domesticated dog and the molar size of a wolf. The fossil was determined to be 35,000–40,000 years old and was identified as a Paleolithic dog (Camarós *et al.*, 2016). A skull found in the Goyet Caves, Mozet, Belgium, was affectionately named the "Goyet dog" and determined to be 36,500 years old (Germonpré *et al.*, 2009). The Goyet dog appears to be an evolutionary dead end; its maternal DNA (i.e. mtDNA, which is usually inherited only from the mother) doesn't correspond to any extant canids (Thalmann *et al.*, 2013). Mandibles, teeth, and a skull found in Razboinichya Cave in Central Asia's Altai Mountains were dubbed the "Altai dog." The remains were believed to be those of a 33,500-year-old Paleolithic dog (Ovodov *et al.*, 2011), a finding supported by DNA analysis (Druzhkova *et al.*, 2013). A follow-up study determined that the Altai dog belonged to a now-extinct lineage. A mandible found at Kostyonki-8 in Voronezh, Russia, was determined to be a Paleolithic dog that was 26,500–33,500 years old (Germonpré *et al.*, 2014). Three canid skulls found in Predmosti, Moravia, in the Czech Republic (Mendel's homeland) were determined to be 31,000-year-old Paleolithic dogs and were discovered in a human burial zone; one of the dogs was buried with a bone placed in his mouth (Germonpré *et al.*, 2014). Pawprints found in the Chauvet Cave of France's Vallon-Point-d'Arc were determined to be 26,000 years old; they are morphologically similar to modern domesticated dogs, but the track line has been argued to be similar to a wolf's (Ledoux and Boudadi-Maligne, 2015). A canid skull found in Ulakhan Sular, northern Yakutia, Siberia, was determined to be a 17,200-year-old Paleolithic dog (Germonpré *et al.*, 2017).

In 2011, mammoth tusk hunters Yuri and Igor Gorokhov found a very familiar face: pushing out of the permafrost was the snout of a puppy. The find wasn't just a skeleton, though; it was a Pleistocene puppy's mummified body, complete with skin and hair. This well-preserved discovery prompted the hunters to contact Sergey Fyodorov, the head of exhibitions at Yakutsk's Mammoth Museum of the North-Eastern Federal University. The brothers Gorokhov donated the first puppy to the Mammoth Museum. There, it was discovered that she was female and had died at 3 months of age, and she was the first—and at that time, only—known puppy of her kind: radiocarbon analysis placed her at 12,400 years old. Fyodorov was richly rewarded for his efforts: 4 years later and 26 feet beyond the first mummified puppy, a second Pleistocene pup was discovered in the Arctic Yakutsk permafrost, 2900 miles outside of Moscow. The pair were dubbed the "Tumat dogs," after the closest settlement. DNA analysis revealed that the puppies, who were likely members of the same litter, were the ancestors of today's dogs (Fig. 1.5). The puppies appeared to have been buried in a landslide, and not by humans, leading to doubt about whether they were wild or domesticated, but Fyodorov noted that ancient human artifacts have historically been found in the vicinity (Harvey, 2016). In addition to these mummified pups, Fyodorov, along with his colleagues, have examined four canine skulls found in this region, with ages dating from 950 to 50,000 years old. Of the

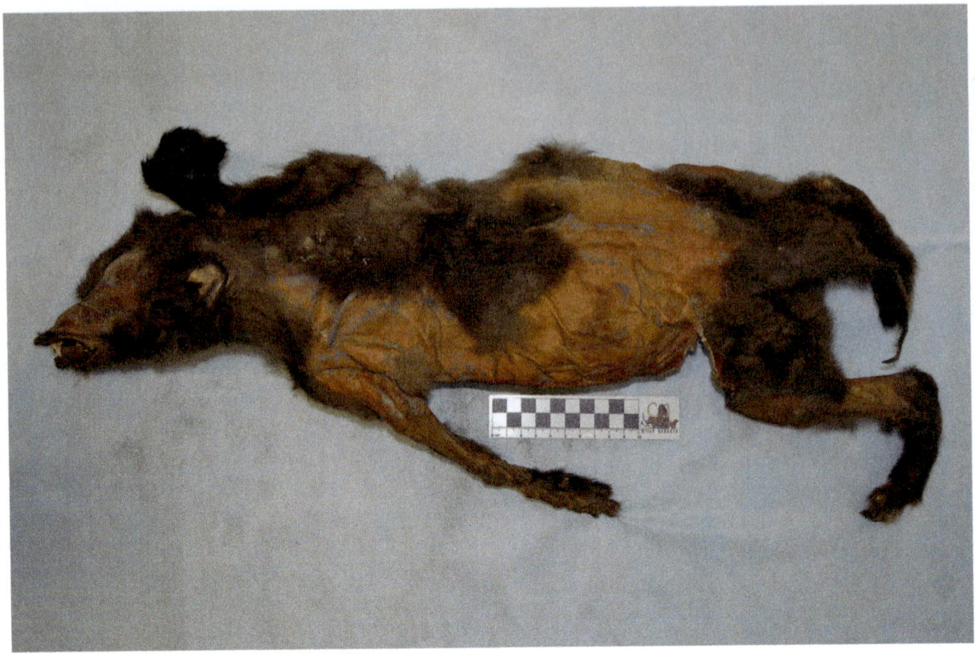

Fig. 1.5. Tumat puppy. Photograph credit to Sergey Fyodorov.

four skulls, one (the "Malyi Lyakhvosky canid"), had a calibrated age of 900 years, and was classified as a Northern dog. A second skull (the "Ulakhan Sular canid") had a calibrated age of 17,200 years and was morphologically similar to the Paleolithic-dog type. A third (the "Badyarikha canid"), was dated at 30,800 years of age, and classified as unknown, while a fourth (the "Tirekhtyakh canid") was dated at 50,000 years old and classified as a Pleistocene wolf (Harvey, 2016).

It shouldn't come as a surprise that many of these finds belonged to now-extinct lineages. The canids have seen many divergences within their family tree, and many extinctions, as well, including the aforementioned Altai dog and dire wolf. There have also been numerous more recent extinctions with wolves and domesticated dogs. Let's start at the "beginning" and work our way from there.

Never Cry Wolf

The descent of dogs from wolves begins with *Canis mosbachensis*, the Mosbach wolf. *Canis mosbachensis* lived during the Early Pleistocene and became extinct approximately 300,000 years ago; it's widely accepted as the ancestor of *Canis lupus*. More recent wolf extinctions include the Falkland Islands wolf (*Dusicyon australis*, previously referred to as *Canis antarcticus*), which became extinct in 1876, the Hokkaido wolf, also known as the Ezo or Sakhalin wolf (*Canis lupus hattai*), which became extinct in 1889, the Japanese wolf, also known as the Honshū wolf (*Nihon ōkami* or *Canis lupus hodophilax*), which became extinct in 1905, the Newfoundland wolf (*Canis lupus beothucus*), which became extinct in 1911, the once-common Kenai Peninsula wolf (*Canis lupus alces*), which became extinct in 1925, and the Sicilian wolf (*Canis lupus cristaldii*), which reportedly became extinct in the 1920s. Wolves used to roam across the Japanese archipelago, but in 1868, Emperor Meiji became the 112th emperor of the Empire of Japan—and he made sweeping changes that consolidated the country's political system under his rule. The time would come to be known as the "Meiji Restoration Period" and saw Japan adopt many of America's agricultural practices. Killing wolves became a matter of national policy.

The Falkland Islands wolf has the rather dubious honor of being the first canid on record to have become extinct during the modern era. While selecting the friendliest of wolves as companions was likely key during the inception of domestication, wolves being "too friendly" with humans could also issue their death sentence. During his 1834 visit to East and West Falkland, Darwin noted that the Falkland Islands wolf was quite tame, and guessed (correctly, unfortunately) that this would lead to their eventual demise.

Death by friendliness is an incident that has occurred in modern times on an individual level, including the case of a wolf who earned the moniker "Romeo." In the winter of 2003, a wolf in Alaska's Mendenhall Valley earned this moniker because of his unusual attraction to domesticated dogs. Romeo, like the Falkland Islands wolves, was comparatively friendly with humans and with dogs, much like the select wolf ancestors of modern dogs likely were. These select ancestors were friendlier, less reactive, more expressive, and more engaging than their aloof canine peers, and it was during this time of mutual gaze holding that dogs with emotion-eliciting expressions would have begun to arise. This facial muscle movement, dubbed "AU101: inner eyebrow raise", didn't just elicit emotional responses from humans, though; it likely aided in canine–human communication, as well (Kaminski *et al.*, 2019). While many wolves will avoid humans or retreat from them, Romeo did neither; he defied all of the typical wolf stereotypes, approaching not only the humans and dogs that he was familiar with, but also showing curiosity (and boldness) toward novel people and their dogs. Romeo first associated with domesticated dogs and their humans in 2003, continuing to do so until 2009. (We'll discuss Romeo at length in Chapter 4). Darwin (1839) later wrote the following about *Canis antarcticus* (now designated as *Dusicyon australis*):

> The only quadruped native to the island, is a large wolf-like fox, which is common to both East and West Falkland. Have no doubt it is a peculiar species, and confined to this archipelago; because many sealers, Gauchos and Indians, who have visited these islands, all maintain that no such animal is found in any part of South America. Molina, from a similarity in habits, thought this was the same with his "*culpeo*"; but I have seen both, and they are quite distinct. These wolves are well known, from Byron's account of their tameness and curiosity; which the sailors, who ran into the water to avoid them, mistook for fierceness. To this day their manners remain the same. They have been observed to enter a tent, and actually pull some meat from beneath the head of a sleeping seaman. The Gauchos, also, have frequently killed them in the evening, by holding out a piece of meat in one hand, and in the other a knife ready to stick them. As far as I am aware, there is no other instance in any part of the world, of so small a mass of broken land, distant from a continent, possessing so large a quadruped peculiar to itself. Their numbers have rapidly decreased; they are already banished from that half of the island which lies to the eastward of the neck of land between Saint Salvador Bay and Berkeley Sound. Within a very few years after these islands shall have become regularly settled, in all probability this fox will be classed with the dodo, as an animal which has perished from the face of the earth. Mr Lowe, an intelligent person who has long been acquainted with these islands, assured me, that all the foxes from the western island were smaller and of a redder colour than those from the eastern. In the four specimens which were brought to England in the Beagle there was some variation, but the difference with respect to the islands could not be perceived. At the same time the fact is far from improbable.

Conclusion

The Falkland Island wolf was indeed the only mammal endemic to the Falklands and it was the only species in the genus *Dusicyon* to endure into the modern era. Like many wolf species, the demise of the Falkland wolf was largely due to persecution from humans, who perceived them to be a threat to their food and livelihood when no real threat likely existed. The plight of these wolves highlights the potential dangers (from a canid's perspective) of commensal living and domestication. How many times did domestication *almost* happen, only to end in the demise of the canid? While we may take our dogs for granted today, a lot of risk, moxie, luck, and serendipity had to be there before we could begin to take the steps to "build" humankind's first friend.

References

Anyonge, W. and Baker, A. (2006) Craniofacial morphology and feeding behavior in *Canis dirus*, the extinct Pleistocene dire wolf. *Journal of Zoology* 269(3), 309–316.

Bardeleben, C., Moore, R.L. and Wayne, R.K. (2005) A molecular phylogeny of the Canidae based on six nuclear loci. *Molecular Phylogenetics and Evolution* 37(3), 815–831.

BiologyOnline (2022) Kingdom Animalia. Content provided and moderated by Biology Online Editors. BiologyOnline.com. Available at: www.biologyonline.com/dictionary/kingdom-animalia (accessed 31 December 2023).

Bower, B. (2018) Dogs lived and died with humans 10,000 years ago in the Americas. *Science News* 193(8), 22.

Callaway, E. (2011) First Aboriginal genome sequenced. *Nature* News. https://doi.org/10.1038/news.2011.551 (accessed 1 November 2023).

Camarós, E., Münzel, S.C., Cueto, M., Rivals, F. and Conard, N.J. (2016) The evolution of Paleolithic hominin–carnivore interaction written in teeth: stories from the Swabian Jura (Germany). *Journal of Archaeological Science* 6, 798–809.

Darwin, C. (1839) Narrative of the surveying voyages of His Majesty's Ships Adventure and Beagle between the years 1826 and 1836, describing their examination of the southern shores of South America, and the Beagle's circumnavigation of the globe. Journal and remarks. 1832–1836. *The Voyage of the Beagle*, Vol. 3. Henry Colburn, London, pp. 149–150.

Darwin, C. (1859) *On the Origin of Species by Means of Natural Selection, or, The Preservation of Favoured Races in the Struggle for Life*. J. Murray, London.

Druzhkova, A.S., Thalmann, O., Trifonov, V.A., Leonard, J.A., Vorobieva, N.V. *et al.* (2013) Ancient DNA analysis affirms the canid from Altai as a primitive dog. *PLoS One* 8(3), e57754.

Dundas, R.G. (1999) Quaternary records of the dire wolf, *Canis dirus*, in North and South America. *Boreas* 28(3), 375–385.

Flynn, J.J., Finarelli, J.A., Zehr, S., Hsu, J. and Nedbal, M.A. (2005) Molecular phylogeny of the Carnivora (Mammalia): assessing the impact of increased sampling on resolving enigmatic relationships. *Systematic Biology* 54(2), 317–337.

Freedman, A.H., Lohmueller, K.E. and Wayne, R.K. (2016) Evolutionary history, selective sweeps, and deleterious variation in the dog. *Annual Review of Ecology, Evolution, and Systematics* 47(1), 73–96.

Germonpré, M., Sablin, M.V., Stevens, R.E., Hedges, R.E.M., Hofreiter, M. *et al.* (2009) Fossil dogs and wolves from Palaeolithic sites in Belgium, the Ukraine and Russia: osteometry, ancient DNA and stable isotopes. *Journal of Archaeological Science* 36(2), 473–490.

Germonpré, M., Lázničková-Galetová, M., Losey, R.J., Räikkönen, J. and Sablin, M.V. (2014) Large canids at the Gravettian Předmostí site, the Czech Republic: the mandible. *Quaternary International* 359–360, 261–279.

Germonpré, M., Fedorov, S., Danilov, P., Galeta, P., Jimenez, E. *et al.* (2017) Palaeolithic and prehistoric dogs and Pleistocene wolves from Yakutia: identification of isolated skulls. *Journal of Archaeological Science* 78, 1–19.

Harvey, I. (2016) Autopsy carried out on the world's oldest dog mummified by ice. *The Vintage News*. Available at: www.thevintagenews.com/2016/08/07/sibling-tumat-dog-discovered (accessed 26 October 2023).

Hawkins, N.J., Bass, C., Dixon, A. and Neve, P. (2019) The evolutionary origins of pesticide resistance. *Biological Reviews of the Cambridge Philosophical Society* 94(1), 135–155.

International Union for Conservation of Nature (2023) The IUCN Red List of Threatened Species. Version 2023-1. Available at: www.iucnredlist.org (accessed 31 December 2023).

Irving-Pease, E.K., Ryan, H., Jamieson, A., Dimopoulos, E.A., Larson, G. *et al.* (2018) Paleogenomics of animal domestication. In: Lindqvist, C. and Rajora, O.P. (eds) *Paleogenomics: Genome-Scale Analysis of Ancient DNA. Population Genomics*. Springer, Cham, Switzerland, pp. 225–272.

Janssens, L., Giemsch, L., Schmitz, R., Street, M., Van Dongen, S. *et al.* (2018) A new look at an old dog: Bonn-Oberkassel reconsidered. *Journal of Archaeological Science* 92, 126–138.

Kaminski, J., Waller, B.M., Diogo, R., Hartstone-Rose, A. and Burrows, A.M. (2019) Evolution of facial muscle anatomy in dogs. *Proceedings of the National Academy of Sciences of the United States of America* 116(29), 14677–14681.

Krause, J., Dear, P.H., Pollack, J.L., Slatkin, M., Spriggs, H. *et al.* (2006) Multiplex amplification of the mammoth mitochondrial genome and the evolution of Elephantidae. *Nature* 439(7077), 724–727.

Kurten, B. (1984) Geographic differentiation in the Rancholabrean dire wolf *Canis dirus* in North America. *Carnegie Museum of Natural History Special Publication* 8, 218–227.

Larson, G. and Bradley, D.G. (2014) How much is that in dog years? The advent of canine population genomics. *PLoS Genetics* 10(1), e1004093.

Ledoux, L. and Boudadi-Maligne, M. (2015) The contribution of geometric morphometric analysis to prehistoric ichnology: the example of large canid tracks and their implication for the debate concerning wolf domestication. *Journal of Archaeological Science* 61, 25–35.

Lindblad-Toh, K., Wade, C.M., Mikkelsen, T.S., Karlsson, E.K., Jaffe, D.B. *et al.* (2005) Genome sequence, comparative analysis and haplotype structure of the domestic dog. *Nature* 438(7069), 803–819.

Linnen, C.R., Kingsley, E.P., Jensen, J.D. and Hoekstra, H.E. (2009) On the origin and spread of an adaptive allele in deer mice. *Science* 325(5944), 1095–1098.

MacHugh, D.E., Larson, G. and Orlando, L. (2016) Taming the past: ancient DNA and the study of animal domestication. *Annual Review of Animal Biosciences* 5, 329–351.

Mayr, E. (1942) *Systematics and the Origin of the Species from the Viewpoint of a Zoologist*. Columbia University Press, New York.

Merriam, J.C. (1918) Note on the systematic position of the wolves of the *Canis dirus* group. *Bulletin of the Department of Geology, University of California* 10, 531–533.

Nowak, R.M. (1995) Another look at wolf taxonomy. In: Carbyn, L.H., Fritts, S.H. and Seip, D.R. (eds) *Ecology and Conservation of Wolves in a Changing World*.

Canadian Circumpolar Institute, Edmonton, Canada, pp. 375–397.

Oetjens, M.T., Martin, A., Veeramah, K.R. and Kidd, J.M. (2018) Analysis of the canid Y-chromosome phylogeny using short-read sequencing data reveals the presence of distinct haplogroups among Neolithic European dogs. *BMC Genomics* 19(1), 350.

Ovodov, N.D., Crockford, S.J., Kuzmin, Y.V., Higham, T.F.G., Hodgins, G.W.L. *et al.* (2011) A 33,000-year-old incipient dog from the Altai Mountains of Siberia: evidence of the earliest domestication disrupted by the Last Glacial Maximum. *PLoS One* 6(7), e22821.

Palkopoulou, E., Lipson, M., Mallick, S., Nielsen, S., Rohland, N. *et al.* (2018) A comprehensive genomic history of extinct and living elephants. *Proceedings of the National Academy of Sciences of the United States of America* 115(11), E2566–E2574.

Perri, A., Widga, C., Lawler, D., Martin, T., Loebel, T. *et al.* (2019) New evidence of the earliest domestic dogs in the Americas. *American Antiquity* 84(1), 68–87.

Perri, A.R., Mitchell, K.J., Mouton, A., Álvarez-Carretero, S., Hulme-Beaman, A. *et al.* (2021) Dire wolves were the last of an ancient new world canid lineage. *Nature* 591(7848), 87–91.

Pilot, M., Greco, C., vonHoldt, B.M., Randi, E., Jędrzejewski, W. *et al.* (2018) Widespread, long-term admixture between grey wolves and domestic dogs across Eurasia and its implications for the conservation status of hybrids. *Evolutionary Applications* 11(5), 662–680.

Tedford, R.H., Wang, X. and Taylor, B.E. (2009) Phylogenetic systematics of the North American fossil Caninae (Carnivora: Canidae). *Bulletin of the American Museum of Natural History* 325, 1–218.

Thalmann, O., Shapiro, B., Cui, P., Schuenemann, V.J., Sawyer, S.K. *et al.* (2013) Complete mitochondrial genomes of ancient canids suggest a European origin of domestic dogs. *Science* 342(6160), 871–874.

The Local SE (2018) Why do today's elephants not interbreed like ancient species? *The Local SE.* Available at: www.thelocal.se/20180227/why-do-todays-elephants-not-breed-like-ancient-species (accessed 26 October 2023).

van't Hof, A.E., Edmonds, N., Dalikova, M., Marec, F. and Saccheri, I.J. (2011) Industrial melanism in British peppered moths has a singular and recent mutational origin. *Science* 332(6032), 958–960.

Vila, C., Savolainen, P., Maldonado, J.E., Amorim, I.R., Rice, J.E. *et al.* (1997) Multiple and ancient origins of the domestic dog. *Science* 276(5319), 1687–1689.

Wayne, R.K. and Ostrander, E.A. (2007) Lessons learned from the dog genome. *Trends in Genetics* 23(11), 557–567.

Wayne, R.K., Benveniste, R.E., Janczewski, D.N. and O'Brien, S.J. (1989) Molecular and biochemical evolution of the Carnivora. In: Gittleman, J.L. (ed.) *Carnivore Behavior, Ecology, and Evolution*. Comstock/Cornell University Press, Ithaca, NY, pp. 465–494.

Wayne, R.K., Geffen, E., Girman, D.J., Koepfli, K.P., Lau, L.M. *et al.* (1997) Molecular systematics of the Canidae. *Systematic Biology* 46(4), 622–653.

Wesley-Hunt, G.D. and Flynn, J.J. (2005) Phylogeny of the carnivora: basal relationships among the carnivoramorphans, and assessment of the position of 'miacoidea' relative to carnivora. *Journal of Systematic Palaeontology* 3(1), 1–28.

2 How to Build a Breed

Abstract

Chapter 2 addresses how breeds of dogs came about and how they proliferated, as well as how breeds are related to one another. There is a review of domestication across species, and for the canine in particular. Artificial selection is discussed, and traits that were valued across historical time are reviewed, including selective breeding for work and appearance of domesticated dogs. Recent advances in genetics allow for a detailed understanding of breed relationships and this is contrasted with historical records of breed relationships.

Hello, Goodbye: Building a Breed

The canines that we know today are but a small representation of the canine species that have ever been on this planet. But dog *breeds*, too, have come and gone due to selective breeding by humans. Before we get into further specifics, though, we'll refer back to what we mean by breed—and why we chose this particular definition. While there are numerous kennel clubs worldwide, for the sake of simplicity, and because your co-authors are US-based, we'll be using the American Kennel Club (AKC) definition: "A domestic race of dogs (selected and maintained by man) with a common gene pool and characterized appearance and function" (AKC, 2023a). But how does a dog become eligible for recognized "breed" status, according to the AKC? To be considered, the following criteria have to be met (AKC, 2023b):

1. A demonstrated following and interest (minimum of 100 active household members) in the breed (in the form of a National Breed Club).
2. A sufficient population in this country (minimum of 300–400 dogs), with a three-generation pedigree. Dogs in that pedigree must all be of the same breed.
3. Geographic distribution of the dogs and people (located in 20 or more states).
4. AKC must review and approve the club's breed standard as well as the club's constitution and by-laws. Breed observations must be completed by AKC Field Staff.

While the long-disappeared dog breeds that we'll now refer to weren't subject to this definition or this set of recognition criteria, this does provide a structural framework for us going forward. Throughout this book, we'll continue to use the AKC definition of breed. Now, on to those who have already come and gone, beginning with the Turnspit dog. The Turnspit dog was bred specifically to run in a wheel called a turnspit that helped meat cook evenly over a fire. With advancing technology, this vocation became obsolete, and people ceased to breed them. The Hare Dog, along with the Salish Wool Dog and Tahltan Bear Dog, were a few of the modern breeds that were endemic to North America. Hare Dogs, which lived in northern Canada, were excellent hunters and once fairly common among the Hare Native Americans. They gradually lost popularity, largely due to their hunting prowess and the fact that they indiscriminately took down both wild animals and livestock belonging to European settlers. Because of this, they eventually died out. The Salish Wool Dog was once popular across what is now modern-day British Columbia. Native American tribes bred them for their "woolly" coats, but with the introduction of sheep from European colonists, people ceased to breed them. The Talbot Dog, which is likely one of the ancestors of today's Bloodhounds and Beagles, became extinct in the late 1700s. The Moscow Water Dog was created by the Russians after World War II. While it was intended for water rescues, the dog preferred to bite people instead. It's easy to see why a tendency like that would be selected against. The Saint John's Water Dog was the founding breed of today's Newfoundlands and Retrievers, but the founding breed is now extinct. The Dogo

DOI: 10.1079/9781800624566.0002

Cubano, or Cuban Mastiff, was endemic to Cuba and bred for guarding and tracking property, both livestock and human. One of the dog's chief jobs was tracking down slaves, but when slavery was abolished, they were no longer bred. Another dog whose extinction came at the hands of important welfare reforms was the Olde English Bulldog: after the UK Cruelty to Animals Act was passed in 1835, these dogs were no longer bred, as they'd been primarily used for dog fighting and bull baiting. The Paisley Terrier had a long, silky coat and helped found the Yorkshire Terrier, but no individuals from this breed exist today. The Tahltan Bear Dog was indigenous to Canada and, while it was bred to hunt bears, it had a gentle temperament with people (Fig. 2.1). The Canadian Kennel Club rescinded recognition of the Tahltan Bear Dog in 1974, after there had been no new registrations for more than a quarter of a century.

Several dog breeds have died out from illness; disease outbreaks in confined areas with a small population of closely related individuals can quickly wipe out the entire group. The Brazilian Tracker became extinct due to a disease outbreak; it had only been recognized by the Brazilian Kennel Club 6 years earlier, in 1967. The Alpine Spaniel, which sported a thick coat, became extinct after a disease outbreak, but some of its ancestors helped found today's Saint Bernard breed. In light of these cases, genetic diversity continues to be an animal welfare concern.

Whether it was for their woolly fur, nimbleness in turning a spit, or ability to track or hunt, many dogs were initially bred for a specific purpose. Researchers have long wondered *what* precipitated their domestication, though. Was it symbiosis? A perfect storm of factors? Hypotheses for the origin of dog domestication include the "Pet Hypothesis," wherein wolf cubs with friendlier temperaments were selected to become part of the family, the "Garbage Dump Hypothesis," wherein wolves chose to spend their time around humans because resources were readily available, and the "Canine Cooperation Hypothesis," wherein wolves became highly cooperative with humans (Range and Virányi, 2015). Perhaps it was some combination of these scenarios. The actual process of domestication comprises the initial taming, followed by selecting individuals with certain traits to breed with one another.

Fig. 2.1. Tahltan Bear Dog. Photography courtesy of Wikipedia (PD-US, unknown author taken prior to 1930). Available at: https://commons.wikimedia.org/wiki/File:Thaltan-bear-dog.jpg.

Small But Mighty

Even tens of thousands of years ago, the morphological and behavioral differences between domesticated dogs and their wild cousins would have already been recognizable, but while dogs exhibit a suite of characteristics not seen in any other species, not all of the changes of domestication are unique to this species.

When a species is domesticated, it undergoes numerous behavioral changes, including a reduction in reactivity to external stimuli and reduced wariness (Price, 1998). Those who witnessed the behaviors of Romeo the wolf (see Chapter 1, and discussed at length in Chapter 4) noted that while he was a wild wolf and wasn't domesticated, his reaction to novelty was unusual for a wild animal. "He was often quite un-shy about unusual stimuli," said writer and photographer Nick Jans. "He was very calm and thoughtful. He would just look, and look, and look, and then sit down."

Many scientists believe that this might be the most important change during domestication, but it isn't the only reduction that domesticated animals have. Whether it's an omnivore, carnivore, ungulate, rodent, or bird, domesticated animals also experience a significantly decreased brain size in comparison to their non-domesticated peers (Kruska, 1988) (Table 2.1). Dogs, in comparison to wolves, have a 20% reduction in their skull size and 30% less brain mass (Serpell, 1995).

Even farmed fish see a reduction in their brain size in comparison to wild fish.

Interestingly, there are other species that have a reduced brain size, in either overall mass or in specific regions, compared to their closest relatives (e.g. the bonobo and the chimpanzee).

Table 2.1. Brain mass reduction as a function of domestication (Kruska, 1988).

Species	Reduction in brain mass compared to their wild counterpart (%)
Cats	27.6
Ducks and geese	16
Ferrets	29.4
Guinea pigs	13.4
Horses	14
Llamas	17.6
Mink	20
Pigs	33.6
Sheep	23.6
Turkeys	29

After adjusting for body size, parts of chimpanzees' brains are larger than bonobos'. Bonobos, compared to chimpanzees, exhibit less aggression, more play behavior, more social tolerance, and a wider range of sexual behaviors. While bonobos' brains are smaller than their chimpanzee cousins, several regions of the bonobo brain are more complex, including the brain's emotional center. Bonobos' behavioral differences, paired with their neural differences, indicate that they excel at regulating their aggressive impulses compared to chimpanzees (Rilling *et al.*, 2012), leading scientists such as Dr Brian Hare to hypothesize that bonobos "self-domesticated" (Hare *et al.*, 2012). Many of the behavioral differences exhibited between bonobos and chimpanzees are comparable to the differences between dogs and wolves.

Wolves and domesticated dogs don't just have a quantitative difference between their skulls and brains, though; the dog's brain also has qualitative differences from its wolf cousins. In fact, there are distinct differences in the brains of different dog breeds; we'll examine this further in Chapter 5. There is a high proportion of differentiation among brain-related genes with the canids. Differences include a highly conserved hypothalamus among wild canids and a high proportion of gene differentiation with domesticated ones (Saetre *et al.*, 2004). The hypothalamus is a small brain region that plays a crucial role in several functions, including thermoregulation and the release of hormones; it also links emotional, endocrinological, and autonomic responses to exploratory behavior. And while dogs and humans have been separated by tens of millions of years, their brains are attuned to many of the same smells and sounds. Scientists at Emory University used magnetic resonance imaging (MRI) and functional MRI (fMRI) machines to create neuroimages of how dogs' brains processed smells. Given their long co-evolution with humans, it makes sense that the odor that triggered the "reward center" (an area called the caudate nucleus) of the brain the most was the smell of their owners.

Many of the sounds that are salient to humans are salient for dogs, as well. When the human brain processes speech, it analyzes both what we say and how we say it. The left hemisphere processes meaning, while the right processes intonation. Andics *et al.* (2014) used comparative neuroimaging of primates and non-primates to discern if there were functionally analogous voice-sensitive regions in the brain. The researchers examined dogs' responses to certain sounds, including canine barks and human barks, sighs, grunts, and speech, finding that dogs, like humans, were sensitive to vocal emotional valence cues. With both domesticated dogs and humans, the auditory cortex lit up when the individual heard "happy" sounds. The lineages of canines and humans split approximately 95 mya, but because of convergent evolution, dogs read humans often better than we read ourselves. Thus, over multiple generations, two species that aren't evolutionarily close co-evolved—and this convergent evolution resulted in dogs having brains that were wired to respond to human smells and sounds.

Focusing again on overall differences in brain mass, we can see that there are key differences in the brain mass of certain primate species, as well. Humans have a reduction of brain size of up to 25% from their extinct hominid cousin, the Neanderthals. Modern humans have an average cranial capacity of 1200–1450 cm, while Neanderthals had a 1600 cm capacity. For most of our evolutionary history, the human brain size to body ratio (called the encephalization quotient) has steadily increased. Modern humans have the highest encephalization quotient at 7.4–7.8, while the chimpanzee's is 2.2–2.5. (Dogs clock in at 1.2, with cats close behind at 1.0.) Approximately 2–2.5 mya, the average human brain was 400–450 g, but by 200,000–400,000 years ago, they weighed 1350–1450 g (Wood and Collard, 1999). Tripling your brain size in a "short" amount of time (relatively speaking, of course) can really differentiate one species from another—and it definitely set the humans apart from the other large-bodied primates. Our closest relatives, the chimpanzees, have brains that are roughly the size of the ancestral human brain, or one-third the mass of the modern human brain. Most of the differentiation with the human brain occurs in the neocortex, which is the brain region responsible for higher cognitive functions associated with self-awareness and language.

Bigger isn't *always* better, though. Recall that dogs have a 20% reduction in their skull size and 30% less brain mass than their wolf cousins (Serpell, 1995). Around the Late Pleistocene (30,000 years ago), there was a reversal with the trend in human brain size: the human brain and body experienced 10% *decreases* in size (Henneberg, 1998). During the past 10,000 years, our species continued to see a slow but steady decrease in brain size. This is likely a result of two factors: our declining body size (most modern humans lead less physically strenuous lives than their ancestors from 10,000 years ago did) and the caloric requirements needed to feed a large brain (*Scientific American*, 2014). Large brains are expensive! What might have been the genetic basis for these trends? Researchers have analyzed the genes controlling brain growth trends and development, with a focus on primary microcephaly (MCPH). This is because microcephalic brains have a volume similar to that of the early hominids. MCPH is associated with several genetic mutations, one of which is called abnormal spindle-like microcephaly (*ASPM*). While *ASPM* is associated with microcephaly, it probably also plays a much larger genetic role in human brain evolution (Zhang, 2003). The *ASPM* gene provides instructions to make a protein that codes for orderly cell division, particularly with neural progenitor cells, the early brain cells that give rise to neurons. Increased neurons, and shorter path lengths between them, are associated with a higher intelligence quotient (IQ). Think of it this way: your brain is a complex, interconnected network that's constantly processing information. Brains affected by the ASPM mutation could be smaller, more connected, faster due to "shortcuts," and less calorically expensive. They would have increased neural efficiency. *ASPM* is a determining factor in brain size and how many neurons the brain has (National Library of Medicine, 2011), and among primates, *ASPM* areas under positive selection are the most highly diverged regions (Kouprina *et al.*, 2004).

Four Tenths of One Percent

Now let's return to our discussion of the differences between bonobos and chimpanzees and dogs and wolves. While domestication is correlated with smaller brain sizes, domestication isn't to be conflated with the gray wolf/domesticated dog date of divergence. This likely occurred shortly after they last shared a common ancestor. Some scientists hypothesize that domestication occurred twice or more during their history with humans. While genome sequencing continues to point to a single domestication event (Botigué *et al.*, 2017) we can look at convergent evolution, where unrelated organisms independently evolve similar traits in the same ecological niche, to see how two separate populations of ancestral wolves may have become domesticated.

A convergent evolution example is that birds, bats, and butterflies have wings to fly and sharks and dolphins have streamlined bodies, fins, and flippers for navigating the water, but neither grouping of species is closely related. The idea of two or more domestication events for dogs isn't without support, given fossil finds of now-extinct lineages of dogs, and it's not without precedent, either; around 9000 years ago, the pig had two domestication events, one in eastern Eurasia (in the Yellow River region, China) (Cucchi *et al.*, 2011) and one in western Eurasia (in Anatolia, Turkey) (Ervynck *et al.*, 2001). Researchers hypothesized that two populations of wolves, one in eastern Eurasia and one in western Eurasia, independently became domesticated shortly after wolves and dogs diverged (Frantz *et al.*, 2016). Some northern dog breeds, such as the Greenland sled dogs and Siberian Huskies, have between 1.4% and 27.3% Taimyr wolf DNA (Skoglund *et al.*, 2015), which could either indicate admixture (which occurs when a previously divergent or isolated genetic lineage is introduced to a novel gene pool, resulting in new genetic lineages) (Rius and Darling, 2014) among the populations, as was the case with *Homo sapiens* and *Homo neanderthalensis*, or multiple domestication events. Regardless of how and when the first wolves joined humans, the genetics of modern dogs reveal that they all descended from a common ancestor of the gray wolf, *Canis lupus*, with domesticated dogs and wolves still sharing 99.6% of their DNA in common. But that 0.4% is crucial in terms of behavior, morphology, nutrition, and development.

Even within the diverse family of Canidae, the domestic dog is unique in the animal kingdom in purpose and in the influence of artificial selection on creating breeds (Wayne and Ostrander, 2007). Wild canids range from the 8-inch-tall fennec fox to the 32-inch-tall gray wolf, but domesticated dogs vary even more, ranging from the 5.9-inch-tall Chihuahua to the massive 44-inch-tall Great Dane (Fig. 2.2). The domesticated dog's unmatched phenotypic and behavioral diversity have made them a model for genetic research: their extreme phenotypic diversification has yielded 400—and by some accounts, 1000 or more!—distinct dog breeds (Mehrkam and Wynne, 2014). Domesticated dogs also have a wider variety of coat lengths and colors, leg lengths, face shapes, tail lengths and curvatures, ear positions, and body types than their wild canid cousins do. Domestic dogs have more phenotypic diversity than all of the other species of carnivore (Freedman *et al.*, 2016). The diversity of dog breeds includes the excessive wrinkling of the Shar-Pei, the chondrodysplasia (disproportionately short legs) of the Pembroke Welsh Corgi, and the distinct brachycephalic skull of the Bulldog. All of these disparate breeds, separated by only 0.4% of their shared genetic makeup.

So what's the significance of 0.4%? vonHoldt *et al.* (2017) delved into some of the structural variants that might lie behind it. Researchers analyzed a genomic region on chromosome 6 in canines. When this region was deleted in humans, it was associated with Williams–Beuren Syndrome, a congenital disorder marked by hypersocial behavior, including being very empathetic and anxious. Hypersociability is one of the key traits that distinguishes dogs from wolves, and a mutation wherein this was selected for would have aided the domestication process. The researchers found four structural variations, including two in genes that were called *GTF2I* and *GTF2IRD1* that were known to cause hypersociability in humans. *GTF2I* has also been linked to hypersociability in mice.

Dr Monique Udell, an experimental psychologist at Oregon State University, studies the behaviors of wolves and dogs. Dr Udell examined the similarities between humans with Williams–Beuren Syndrome and dogs, noting that humans with this syndrome were overly friendly with previously unknown persons (as dogs often are) and didn't persevere on cognitive tests. This was similar to what researchers found when comparing dogs to wolves in problem-solving experiments: wolves would be persistent, while dogs would look to the human experimenters for help. Among humans, Williams–Beuren Syndrome marks developmental delays; interestingly, domesticated dogs appear to also have a period of delayed development compared to wolves, as they develop at slower rates than their wolf cousins do across all stages.

This is similar to the differences between humans and chimpanzees. Humans tend to be more altricial (meaning that they have increased periods of dependence during infancy and adolescence) than their chimpanzee relatives, who reach all of their developmental landmarks at an earlier age than we do. But premature human infants have a better "cling" reflex than full-term babies—a cling reflex that's very similar to the one shown by full-term chimpanzee infants. Explanations for why human babies are so helpless at birth include: (i) limitations of the birth canal which constrain prenatal

Fig. 2.2. Great Dane and Chihuahua skeletons. Photography courtesy of Wikipedia under Creative Commons CC0 1.0 Universal Public Domain Dedication. Available at: https://commons.wikimedia.org/wiki/File:Great_Dane_and_Chihuahua_Skeletons.jpg.

brain development, (ii) the metabolic cost of carrying a fetus for longer than 9 months, and (iii) the importance of culture and the environment on shaping brain development (Dunsworth *et al.*, 2012). Researchers have recently debunked the hypothesis about the birth canal, leaving the metabolic and developmental hypotheses still under debate. Regardless of the cause of altriciality in human infants, having babies that cannot cling to their mothers selects for pair-bonding, grandparenting, and alloparenting (care by individuals other than the biological parents) (Dunsworth *et al.*, 2012). While delayed development takes more investment from the parent or parents, there are benefits to this delay, particularly with emotional and social behaviors such as the development of language and comprehending complex social cues.

Consider, now, how dependent dogs are on their humans, from infancy onward. The domesticated dog's extended adolescence and increased dependence, in comparison to their wolf cousins, are the perfect recipe for interspecies bonding. This is evident in the cognitive work performed by Dr Udell, who has found that dogs will "look" to their humans for help during a problem-solving task rather than persevering (Udell, 2015). Dogs, like human children, are inherently dependent upon us.

Dr Udell recalled the discovery of the Williams–Beuren link. "I was working with Clive Wynne when we ran into Williams–Beuren Syndrome. Now, we know that dogs have a developmental delay in the social realm, in comparison to wolves. In cognitive tests, dogs will find the person so distracting that they focus only on the person and not the task. Even when dogs can see that the object that they want is in the apparatus on the left, if the human points to the right, the dogs go to the right. These overriding social cues happen with the dogs, but not with wolves. The wolves were cognitively aware and used the social cue when it was the right answer and the physical cue when it was the right answer."

"We wondered: instead of being socially advanced, maybe dogs are actually having a hard time. Maybe instead the wolves are acting how a normal canine *should* act."

Dr Udell was presenting on this topic when a faculty member at Princeton asked her: "Do you know Bridget von Holdt?" von Holdt was a geneticist at Princeton and she was the first to discover the Williams-Beuren region.

"We sat down for coffee," Dr Udell recalled, "and it was so serendipitous. We moved Clive into the conversation, too—he'd said that we should find a geneticist for this project—and Bridget was the best possible person to have collaborated with. We were able to get the samples, test the animals, pull in some of her collaborators ... And we received some pretty straightforward outcomes. And it's not even differences in a gene; it's differences in SNPs [single nucleotide polymorphisms; we'll get to this in Chapter 3] in a region on the gene. It looks like a whole host of changes likely had to have happened around the same time. But did one lead to another, or did they happen together?"

"Anna Kukova's laboratory is looking at a similar region in the brain, and she's examining the factors that go into fear reduction."

The dog's developmental trajectory is distinct from the wolf and plays an important part in their relationship with us. The dog's development begins with the prenatal period, when their mother is still pregnant. After birth, the dog's life stages are divided into the neonatal period (from birth to 2 weeks), the transition period (the following week), the awareness period (the following 3 days), the socialization period (3–12 weeks), and the juvenile period (approximately 12 weeks to 6 months or later) (Scott and Fuller, 1965). During the neonatal period, puppies are unable to thermoregulate (maintain their body temperature), defecate, or urinate on their own; they're blind, deaf, and wholly dependent upon their mothers. During their transition period, their eyes and ears open and their senses of smell and taste kick in, as well. During the awareness period, puppies begin to learn—and their brainwaves exhibit qualitative changes. During the socialization period, puppies learn the language of their species, including species-specific behaviors, such as context-specific reactions, body postures, and vocalizations, that will aid them during social encounters. During the socialization period, puppies learn at a rapid rate and begin bonding with their humans, as well.

Dogs, like humans, have sensitive and critical periods of their development. Being exposed to or socializing with humans outside of these times decreases their likelihood to bond with a human partner. A critical period of development is when the nervous system is particularly sensitive, and an individual cannot learn something beyond this time frame. A sensitive period is a time during which an individual can learn rapidly. It's in a dog's genes to bond with humans, but timing is everything, and dogs' tendency to be more altricial compared to wolves plays a crucial role in this.

Some species tend to be more precocial (meaning that they're active and capable of moving about freely shortly after birth, with little or limited parental care and high levels of independence) while others tend to be more altricial. Prey animals, such as horses and deer, are particularly precocial; they're at the farthest end of the mammalian precocity spectrum. They can stand, walk, and even run within an hour of birth. Humans are on the opposite end of that spectrum and are one of the most (if not the most) altricial species on the planet; domesticated dogs and cats are also considered to be particularly altricial, as they are both born deaf, blind, and unable to thermoregulate, defecate, or urinate on their own (Arterberry, 2000). Wolves are also both born deaf and blind, but across all major developmental milestones, they're more precocious than dogs: their eyes open earlier, they hear and walk sooner, and they are also able to manipulate objects with their mouths at an earlier age than domesticated dogs.

This comparative developmental delay could have been the key to domestication. Domesticated dogs have a considerably shorter period of maternal investment than wolves do (and typically no paternal investment). While mother dogs will nurse their pups for 8 or more weeks, investment tends to tail off between 6 and 11 weeks of age. Perhaps that's because that's when humans have stepped in, throughout their evolutionary histories—or perhaps people stepped in because this was when dog mothers began to decrease their investment. Wolf pups, in contrast, are raised by both parents well into their second years, even though they're far more precocial than dogs are. Thus, wolves are more developmentally advanced, in comparison to dogs, but wolves' time with their parents is often tenfold what domesticated dogs would spend with their mothers. Dogs are both altricial and highly dependent upon humans. Unsurprisingly, domesticated dogs that live in feral situations have higher rates of infant mortality than their wolf cousins do. Rates of mortality prior to reaching the first year of age range from 45% (Beck, 1973) to 95% for feral dogs (Boitani and Ciucci, 1995), while wolf pup mortality rates are highly variable, but are estimated at approximately 40–60% (International Wolf Center, 2023) (and largely due to human causes).

So what's the mechanism behind this difference? Well, it's us—you, me, and everyone who has ever had a dog, for generations and generations. One of the most popular hypotheses behind this differential investment is the attractiveness of puppies, which have the retention of neotenous characteristics. Neoteny, also referred to as paedomorphosis or juvenilization, is the retention of infantile characteristics—physical traits one would find in the youngsters of a species. Ethologist Konrad Lorenz was the first to identify these characteristics—and humans' propensity to like species that had them. Lorenz called these traits "Kindchenschema" (infant or baby schema) and they include a round face, large eyes and cheeks, a small mouth and nose, and a protruding forehead (Lorenz, 1943) (Fig. 2.3). These traits are hypothesized to inspire humans to be nurturing. Researchers believe that these physical traits, paired with particularly "endearing" behaviors like "puppy dog eyes," developed during the domestication process (Kaminski *et al.*, 2017). Now picture a baby human, chimpanzee, cat, or dog: these are all species that have a high degree of "Kindchenschema". While wolves do have these traits to a certain extent, as well, dogs have more exaggerated and prolonged neotenous characteristics. These are thought to enable us to deal with any behavioral challenges as they grow. Between their mother's decreased investment and dogs' typically difficult (think teenage) stage of 18–24 months, it's a good thing that dogs, like many domesticated animals, retain highly attractive characteristics throughout their lifetimes.

So how do you "measure" cute? Certain breeds (especially those with brachycephalic faces, like the Cavalier King Charles Spaniel and the Pomeranian) are said to be more neotenous than others. A growing body of research supports the hypothesis that neotenous traits among human infants elicit caregiving responses (Glocker *et al.*, 2009). Researchers tested this hypothesis with dogs, as well, rating the "attractiveness" of dogs from birth to 7 months of age (Chersini *et al.*, 2018). They selected three breeds (White Shepherd, Cane Corso, and Jack Russell Terrier). The dogs had similar "peaks" of attractiveness, with the Cane Corso's peaking at 6.3 weeks, the Jack Russell Terrier at 7.7 weeks, and the White Shepherd at 8.3 weeks. This peak attractiveness corresponded with weaning and when humans would have likely stepped in to take over the "parental" role. Humans weren't the only species interested in faces, though; dogs were closely examining ours, as well. Recent research has shown that dogs can discriminate the emotional expressions of human faces—a skill likely developed early on in their domestication (Müller *et al.*, 2015).

The Origin of Breeds

We have stated that for the sake of convenience and simplicity that we will be using the AKC definition of breed, but the AKC was first founded in 1884, and dogs were first domesticated tens of thousands of years ago. We concede that the origin of dog breeds—and thus, quite often, how a breed should be defined—remains under debate, not because we don't know when particular traits arose and were identified (kennel clubs help with this immensely), but because breed definitions and groups vary from club to club. Again, we're still arguing over their Latin binomial (*Canis familiaris* or *Canis lupus familiaris*); breed differences are even harder to define.

Wade into the murk with us. Modern dogs currently vary from the 1-pound Teacup Poodle and Chihuahua to the 200-pound Giant Mastiff, but the first members of *Canis familiaris* were likely very wolf-like in behavior and appearance. These first dogs likely exhibited minimal phenotypic and genotypic variation compared to modern dogs. Wolves' phenotypic variation, including coat color and skull size (Jolicoeur, 1959), height, and weight, however, provided the foundation for the variety that we see with modern dogs. This variation led Scott and Fuller (1965) to refer to wolves as polymorphic, concluding that the ancestors of modern domesticated dogs and wolves also had a high level of variability, excepting their fairly uniform skeletal proportions (Hildebrand, 1954).

Ancient dog breeds predate their documentation, but the first historical record of distinctive dog breeds appeared around 3000–4000 years ago (Clutton-Brock,

Fig. 2.3. Diagram demonstrating neoteny on the left, with large eyes and short snouts, and elongated snouts with small eyes on the right. Creating author: Ephert, 2014. Permission to use based on Creative Commons Attribution-Share Alike 4.0 International license. Available at: https://commons.wikimedia.org/wiki/File:Animal_human_growth_skull_neoteny_cuteness_maturation.png.

1995; Mehrkam and Wynne, 2014). Many of the breed groups we know today had been identified by the Roman period, and Europe saw a sizable propagation of breeds during the Middle Ages (Mehrkam and Wynne, 2014). The Molosser dog (or Molossus) belonged to the Molossians, a tribe of ancient Greece. These dogs were robust, brave, and loyal. Philosophers, including Aristophanes, Aristotle, Horace, and Virgil wrote about the Molossus. "Never, with them on guard," says Virgil in his poem *Georgics*, "need you fear for your stalls a midnight thief, or onslaught of wolves, or Iberian brigands at your back." There were reportedly two types of Molossus: a broad-muzzled hunting dog that was the likely progenitor of modern-day Mastiffs and a second, larger livestock guardian dog.

The proliferation of most dog breeds, however, didn't occur until a few hundred years ago, when selective breeding really took hold among dog owners. That doesn't mean that "breed standards" weren't a concept yet, though. Pitulko and Kasparov (2017) found that even as far back as 9000 years ago, dogs on Zhokov Island, Siberia were being selected for different roles, much like today's Golden Retrievers can be bred for field or show. The Zhokov dogs were either bred to be sled dogs, with a weight range of 35–55 pounds, or bred to help hunters, with an average weight of 66 pounds. Modern sled dogs weigh 44–55 pounds, a size that provides enough power to pull while still allowing them to thermoregulate efficiently (Pitulko and Kasparov, 2017).

Estimates on the origin of dog breeds and breed types range from ancient times (breeds that are thousands of years old) to more modern ones, such as those created during the Victorian era (1830–1900). Dogs have a unique demographic history and have experienced multiple breed-specific bottlenecks, which in turn has left genetic signatures in their genomes: many modern breeds have genomic signatures from early ancestors (Ostrander *et al.*, 2017). Parker *et al.* (2004) examined the microsatellites (repetitive tracts of DNA, ranging from one to six or more repeated base pairs) of 85 dog breeds, finding that each breed was so genetically distinct that 410 of the 414 (99%) purebred dogs in the study could be accurately placed with their breed based upon their genotypes. This demonstrates that dog breeds can be considered distinct genetic units. Four dog breeds, including the Australian Shepherd, Chihuahua, German Shorthaired Pointer, and Perro de Presa, failed to cluster with other members of their breed, while six breed pairs, including the Alaskan Malamute (Fig. 2.4) and Siberian Husky, Belgian Sheepdog and Belgian Tervuren, Bernese Mountain Dog and Greater Swiss Mountain Dog, Bullmastiff and Mastiff, Collie and Shetland Sheepdog (also known as the Sheltie), and Greyhound and Whippet, primarily clustered together. Nine breeds could be isolated from those with European origins: the Chinese Shar-Pei, Shiba Inu, Chow Chow, Akita, Basenji, Siberian Husky, Alaskan Malamute, Afghan Hound, and Saluki.

Akey *et al.* (2010) tracked the footprints of artificial selection in the dog genome, identifying 155 genomic regions with strong signatures of recent selection and containing genes for highly varying phenotypes, including behavior, body size, skeletal morphology, physiology, and coat color and texture. Another 2010 study examined 48,000 SNPs (variations in a single nucleotide in a DNA sequence; we'll discuss these further later on) with 912 dogs in 85 breeds. The researchers found that modern dogs, including the Herding Dogs, Mastiff-type breeds, Retrievers, Scenthounds, Sighthounds, Small Terriers, Spaniels, Spitz breeds, and Toy Dogs, had distinct genetic clusters that corresponded to

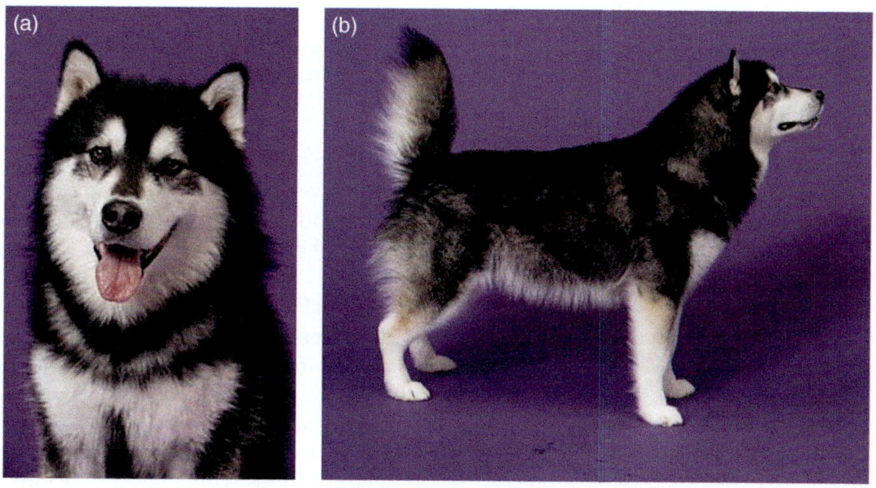

Fig. 2.4. Alaskan Malamute, Dubs, the mascot of the University of Washington: (a) front-on view and (b) stacked position. Photos courtesy of the University of Washington.

their function or phenotype (vonHoldt *et al.*, 2010). Sighthounds were bred to hunt independent from their people. Spitz types, which include some of the oldest breeds, originated in North America, Russia, and Scandinavia. Spitz means "pointed" in German and Spitz-type dogs typically have a wedge-shaped head. According to the AKC, there are currently 50–70 different Spitz breeds. Spitzes typically have a "wolf-like appearance, [including] pointy, pricked ears, almond-shaped eyes, a heavy, double coat, and a feathery tail carried over the back" (Johnstone, 2021).

vonHoldt *et al.* (2010) also found 13 breeds that genetically diverged from modern breeds: the Afghan Hound, Akita, Alaskan Malamute, American Eskimo Dog, Basenji, Canaan Dog, Chinese Shar-Pei, Chow Chow, Dingo, New Guinea Singing Dog, Saluki, Samoyed, and Siberian Husky. The researchers found evidence of three dog groups that were distinct and divergent from modern ones: an Asian group, which included admixture with Chinese wolves, and included the Akita, Chow Chow, Dingo, New Guinea Singing Dog, and Shar-Pei; a northern group, which included the Alaskan Malamute and Siberian Husky; and a Middle Eastern group, which included the Afghan Hound and Saluki.

The preponderance of genomic studies with dogs thus far have been single-gene analyses, arising from questions based upon these phenotypic traits. Some breeds, such as the Beagle, German Shepherd Dog, Greyhound, and Shar-Pei, are particularly genetically distinct from one another (Akey *et al.*, 2010), with the Beagle being traced back to the fifth century BCE in ancient Greece, the Greyhound originating 4000 years ago in Greece and the fertile crescent (Persia and Egypt), and having biblical references, and the Shar-Pei originating around 200 BCE in China. German Shepherd Dogs (also known as Alsatians), conversely, are recognized relatively more recently, descending from a herding dog of unknown origin and being recognized as a distinct breed in Germany in 1899.

The AKC began keeping track of breeds in 1885. Their archival records show that during that year, there were 24 registered breeds. The most common breeds by number were the English Setter, with 834 registered, the red Irish Setter, with 390, Irish Water Spaniels, with 341, Pointers, 285, Spaniels, 219, Black and Tan Setters (Gordon Setters), 121, a mysterious category called "Crossbreds (setters/pointers) no longer registered," 77, Beagles, 35, Collies, 22, Fox Terriers, 17, Dachshunds, 11, Mastiffs, nine, and Greyhounds, eight. Within only 10 years, the number of registered breeds had doubled, with Rough-coated Saint Bernards taking the lead as the most popular breed, with 768 registered. By 1915, three decades after the AKC began tracking breeds, the original breed registry had tripled, with 72 recognized breeds. Boston Terriers had taken the lead as the most popular breed in the US, with 4901 registered (AKC, 1945).

Exactly *where* the first dog originated is still being debated. Current hypotheses are that they originated from multiple sites, including Eastern Asia (Wang et al., 2016), Central Asia (Shannon, 2015), and Western Europe (Thalmann *et al.*, 2013). Dogs have been in the New World for at least 10,000 years, with genetic analyses revealing that some ancestral dogs followed travel routes with humans from East Asia and that dogs originating from southern East Asia had significantly higher rates of genetic diversity in comparison to dogs from other locations (Wang *et al.*, 2016). While it's still cloudy *where* dogs first originated, it's clear that ancient dogs were thriving on multiple continents. Over time, ancient dogs in these diverse areas became isolated islands of different breeds. An examination of 161 dog breeds established 23 clades of dogs (Parker *et al.*, 2017) (Fig. 2.5). This genetic cladogram (a branching diagram showing the cladistic relationship between a number of species) predates the phenotypic breed groups.

While kennel clubs worldwide vary in the number of breed groups recognized, with the Canadian Kennel Club and UK Kennel Club recognizing seven, the AKC and UKC recognizing eight, and the FCI recognizing ten, this cladogram finds almost two dozen distinct clades.

Fit For a Queen

Victoria gazed lovingly at her red-headed Italian companion, her blue eyes darkening as her pupils dilated. She placed one hand on each of his cheeks, pulling his face closer to hers. The woman who had ruled England alone for 27 years hadn't found a replacement for her beloved Prince Albert, but she was definitely smitten with the diminutive Volpino Italiano, a Spitz-type dog that originated in the Pomerania region near the Baltic Sea. Queen Victoria, the granddaughter of King George III of England and his wife, Queen Charlotte, was just as enthusiastic about the fledgling "Pomeranian" breed as her grandparents were. While the breed

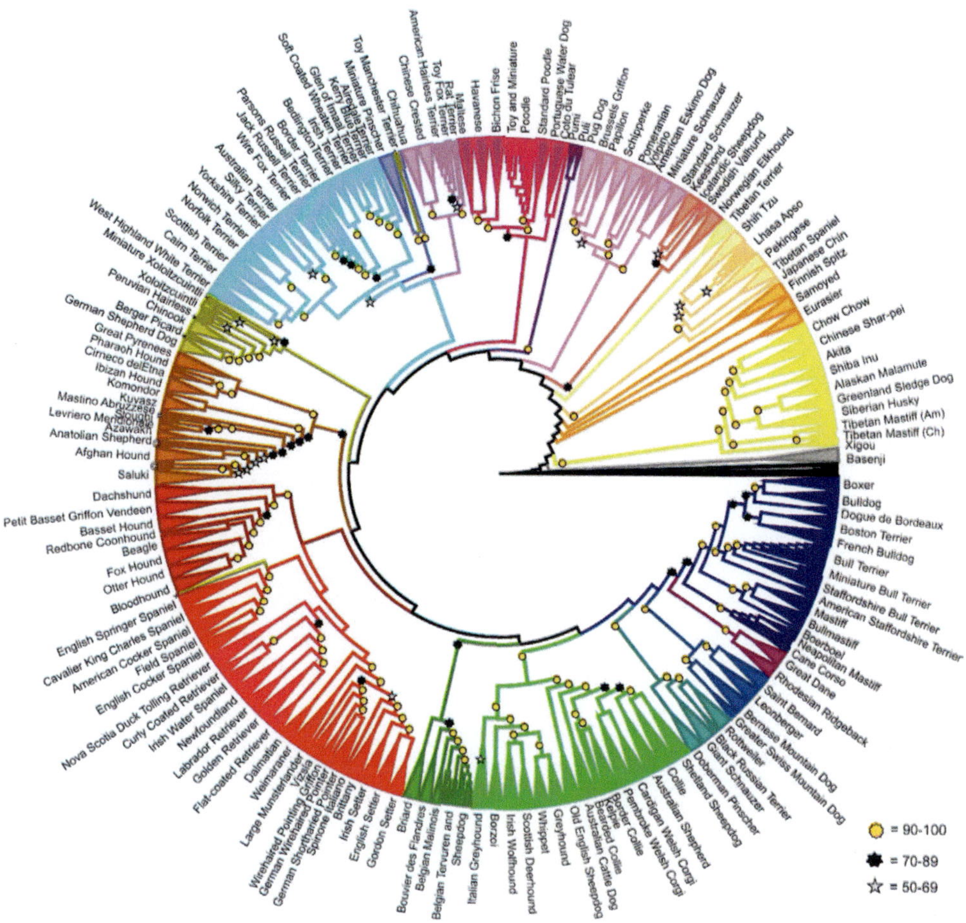

Fig. 2.5. Cladogram of 161 domestic dog breeds (Parker *et al.*, 2017). Breeds that form unique clades supported by 100% of bootstraps are combined into triangles. For all other branches, a gold star indicates 90% or better, black star 70–89%, and silver star 50–69% bootstrap support. Breeds are listed on the perimeter of the circle. A small number of dogs do not cluster with the rest of their breed, indicated as follows: *Cane Paratore, +Peruvian Hairless Dog, #Sloughi, @Country-of-origin Salukis, and ^Miniature Xoloitzcuintli. This work is licensed under license: http://creativecommons.org/licenses/by/4.0/ Creative Commons Attribution 4.0 International License.

standard of the time was approximately 30–50 pounds, Victoria's favorite dog was a diminutive redhead she named "Windsor's Marco" (Fig. 2.6). Marco tipped the scales at only 12 pounds, and shortly after she began to exhibit him in 1891, the breed standard shifted.

Dog breeds like the Pomeranian, the breed of dog that Jeff and Julia were dealing with in the Prologue of our book, are a relatively modern innovation, stemming primarily from Victorian-era efforts to create more "perfect" specimens. During her lifetime, the size of the Pomeranian decreased by 50% as Victoria selected those dogs who would produce her preferred, smaller-statured dogs. This revealed just how rapidly a breed could be reshaped, given preferable mating opportunities for only those individuals chosen to reproduce. Essentially, Victoria took the bowl of paint and added only blue to it. Other Pomeranian owners, wanting to follow this new trend, did the same.

While the Pomeranian's physical changes were particularly fast, all domesticated dogs have been shaped by artificial selection in a relatively short amount of time, evolutionarily speaking. We share our lives with dogs, but compared to so many other species, we still know relatively little about the

Fig. 2.6. Pencil and watercolor portrait painting by Frances C. Fairman (1836–1923), of Marco, a brown Pomeranian Terrier. It hangs in the Durbar Corridor at Osborne House, Isle of Wight, UK, and the dog belonged to Queen Victoria. This work is in the public domain in its country of origin and other countries and areas where the copyright term is the author's life plus 70 years or fewer. This work is in the public domain in the US because it was published (or registered with the US Copyright Office) before January 1, 1928. Available at: https://commons.wikimedia.org/wiki/File:Marco_by_Frances_C._Fairman.jpg.

genetics of dog breeds. The scientific study of domesticated dogs is still in its infancy, but public demand for information pertaining to their origins and behaviors associated with their genes is at a record high. While other animals, including horses, cows, and cats, have experienced selective pressures, as well, their selection has been both more recent and less pressured than the selection with domesticated dogs. Researchers have just begun to delve into the specific behavioral differences between breeds among several species, including horses, cats, chickens, and dogs. Patterns and propensities have been found among certain breeds, and while generalizations on breed differences can be made, the topic remains controversial due to potential misuse with breed bans.

The Victorian era prompted a paradigm shift in purposefully breeding dogs to enhance or remove certain physical and behavioral characteristics. The domesticated dog's divergence from the rest of its more closely related canid cousins occurred more quickly and more dramatically than that of the peppered moth, and in widely varied phenotypical iterations. Frequently, these iterations have yielded unexpected results: efforts to enhance or remove a trait have also resulted in some unintended side effects being increased or decreased. Manipulating the intermixing of the genetics of two animals might have advantageous side effects for the breeder, such as a highly desired coat color, or they may be disadvantageous, such as hip dysplasia. For decades, animal behavior researchers preferred to work with more exotic wild species rather than dogs, as dogs were artificially created, but this swift, dramatic change in phenotype among the breeds over only 200 years makes them an even more compelling species to study. When Darwin (1859) wrote, "Man selects only for his own good: Nature only for that of the being which she tends," there was no better example than the dog.

The Brothers Belyaev

Resting prone against the cold floor, Nikolai Ivanovich Vavilov stared at the bars of his cell, his eyes struggling to bring them into focus. His trademark mustache trembled slightly as he drew a few more shaky breaths, and then he was still. Vavilov had dedicated his life to cultivating and improving food crops, but he starved to death in a Russian prison in 1943. His crime? Science. Specifically, the "wrong" kind of science for the political climate.

Vavilov's death was a stark warning for other scientists. Dmitry Constantich Belyaev and his brother, Nikolai, were Darwinists in the Soviet Union during the Cold War. Nikolai, 18 years Dmitry's senior, was already an established geneticist before his younger brother entered the field. The Belyaev brothers believed in the principles of Mendelian genetics, but in Russia during Stalin's regime, this was dangerous to admit. Stalin was a staunch supporter of Trofim Lysenko, who supported a Lamarckian viewpoint that was creatively re-packaged as "Lysenkoism." Stalin outlawed Mendel's works. Those who continued to adhere to Darwinism and Mendelian genetics were labeled as enemies of the state—a crime punishable by death, as leading Russian geneticist Nikolai Vavilov had discovered. Vavilov had received a death sentence for his scientific viewpoints; the sentence was later changed to 20 years' imprisonment, but he died of starvation under questionable conditions. Despite the inherent dangers of their science, the Belyaev brothers persisted in their field. Dmitry's dissertation was: "The variation and inheritance of silver-colored fur in silver-black foxes," and even though he disguised his work as "animal physiology" rather than genetics, he lost his position managing the Department of Fur Animal Breeding. At least he only lost his job. And thus, the stage was set for Belyaev's subversive scientific triumph.

Cold War Canines

During the 1940s and early 1950s, the Cold War between the US and the USSR continued to heat up—and so was government-funded science and the space race. Fueled by fear, espionage, and an arms race with the specter of nuclear retaliation, science was particularly volatile during this time. Stalin's death in 1953 provided some relief for geneticists. In the US, the fear of Communist threats was dubbed the "Red Scare," and in 1954, the words "Under God" were added to the Pledge of Allegiance as an anti-Soviet proclamation. In Russia, scientists that were interested in work that didn't apply to the Cold War and *did* apply to genetics continued to be a bit duplicitous about the aims of their research. Dmitri Belyaev partnered with fellow scientist Nina Sorokina to reproduce the effects of canine domestication—but they wanted to do so not with the gray wolf, but on a new model: the silver fox (*Vulpes vulpes*). Securing funding for work of this kind was tricky, so Belyaev and Sorokina said that they needed funding for a study on how to multiply the lucrativeness of the fox fur trade: namely, increasing the number of offspring born and the amount of fur that each fox produced.

Five years after Stalin's death, Belyaev secured his funding. He asked his assistant, Lyudmila Trut, to select the calmest foxes that she could find. Belyaev and Sorokina started with a population of 130 foxes, continuing to select those that were less fearful of humans to reproduce. As friendly individuals were mated with friendly individuals, each succeeding generation of foxes became less fearful of humans—and along the way, they lost a lot of their "foxy" physical characteristics, as well. From 1 month of age onward in 1-month intervals, the experimenters tested the fox pups to determine whether they tolerated being handled by humans and whether they preferred spending time with their peers or with human companions. At 7 or 8 months of age, when they reached sexual maturity, the foxes were given a "tameness score." Class I comprised individuals that were the friendliest with humans, Class II allowed handling, but weren't friendly, and Class III were the least domesticated individuals. The scores were based on several factors, including whether they bit the experimenter and whether they approached or avoided them. Over the generations, the foxes began to exhibit not larger litter sizes and more fur production, as Belyaev and Sorokina cleverly claimed to their benefactors, but floppy ears, mottled coats, and tails that were markedly less fluffy and more curled. Within several generations, the foxes also lost much of their "foxy" behavior, and they began to wag their tails and lick their human handlers. The foxes had been artificially selected to "become" dogs; they had, as Lee Dugatkin so aptly put it in his scientific best-seller, tamed a fox and built a dog (Dugatkin and Trut, 2017). After only four generations, the first tail wag occurred. Soon afterward, the foxes were so friendly that they would leap into the experimenters' arms—much like a domesticated dog would. Six generations into the experiment, Belyaev amended the classes, adding Class IE, indicating "domesticated elite." Individuals with this rating eagerly established contact with humans, acting much like domesticated dogs would, including licking the experimenters. By ten generations, 18% of the fox pups were in the IE category, and ten generations after that, 35% had that designation. In 2009, the IE category comprised 70–80% of the foxes. While Belyaev died in 1985, their research continues today: more than 40 generations of foxes have been bred (Trut, 1999) (Fig. 2.7). Along with their behavioral and morphological changes, the foxes also experienced physiological ones: their reproduction was no longer seasonal and they experienced changes in their pituitary–adrenal function, as well. The latter indicated that the tamed foxes had decreased activity in their hypothalamus–pituitary–adrenal (HPA) axis, the body's central stress response system, a connection between the brain and the endocrine system. This is similar to the changes seen in the highly differentiated

Fig. 2.7. Lyudmila Trut with domesticated fox. This photograph is courtesy of Svetlana Argutinskaya (author, 1974). Available at: https://commons.wikimedia.org/wiki/File:L._Trut_and_domestic_Fox-1974.jpg, licensed under CC BY 4.0.

hypothalamus of domesticated dogs, indicating that this area was important during the domestication of both dogs and foxes (Saetre *et al.*, 2004).

Researchers have sequenced the genes of ten foxes whose temperaments ranged from aggressive to tame. Researcher Anna Kukekova and her team assembled a genome of the silver fox, focusing on the 103 genomic regions that differed between the two temperaments. The tamest foxes had a version of a gene called *SORCS1* that was not present in aggressive or conventionally bred foxes. A different *SORCS1* version was found in the aggressive foxes, but absent in conventionally bred ones. Interestingly, the aggressive foxes, but not the tame ones, had the "Williams–Beuren region," which is associated with hypersociality, but also increased rates of anxiety, which the aggressive foxes did display. When confronted with novel objects and unfamiliar people, domesticated animals appear to experience less stress than their wild peers do, and Kukekova's team found that this may be due to the "blunted response" in the HPA axis. Thus, while there are many similarities in the neural changes between dogs and foxes, there appeared to be a different genetic mechanism in foxes as it pertains to the Williams–Beuren region.

Conclusion

The four-legged, omnivorous "formula" for a canine might always be the same, but researchers want to know what, when, where, why, and how this recipe for our best friends came together. While we've closed in on the answers to many of the questions about dog domestication, including *who* (the canines that share an ancestor with *Canis lupus*) and *when* (approximately 40,000 years ago), we cannot say for sure *which* specific canines took the plunge toward domestication, *where* the first domestication occurred, *why*, or *how*, although we're closing in on this—and we have a rather provocative hypothesis that's driving these questions. We assert that breed and pedigree are just as important as the environment, including epigenetics and early life experiences, in shaping a dog's behavior. While this might sound like a bold claim, throughout these chapters, we will provide ample evidence of the interrelationship of these factors.

References

AKC (1945) Breed popularity AKC registrations 1885–1945. Available at: http://images.akc.org/pdf/archives/AKCregstats_1885-1945.pdf (accessed 27 October 2023).

AKC (2023a). Available at: www.akc.org/about/glossary (accessed 25 October 2023).

AKC (2023b). Available at: www.akc.org/press-center/facts-and-stats/becoming-recognized (accessed 25 October 2023).

Akey, J.M., Ruhe, A.L., Akey, D.T., Wong, A.K., Connelly, C.F. *et al.* (2010) Tracking footprints of artificial selection in the dog genome. *Proceedings of the National Academy of Sciences of the United States of America* 107(3), 1160–1165.

Andics, A., Gácsi, M., Faragó, T., Kis, A. and Miklósi, A. (2014) Voice-sensitive regions in the dog and human brain are revealed by comparative fMRI. *Current Biology* 24(5), 574–578.

Arterberry, M.E. (2000) *The Cradle of Knowledge. Development of Perception in Infancy*. MIT Press, Cambridge, MA, p. 27.

Beck, A. (1973) *The Ecology of Stray Dogs*. York Press, Baltimore, MD.

Boitani, L. and Ciucci, P. (1995) Comparative social ecology of feral dogs and wolves. *Ethology Ecology & Evolution* 7(1), 49–72.

Botigué, L.R., Song, S., Scheu, A., Gopalan, S., Pendleton, A.L. *et al.* (2017) Ancient European dog genomes reveal continuity since the early Neolithic. *Nature Communications* 8, 16082.

Chersini, N., Hall, N.J. and Wynne, C.D.L. (2018) Dog pups' attractiveness to humans peaks at weaning age. *Anthrozoös* 31(3), 309–318.

Clutton-Brock, J. (1995) Origins of the dog: domestication and early history. In: Serpell, J. (ed.) *The Domestic Dog, Its Evolution, Behaviour and Interactions with People*. Cambridge University Press, Cambridge, MA, pp. 16–20.

Cucchi, T., Hulme-Beaman, A., Yuan, J. and Dobney, K. (2011) Early Neolithic pig domestication at Jiahu, Henan Province, China: clues from molar shape analyses using geometric morphometric approaches. *Journal of Archaeological Science* 38(1), 11–22.

Darwin, C. (1859) *On the Origin of Species by Means of Natural Selection, or, The Preservation of Favoured Races in the Struggle for Life*. J. Murray, London.

Dugatkin, L. and Trut, L. (2017) *How to Tame a Fox (and Build a Dog)*. The University of Chicago Press, Chicago, IL.

Dunsworth, H.M., Warrener, A.G., Deacon, T., Ellison, P.T. and Pontzer, H. (2012) Metabolic hypothesis for human altriciality. *Proceedings of the National Academy of Sciences of the United States of America* 109(38), 15212–15216.

Ervynck, A., Hongo, H., Dobney, K. and Meadow, R. (2001) Born free? New evidence for the status of *Sus scrofa* at Neolithic Çayönü Tepesi (Southeastern Anatolia, Turkey). *Paléorient* 27(2), 47–73.

Frantz, L.A.F., Mullin, V.E., Pionnier-Capitan, M., Lebrasseur, O., Ollivier, M. *et al.* (2016) Genomic and archaeological evidence suggest a dual origin of domestic dogs. *Science* 352(6290), 1228–1231.

Freedman, A.H., Lohmueller, K.E. and Wayne, R.K. (2016) Evolutionary history, selective sweeps, and deleterious variation in the dog. *Annual Review of Ecology, Evolution, and Systematics* 47(1), 73–96.

Glocker, M.L., Langleben, D.D., Ruparel, K., Loughead, J.W., Gur, R.C. *et al.* (2009) Baby schema in infant faces induces cuteness perception and motivation for caretaking in adults. *Ethology: Formerly Zeitschrift fur Tierpsychologie* 115(3), 257–263.

Hare, B., Wobber, V. and Wrangham, R. (2012) The self-domestication hypothesis: evolution of bonobo psychology is due to selection against aggression. *Animal Behaviour* 83(3), 573–585.

Henneberg, M. (1998) Evolution of the human brain: is bigger better? *Clinical and Experimental Pharmacology & Physiology* 25(9), 745–749.

Hildebrand, M. (1954) Comparative morphology of the body skeleton in recent Canidae. *University of California Publications in Zoology*, 52, 399–470.

International Wolf Center (2023) Wolf FAQs. Available at: www.wolf.org/wolf-info/basic-wolf-info/wolf-faqs (accessed 27 October 2023).

Johnstone, G. (2021) Get to know the spectacular spitz breeds. American Kennel Club. Available at: www.akc.org/expert-advice/lifestyle/spitz-dog-breeds (accessed 27 October 2023).

Jolicoeur, P. (1959) Multivariate geographical variation in the wolf *Canis lupus*. *Evolution* 13, 283–299.

Kaminski, J., Hynds, J., Morris, P. and Waller, B.M. (2017) Human attention affects facial expressions in domestic dogs. *Scientific Reports* 7(1), 12914.

Kouprina, N., Pavlicek, A., Mochida, G.H., Solomon, G., Gersch, W. *et al.* (2004) Accelerated evolution of the *ASPM* gene controlling brain size begins prior to human brain expansion. *PLoS Biology* 2(5), E126.

Kruska, D. (1988) Mammalian domestication and its effect on brain structure and behavior. In: Jerison, H.J. and Jerison, I. (eds) *Intelligence and Evolutionary Biology*. Springer-Verlag, New York, pp. 211–50.

Lorenz, K. (1943) Die angeborenen formen möglicher erfahrung. *Zeitschrift für Tierpsychologie* 5(2), 235–409.

Mehrkam, L.R. and Wynne, C.D.L. (2014) Behavioral differences among breeds of domestic dogs (*Canis lupus familiaris*): current status of the science. *Applied Animal Behaviour Science* 155, 12–27.

Müller, C.A., Schmitt, K., Barber, A.L.A. and Huber, L. (2015) Dogs can discriminate emotional expressions of human faces. *Current Biology* 25(5), 601–605.

National Library of Medicine (2011) *ASPM* gene: assembly factor for spindle microtubules. Available at: https://ghr.nlm.nih.gov/gene/ASPM (accessed 27 October 2023).

Ostrander, E.A., Wayne, R.K., Freedman, A.H. and Davis, B.W. (2017) Demographic history, selection and functional diversity of the canine genome. *Nature Reviews Genetics* 18(12), 705–720.

Parker, H.G., Kim, L.V., Sutter, N.B., Carlson, S., Lorentzen, T.D. *et al.* (2004) Genetic structure of the purebred domestic dog. *Science* 304(5674), 1160–1164.

Parker, H.G., Dreger, D.L., Rimbault, M., Davis, B.W., Mullen, A.B. *et al.* (2017) Genomic analyses reveal the influence of geographic origin, migration, and hybridization on modern dog breed development. *Cell Reports* 19(4), 697–708.

Pitulko, V.V. and Kasparov, A.K. (2017) Archaeological dogs from the early Holocene Zhokhov site in the Eastern Siberian Arctic. *Journal of Archaeological Science: Reports* 13, 491–515.

Price, E.O. (1998) Behavioral genetics and the process of animal domestication. In: Grandin, T. (ed.) *Genetics and the Behavior of Domestic Animals*. Academic Press, San Diego, CA, pp. 31–66.

Range, F. and Virányi, Z. (2015) Tracking the evolutionary origins of dog-human cooperation: the "Canine Cooperation Hypothesis." *Frontiers in Psychology* 5, 1582.

Rilling, J.K., Scholz, J., Preuss, T.M., Glasser, M.F., Errangi, B.K. *et al.* (2012) Differences between chimpanzees and bonobos in neural systems supporting social cognition. *Social Cognitive and Affective Neuroscience* 7(4), 369–379.

Rius, M. and Darling, J.A. (2014) How important is intraspecific genetic admixture to the success of colonising populations? *Trends in Ecology & Evolution* 29(4), 233–242.

Saetre, P., Lindberg, J., Leonard, J.A., Olsson, K., Pettersson, U. *et al.* (2004) From wild wolf to domestic dog: gene expression changes in the brain. *Molecular Brain Research* 126(2), 198–206.

Scientific American (2014) Why have our brains started to shrink? *Scientific American*. Available at: www.scientificamerican.com/article/why-have-our-brains-started-to-shrink (accessed 27 October 2023).

Scott, J.P. and Fuller, J.L. (1965) *Genetics and the Social Behavior of the Dog*. The University of Chicago Press, Chicago, IL.

Serpell, J. (1995) *Origins of the Dog: Domestication and Early History. The Domestic Dog*. Cambridge University Press, Cambridge.

Shannon, L. (2015) Genetic structure in village dogs reveals a Central Asian domestication origin. *Proceedings of the National Academy of Sciences of the United States of America* 112(44), 13639–13644.

Skoglund, P., Ersmark, E., Palkopoulou, E. and Dalen, L. (2015) Ancient wolf genome reveals an early divergence of domestic dog ancestors and admixture into high-latitude breeds. *Current Biology* 25(11), 1515–1519.

Thalmann, O., Shapiro, B., Cui, P., Schuenemann, V.J., Sawyer, S.K. *et al.* (2013) Complete mitochondrial genomes of ancient canids suggest a European origin of domestic dogs. *Science* 342(6160), 871–874.

Trut, L. (1999) Early canid domestication: the farm-fox experiment. *American Scientist* 87(2), 160–169.

Udell, M.A.R. (2015) When dogs look back: inhibition of independent problem-solving behaviour in domestic dogs (*Canis lupus familiaris*) compared with wolves (*Canis lupus*). *Biology Letters* 11(9), 20150489.

vonHoldt, B.M., Lohmueller, K.E., Han, E., Parker, H.G., Quignon, P. *et al.* (2010) Genome-wide SNP and haplotype analyses reveal a rich history underlying dog domestication. *Nature* 464(7290), 898–902.

vonHoldt, B.M., Shuldiner, E., Koch, I.J., Kartzinel, R.Y., Hogan, A. *et al.* (2017) Structural variants in genes associated with human Williams–Beuren syndrome underlie stereotypical hypersociability in domestic dogs. *Science Advances* 3(7), e1700398.

Wang, G.-D., Zhai, W., Yang, H.-C., Wang, L., Zhong, L. *et al.* (2016) Out of southern East Asia: the natural history of domestic dogs across the world. *Cell Research* 26(1), 21–33.

Wayne, R.K. and Ostrander, E.A. (2007) Lessons learned from the dog genome. *Trends in Genetics* 23(11), 557–567.

Wood, B. and Collard, M. (1999) The human genus. *Science* 284(5411), 65–71.

Zhang, J. (2003) Evolution of the human *ASPM* gene, a major determinant of brain size. *Genetics* 165(4), 2063–2070.

3 A Crash Course in Genetics

Abstract

Chapter 3 provides a background on genetics to achieve an understanding of the influence of genetics on the creation of breed differences, and in driving variation amongst individuals of the same breed. This includes the definition of some common terms and provides examples of these, including single-gene effects, linked genes, multiplicative effects, genetic variability, mutations, and epistasis as they pertain to dogs.

Better Late Than Never

Eleven-year-old Johann Mendel gazed across the rolling landscape of his hometown, trying to memorize each hillock, swale, and seedling before he left. Northern Moravia, in what is now the Czech Republic, was a good place to put down roots if you wanted to be a farmer, but not if you were an aspiring scientist. Anton and Rosine Mendel valued education and recognized that their son was exceptionally bright, but as peasant farmers, funds were scarce for schooling; farming was a far more likely vocation for Johann than a career in the sciences. When a local schoolmaster recognized Johann's potential and suggested that he go to secondary school in nearby Troppau though, his family made some sacrifices: his sister, Theresia, donated part of her dowry. Johann honored his family's contribution by graduating with honors in 1840. He never married or had a family of his own; instead, he chose the monastic life to further his education without additional cost. It was there that he received the moniker "Gregor" when he became an Augustinian friar. It was with this name that people would recall him from then onward. He remembered Theresia's generosity, as well: when his education paid off, he supported her three sons, two of which went on to become doctors.

Perhaps it was his early years spent in the lush Moravian countryside that inspired Mendel to focus his life's work on botanical studies. He was particularly interested in the inheritance of traits and chose pea plants as the model for his research because they were diverse and produced easily. Using more than 10,000 pea plants, he examined seven traits: flower position and color, plant height, seed shape and color, and pod shape and color. Mendel cross-fertilized "purebred" pea plants with opposing characteristics, quickly finding that he could manipulate the traits of succeeding generations of plants. Finding mathematical patterns of inheritance, he deduced that each parent contributed one gene to its offspring, with the offspring then having a pair of genes that acted together as one distinct unit with physical appearances that hinted at the source genes. For example, when a purebred green pea plant and a purebred yellow pea plant (the parental, or P generation) were crossed, the resulting plants in that first generation (known as the F_1 generation, where the "F" stands for "filial", or offspring) would always have yellow seeds. In the succeeding generation (known as the F_2 generation), however, there would be one green seed for every three yellow ones. In a similar experiment, Mendel crossed a purebred white flowered pea plant with a purebred purple flowered pea plant. All of the F_1 generation offspring had purple flowers, while the F_2 generation again had a ratio of three purple flowers for every one white flower (Fig. 3.1). Mendel discovered that not all genes were created equally: some were dominant, while others were recessive.

To understand the principles that guided Mendel and so many other later scientists, it's important to understand what genotype and phenotype mean, and what dominant and recessive mean. Genotype refers to the genes in an individual's DNA that are responsible for a particular trait. Phenotype refers to the physical expression of that trait. When a

DOI: 10.1079/9781800624566.0003

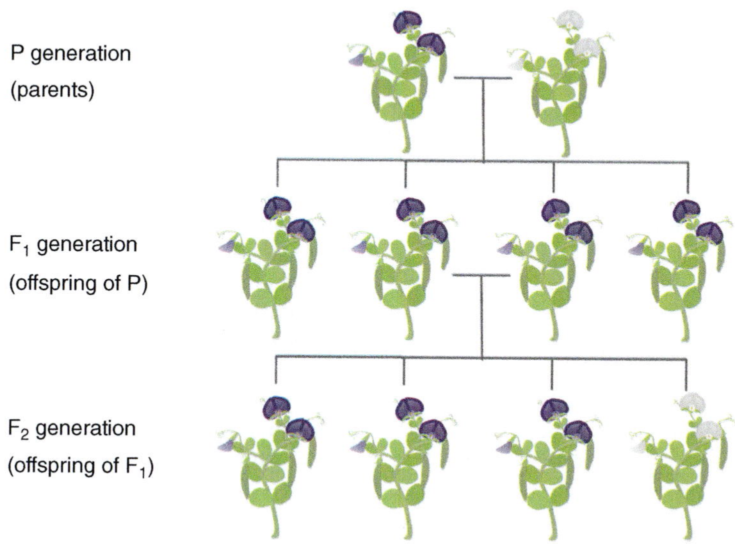

Fig. 3.1. Results of Mendel's cross for true breeding purple and white pea plants. Permission to use based on Creative Commons Attribution-Share Alike 4.0 International license. Available at: https://commons.wikimedia.org/wiki/File:Mendels_True_Breeding_Cross.png.

gene is dominant, the effect of the phenotype of that gene masks the contribution of a second gene at the same locus, or particular point, on the DNA strand. In contrast, a recessive gene is one that can be masked by a dominant gene. Each alternative form of a gene at the same place on the DNA strand is referred to as an allele. To have a trait that's expressed by a recessive allele, such as blue eyes, you must get the allele for blue eyes from both of your parents. To summarize this, Mendel wrote three laws to outline his research:

- *The Law of Segregation of Genes (the "First Law")*: Each inherited trait is defined by a gene pair. Offspring inherit one genetic allele from each parent when sex cells unite in fertilization.
- *The Law of Independent Assortment (the "Second Law")*: Genes for different traits are sorted separately from one another so that the inheritance of one trait is not dependent on the inheritance of another.
- *The Law of Dominance (the "Third Law")*: An organism with alternate forms of a gene will express the form that is dominant.

After 8 years of research, Mendel finally published the results of his botanical study. That same year, a skull discovered in Forbes' Quarry, Gibraltar, was finally announced to the scientific community, 17 years after it was unearthed by Captain Edmund Flint of the British Royal Navy. The skull, which would be dubbed "Gibraltar 1," belonged to a female with swept-back cheeks, a prominent brow ridge, and a projecting nose—physical traits associated with Neanderthals. When she was first discovered, "Neanderthals" were unknown to science. Her remains were initially given to the Gibraltar Scientific Society. This mystery woman also had worn teeth and bony growths within her skull, just inside her forehead. With modern humans, this is associated with menopause; had they found more of Gibraltar 1's skeleton, they would have seen other telltale signs in her bones, such as the grooved scrape marks of her pelvis, indicative of birthing several children. While her remains were dated at 30,000–50,000 years ago, yet to be discovered were the remains of her descendants, born some ten thousand years later, yet still inhabiting one of the last strongholds for Neanderthals as *Homo sapiens* continued to spread across what is now the European continent.

The year 1865 was full of promise and scientific discovery: life, in the promise borne from blossoming pea plants, and death, in the discovery of those first identified Neanderthal remains. Mendel's paper, which was published in *The Proceedings of the Brünn Natural Science Society*, was shared widely among scientific circles. Mendel forwarded 40 copies of the study to leading scientists, including Charles Darwin and a botanist named Carl Nägeli. There was no evidence that Darwin read Mendel's work, but Nägeli did—and he was intrigued. While Mendel had been working with pea plants, Nägeli had been conducting studies with hawkweed; he wanted Mendel to replicate his published findings with this plant, as well. Unbeknownst to either scientist, hawkweed reproduces both sexually and asexually, unlike pea plants, which only reproduce sexually. Thus, Mendel was unable to find support for his laws of inheritance using this plant. This failure suddenly filled Mendel with doubt; the climate was perfect for presenting new scientific concepts, and he wanted to be a part of that—but first he had to replicate his findings. Darwin's provocative *On the Origin of Species* had been published 6 years prior and was still causing a stir. Mendel himself purchased a second edition of Darwin's tome in 1863 in the German translation, *Uber die Entstehung der Arten*. He'd even underlined the parts of the book that were most salient to him, peppering the pages with notes and exclamation marks (Galton, 2009). Given the debacle with the hawkweed and the controversy surrounding *On the Origin of Species*, Mendel's discoveries remained in Darwin's shadow until 1900, when evolutionary scientists Carl Correns, Hugo de Vries, and Erich von Tschermak were examining the inheritance of variation. They rediscovered Mendel's work—but even more importantly, they finally understood its importance. Mendel was posthumously recognized as the father of the science of genetics. (Better late than never.)

Carl Linnaeus' work inspired Darwin, who, in turn, inspired Mendel. Darwin and Mendel were both fascinated with genetics and the driving forces behind population changes. But what, exactly, is genetics, and how does it work? The word *genetics* comes from the ancient Greek word *genetikos*, meaning "generative." *Genetikos* derives from *genesis*, meaning "origin." Genetics is a branch of biology that studies heredity and the variation of inherited characteristics among organisms. As we saw in Chapter 1, genes comprise DNA, which consists of three parts: a sugar group, a phosphate group, and one of four different nitrogen bases: adenine (A), cytosine (C), guanine (G), and thymine (T). DNA provides the instructions for making molecules and controlling the chemical reaction of life. These directions are inheritable, as they can be passed down from parent to offspring. Mendel studied this trait inheritance, noting the patterns in how traits were passed from parents to their offspring.

A Brief History of Genetics

Mendel wasn't the first scientist to study or make predictions about genetics; its history predates him by thousands of years, led by Hippocrates (460–370 BCE) and Aristotle (384–322 BCE). Hippocrates believed that characteristics were inherited from parents because hereditary material was collected from throughout the body, while Aristotle thought that the non-physical, form-giving principle of an organism was transmitted through semen and the mother's menstrual blood, which interacted in the womb to direct an organism's early development. Both believed, along with the prevailing scientists of their time, that individuals inherited acquired traits and that individual species had a fixed essence.

The Athenian philosopher Epicurus (341–270 BCE) proposed that males and females both provided hereditary contributions, which he referred to as "sperm atoms." He noted that there were dominant and recessive types of inheritance, but not what caused them; he did, however, describe segregation and the independent assortment of sperm atoms. In 300 CE, Indian medical writers thought that four factors determined a child's characteristics:

1. Input from the mother's reproductive material
2. Input from the father's sperm
3. Factors due to the mother's diet during pregnancy
4. Input from the soul entering the fetus

So, while early philosophers and philosopher-scientists had a few of the right ideas, it wasn't until Mendelian genetics hit the scene that the math and science of genetics really started to come together.

Let's review a few of the key concepts in genetics. We'll start with mutations, which are the permanent alteration of the nucleotide sequence of the genome. Mutations are the movers and shakers of evolution—the ultimate source of genetic variation, but by artificially selecting for a mutation that appears to be "desirable," other traits can be inadvertently selected for, as well. This occurs because of the proximity of these "other traits" to the locus that's being artificially selected. If they are near the area of selection, they are likely to be retained, as well. Unfortunately, this means that Mendel's Law of Independent Assortment isn't as "independent" or random as was originally thought. Over the decades, as mutations have arisen in the lineage of different dog breeds, they have either been selected for or against by breeders, and with mixed results.

For example, in the 1800s, the Bull Terrier was created by breeding Bulldogs with English Terriers. The resulting dog was light-bodied with a delicate-to-average head shape. Over the decades, when mutations such as a slightly Roman nose arose within the lineage, breeders chose those Bull Terriers with those more exaggerated features to breed with one another. As a result, today's Bull Terrier looks nothing like the Bull Terrier of yesteryear: they're stocky and Roman-nosed (Fig. 3.2). But the mutations weren't morphological alone. In addition to the physical changes from the founders of the breed, Bull Terriers also inherited a high rate of congenital issues. Approximately 18% of all Bull Terriers are born with hearing problems, compared to the American Kennel Club (AKC)'s estimate of 5–10% of dogs in the general population having some form of hearing loss (these estimates include both inherited and acquired deafness). Bull Terriers have a large percentage of white in their coats; similarly, Dalmatians were aggressively selected to have a larger amount of white coat pigmentation and more pronounced spots on their coat. Dalmatians with "louder" patterns were bred with similarly colored Dalmatians, yielding offspring that Cruella de Ville would have drooled over. But coat color can't be selected for alone. Dalmatians, like Bull Terriers, also inherited a higher than expected rate of deafness, and this is due to the pursuit of that loud coat color: Dalmatians' predominantly white coat is caused by the extreme piebald allele. Dalmatians, Bull Terriers, and all other white dogs are homozygous (identical alleles on both chromosomes) for the piebald gene (designated by s^ps^p) and have a coat that's classified as "extreme white." And it's "extreme" in its genes, as well.

Let's clarify the difference between homozygous and its sister classification, heterozygous. "Zygosity" refers to the degree of similarity in an organism's genetic sequence. Homozygosity refers to having two identical alleles on one or more genes, referenced as XX with dominant traits and xx for recessive ones. Heterozygosity refers to having two different alleles on one or more genes. Recall Mendel's pea plants. A pea plant with a particular flower could be homozygous dominant for that color (purple–purple, white–white, or red–red), or they could be heterozygous for it (purple–white, white–purple, or red–white).

There are two known alleles on the S locus: S, which codes for "no white," and s^p, which codes for piebald. It has been proposed that there is a third allele for "extreme white," coded as s^w (Płonek *et al.*, 2016). The extreme white coat color is correlated with deafness, due in large part to a lack of skin pigmentation in the inner ear. The more extreme the coat color, the higher the likelihood

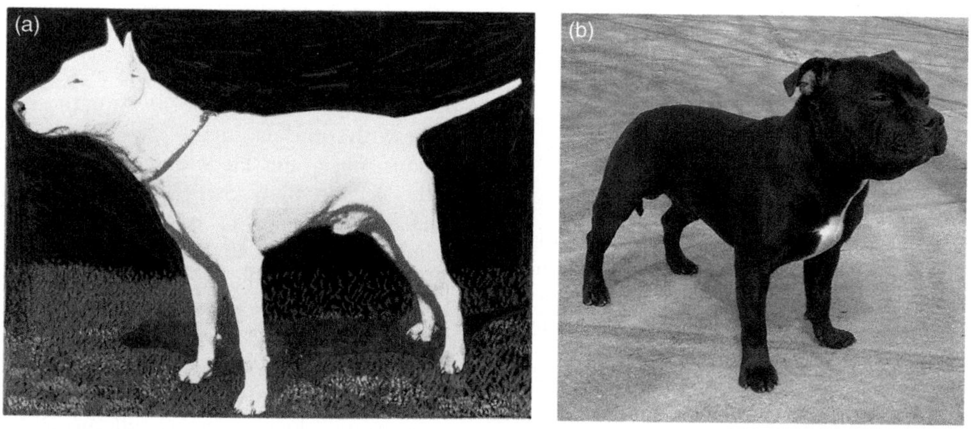

Fig. 3.2. (a) Bull Terrier (exact date unknown but prior to 1935). Permission to use based on Creative Commons license, CC-PD-Mark and access in the public domain (National Archives). Available at: https://commons.wikimedia.org/wiki/File:Photograph_of_a_Bull_Terrier_-_NARA_-_34929528.jpg. (b) Staffordshire Bull Terrier (2003). Permission to use licensed under CC BY 3.0, based on GNU Free Documentation License. Available at: https://commons.wikimedia.org/wiki/File:Staffordshire_Bull_Terrier_600.jpg.

that the dogs would also have blue irises or a lack of pigmentation in their inner ears; these traits increase the likelihood of hearing loss. Unsurprisingly, approximately 25–30% of Dalmatians have either bilateral or unilateral deafness (Strain, 2004). The only likely way to resolve this would be to go against breed standards and seek higher percentages of black in the coat. Back breeding, or breeding individuals with those who have a more "ancestral" look (and more genetic diversity), has been performed for both aesthetic and health-related issues. For Dalmatians, this has been done before to rectify other health issues, as we'll address shortly; perhaps this practice needs to be revisited. Many other breeds with high-white coats, such as Boxers, Australian Cattle Dogs, and Jack Russell Terriers, also have higher than expected rates of deafness.

Dalmatians aren't just prone to hearing loss; they're plagued by other genetic issues, as well. Over the course of artificially selecting the "perfect" Dalmatian, many of the preferred dogs were closely related, increasing the chance of rare but harmful gene mutations arising in subsequent generations—and that's exactly what happened. Over multiple generations, the resultant offspring were homozygous for two copies of the *SLC2A9* gene, which codes for creating a protein called glucose transporter 9 (GLUT9). This protein helps the body reabsorb water and nutrients. Having two copies of *SLC2A9* is associated with hyperuricosuria and hyperuricemia (HUU), a syndrome that causes elevated levels of uric acid in the urine and blood (Zierath *et al.*, 2017). All Dalmatians are homozygous for HUU, although efforts to remedy this have been made with outcrossing experiments.

Outcrossing, also referred to as "outbreeding," is the opposite of inbreeding. While inbreeding involves the mating of closely related individuals with one another, outcrossing involves breeding individuals that aren't closely related. Pairing individuals that don't have recent common ancestors can increase genetic variation, reducing the likelihood that offspring will have genetic abnormalities. With dog breeding, outcrossing involves mating two dogs within the same breed that aren't closely related, often to introduce desired characteristics from a different bloodline. For some breeders, outcrossing comprises breeding dogs that don't share a common ancestor within the last four generations.

Whippet ... Into Shape

The AKC recognized the Whippet as a breed in 1889. The Whippet is lithe, lean, and fast, and typically has a small head, pointed snout, and long neck. They were originally developed for chasing small game. Now, they are often bred for the races. They have a delicate appearance and usually tip the scales at 25–40 pounds. But now there's a new Whippet morph—the "Bully Whippet"—which has a mutation called "double muscling." The Bully Whippet is almost unrecognizable compared to the original. This mutation for double muscling, referred to as muscular hypertrophy (*mh*), is associated with a protein called myostatin, also known as growth differentiation factor 8 (GDF-8). GDF-8 acts on the growth and differentiation of muscle cells. In 2017, Wendy, the most famous of the Bully Whippets, passed away just shy of 14 years of age. Despite her unusual appearance, she was otherwise reasonably healthy up until her death. The average lifespan of a Whippet is 12–15 years, putting Wendy squarely in the middle of the breed's typical life expectancy. At 60 pounds, she was approximately twice the weight of an average Whippet. Her owner chose to euthanize her after she began showing signs of aging, including weight loss and mobility issues. Before her passing, she had been included in a genetics study on Whippets that examined this myostatin mutation and how it occurred. While most Whippets with double muscling don't appear to have life-threatening health issues, they are unable to race due to muscle spasms. Approximately 50% of them also have a pronounced overbite (Generatio: Center for Animal Genetics, 2023).

While Wendy was perhaps the most famous for muscular hypertrophy, the *mh* mutation happens often with other animals, including cats, sheep, humans, mice, and Piedmontese and Belgian Blue cattle. Belgian Blue cattle with this mutation are homozygous for an 11 base pair deletion in this coding region; this deletion removes the portion of myostatin protein that codes for conservation of muscle tissue and is not seen in normally muscled animals. Mice with double muscling share this same mutation, while double-muscled Piedmontese cattle have a mutation that changes a cysteine residue to a tyrosine one, which interferes with the normal function of the protein (Kambadur *et al.*, 1997).

Whippets with double muscling are homozygous for the mutation and, instead of having that slim build that the breed is well known for, they are heavily muscled, with thick legs, broad chests, and wide necks. They look like photoshopped iterations of a dog—the body builders of the canine world. With Whippets, double muscling comes from a 2 base pair deletion in their myostatin, located on canine chromosome 37. Because myostatin acts as a delimiting factor in the growth of muscle mass, this deletion codes for a premature stop codon, and the end result is a larger number of muscle fibers (Mosher *et al.*, 2007). If you're familiar with HTML, the coding language used to create web pages, think of this as a command that's missing the end "slash" that tells it when to stop the command. Instead of ending the instructions for muscle fiber growth, the command continues. While the difference in body types between Whippets and Bully Whippets is perhaps the most extreme, scientists have engineered another breed that can have this mutation, as well. A study out of the Guangzhou Institutes of Biomedicine and Health edited the myostatin (*MSTN*) genes in Beagles, resulting in double muscling in one of the two test subjects (Zou *et al.*, 2015).

While this mutation could be considered harmful (due to the muscle spasms and limited mobility that mh animals typically have) or neutral (due to the lack of other major associated health issues), multitude of known deleterious mutations are identified in purebred dogs. These include congenital hypothyroidism, skeletal dysplasia 2 (SD2, also known as dwarfism), and canine cyclic neutropenia (CCN, also known as Grey Collie Syndrome). CCN is a mutation of a single base pair in the *AP3B1* gene. Dogs that have two copies of this allele are affected with diluted coat pigmentation and have 2-week white blood cell fluctuations that are often accompanied by fevers and life-threatening infections. CCN affects Rough and Smooth Collies (Benson *et al.*, 2003). Congenital hypothyroidism involves a single base pair insertion in the *TPO* gene, with the Spanish Water Dog (Fyfe *et al.*, 2013), or a single base pair change in the *TPO* gene, with the Rat (Pettigrew *et al.*, 2007), Toy Fox (Fyfe *et al.*, 2003), and Tenterfield Terriers (Dodgson *et al.*, 2012). Congenital hypothyroidism, which can be fatal without early intervention, presents with growth retardation, delayed ear and eye opening, delayed tooth eruption, thickened skin, coat abnormalities, and goiter (Fyfe *et al.*, 2013). SD2 occurs when there is a single base change in the gene *COL11A2*. This form of dwarfism presents with typical body size and legs that are approximately 6 cm shorter than breed standard (Frischknecht *et al.*, 2013). SD2 affects Labrador Retrievers; breed variation can lead to difficulty in determining whether dogs have the mutation. While the physical form of dwarfism isn't harmful, dwarfism often has a suite of associated health issues. It's recommended that dogs affected with SD2 not breed to avoid the continuation of dwarfism in the population.

Hey Bulldog

Some harmful mutations are lethal, but others only deliver a selective disadvantage. In free-living populations, harmful or mildly disadvantageous mutations would be culled through natural selection, but with artificial selection, many of these traits can persist due to human intervention. Humans have been breeding dogs for increasingly extreme traits, like coat, height, and face shape, for centuries, often with little knowledge, and later, little regard, for the genetic consequences. The English Bulldog, for example, has low genetic diversity due to artificial bottlenecks and a small founder population of only 68 dogs after 1835. That's in large part due to the desire to breed the English Bulldog for exaggerated characteristics, including an enlarged head; many have pelvic abnormalities, which typically necessitates birth via Caesarian. In the wild, these traits would disappear from the gene pool within one generation, with only those Bulldogs that could give birth naturally surviving. If all English Bulldogs were released into the wild (which is unadvisable) and had only had mating opportunities with other English Bulldogs, they would either completely die out or those with the least exaggerated traits would be selected for, creating a new population. The plight of the Bulldog has geneticists wondering if this breed has gone beyond the point of saving; with such limited diversity, it's unlikely that even "reverse selection" could decrease the number of recessive deleterious traits in the breed (Pedersen *et al.*, 2016). This breed, like the Dalmatian, is a candidate for outcrossing for the welfare of future generations.

As discussed previously, mutations aren't always harmful. They also may have led to the dog's friendly, hypersocial temperament. Dr Monique Udell examined the similarities between humans with Williams–Beuren Syndrome and dogs. Dr Udell found that humans with this syndrome were overly friendly with previously unknown persons (as many, but not all, dogs often are) and didn't persevere on cognitive tests. A similar pattern was found with wolves, who will persevere during cognitive tests, while dogs will either quit or look to their humans for assistance. Dr Udell examined DNA taken from wolves and dogs, finding that four mutations influenced the behavior of domesticated dogs. The mutations on one gene, *WBSCR17*, were particularly important; in humans, that same gene is associated with Williams–Beuren Syndrome.

It's highly likely that a mutation such as the one associated with *WBSCR17* was selected for very early on during the domestication of the ancestors of modern dogs. This wasn't "just" friendly dogs being bred to friendly dogs, though; they weren't selecting each other for these traits. Had Lamarck researched these changes, or seen what's happening with many of modern-day Mozambique's elephants, he may have felt that it was evidence for his heritability of acquired characteristics, where changes over time resulted in different forms in subsequent generations, such as giraffes having extra long necks due to continued stretching. While it isn't Lamarckism, strong selection pressure on elephants has resulted in those elephants without tusks becoming increasingly common in the gene pool in some geographic areas. Elephants are known for their tusks and trunks, but perhaps not for long. Researchers have found that the selective pressure of hunters seeking out elephants with tusks for their ivory has provided an advantage to those elephants without tusks (Okane, 2019). A growing number of elephants in the Gorongosa National Park in Mozambique, Africa, are being born sans tusks (Fig. 3.3). Now this distinguishable feature is very important to elephants, particularly to males, who use them to fight with one another over breeding access to females. Under typical environmental conditions, approximately 2–4% of female elephants are born without tusks (Maron, 2018), but males typically always have them; in the absence of their tusks, what will be the new determining factor for breeding rights?

Fig. 3.3. A bull elephant born without tusks. Credit to Steve Garvie. This file is licensed under the Creative Commons Attribution-Share Alike 2.0 Generic license. Provided by Wikipedia Commons and available at: https://commons.wikimedia.org/wiki/File:Flickr_-_Rainbirder_-_Asian_Elephant_(Elephas_maximus_maximus).jpg.

The Effects of Domestication

As our oldest domesticate, dogs have been subject to the effects of our selective breeding for centuries, with breeds experiencing population bottlenecks at the whims of our aesthetics and opinions on their utilitarian value. The domestication process yields both positive (in the view of humans, at least) and negative effects. There are four ways that domestication can affect deleterious variation. The first way is a shift in environment due to living with humans, which could allow mutations that change amino acids, and usually proteins, to increase in frequency because they are less harmful in the new environment. The second is population bottlenecks that allow the increase of deleterious variants by random chance due to the small number of breeding individuals. The third is increased inbreeding, particularly between very close relatives. And the fourth is the strong artificial selection for traits that are desirable to humans, but which may have been disadvantageous in their ancestral environment (Freedman *et al.*, 2016).

The interplay between mutations and selective pressures is critical to the direction of a species' evolution. With the selective pressure of hunting, the elephant's once-important morphological trait became a liability. While researchers are still trying to determine the genetics behind this mutation, the gene for tusks continues to be selected against in this population, with tuskless elephants contributing to the gene pool at a higher rate than their tusked peers. In a short amount of time, evolutionarily speaking, the population of tuskless elephants has increased rapidly. According to elephant behavioral researcher Joyce Poole, this site used to be the home of 4000 elephants, but after civil war ravaged their country, more than 90% of these iconic animals were killed. There are currently only 200 female elephants at Gorongosa National Park. Of these, 51% aged 25 years or older are tuskless, and 32% of the females born after 1992 don't have tusks (Maron, 2018). With continued poaching pressure, the trend looks to continue.

Consider the trend with domesticated dogs and the interplay of mutations and selective pressures. With dogs, the selective pressure was again humans, but this time, it wasn't pressure from hunting or poaching, but from choosing dogs with friendly, hypersocial temperaments—dogs, for example, that had mutations on *WBSCR17*. All dogs that didn't have these traits would alternately be denied human interaction or, in some cases, hunted; it's easy to see, given the extreme change from tusked to tuskless elephants in only a few decades, how quickly dogs, too, would have been changed by this pressure.

True Blue

Many other traits are the result of genetic mutations. Approximately one in every 4 million lobsters is born with a genetic mutation that causes blue coloration. While a cobalt crustacean is certainly novel and striking, the bright color isn't beneficial for the lobster: it makes them stand out to predators and they often die young. Blue used to be an uncommon color for humans' eyes, too. People with blue eyes all have a single common ancestor who had a genetic mutation between 10,000 and 6000 years ago. That mutation occurred within the *HERC2* gene, which regulates *OCA2* expression. The *OCA2* gene provides instructions for making a protein that's located in melanocytes, specialized cells that produce melanin, which provides skin, hair, and eyes with their pigmentation (National Library of Medicine, 2022). Variations in the *HERC2* gene are associated with variability in skin, hair, and eye pigmentation (National Library of Medicine, 2023). Today, approximately 17% of the US population and 22.3% of the Caucasian population has blue eyes (Belkin, 2006). The recessive mutation for blue eyes, which was once common in the US, is decreasing from 57.4% of the Caucasian population born between 1899 and 1905 to 33.8% born between 1936 and 1951 (Grant and Lauderdale, 2002).

While blue eyes are on the decline, there are even rarer mutations, some of which have questionable selective advantages and have only recently been discovered. In 2019, scientists reported on a Scottish woman who "never felt pain" and has a rare genetic mutation that inhibits the feeling of anxiety or pain (Habib *et al.*, 2019). While scientists have worked on cases of patients who had little to no pain for a century, Jo Cameron of Inverness, Scotland is the first person with this specific condition. It's attributed to a genetic mutation in a previously unidentified gene. Not only does she feel no pain or anxiety, but she heals quicker than the average human, as well.

There's also a genetic mutation that dramatically increases bone density. Depending upon whether you aspire to be an Olympic swimmer (then this would be a bad mutation for you, as you'd probably sink like a stone) or a human crash test dummy (then this would be a rather serendipitous mutation for you), this mutation could have very different effects on your life. The mutation, called "sclerosteosis," is on the *LRP5* gene. It causes people to be born with bones that are several times denser than average.

Sickle cell anemia is another well-known mutation. When people who are living in malaria-infested areas have one of the two alleles of sickle cell disease, it provides an advantage: they cannot become sick from malaria. Humans possessing two alleles, however, have an incurable condition called sickle cell anemia, where their blood cells are sickle-shaped, sticky, and rigid, instead of round and flexible. These crescent-shaped cells can lodge in small blood vessels, diminishing oxygen and blood flow. Only 50% of those diagnosed with sickle cell anemia live beyond 50 years old (Platt *et al.*, 1994).

Sickle cell anemia exemplifies a single gene disorder, sometimes called a Mendelian disorder. More generally, a single gene effect is when each allele has a specific biological effect and leads to the expression of a certain trait with two distinct phenotypes (like the presence or absence of sickle cell anemia). Among humans, there's evidence for a single gene effect with several conditions, including male pattern baldness and polycystic ovaries (Carey *et al.*, 1993), as well as uncommon disorders such as

Huntington disease, muscular dystrophy, cystic fibrosis, and Fragile X syndrome. Recall that dominance is when the effect of the phenotype of that allele masks the contribution of a second allele at the same locus. Among dogs, the white spotting in the Dutch Boxer's coat is from a single gene effect with "incomplete dominance" (van Hagen *et al.*, 2004), which happens when one allele for a trait isn't completely expressed over its paired allele. Thus, the expressed physical trait is neither the dominant nor the recessive phenotype, but a third phenotype with a combination of the phenotypes of both alleles. The simple case of a single gene affecting one trait are less common than most people think, and some of them have been debunked as myths, such as the one about ear lobe attachment (Shaffer *et al.*, 2017). For example, some genes affect multiple phenotypic traits (this is called pleiotropy). Pleiotropy derives from the Greek words *pleion* ("many") and *tropos* ("affecting").

Examples of pleiotropy include pigmentation and deafness with cats, chickens and the "frizzle" trait, and fruit flies and the *vestigial* gene. White cats can have a range of eye colors, including yellow, green, orange, blue, and mixed colored (e.g. one green and one blue), they can also have a range of hearing, including congenital sensorineural deafness (CSD). CSD is deafness that occurs from birth, and is caused by a disease or lesion of the auditory nerve or inner ear. It's common among white cats, and while it can occur with any of the above eye colors, it's most common with blue-eyed cats. CSD happens when there is degeneration in the inner ear; among cats with two different colored eyes, deafness is more common on the side of the head that has the blue eye. Some estimates find that as many as 40% of cats who have white fur and blue eyes are deaf (Hartl and Jones, 2005). Cats aren't the only species afflicted with an auditory impairment that's linked to pigmentation; a disease called Waardenburg syndrome acts similarly on humans. Waardenburg syndrome is associated with changes in pigmentation and hearing loss; those who have the syndrome often have very light blue eyes or mixed colored eyes along with congenital hearing loss and commonly, unusual hair coloring such as prematurely gray hair or hair with white patches. The gene associated with these phenotypes is pleiotropic, as it affects both hearing and pigmentation. Among mice, research has shown that pigmentation is associated with the maintenance of the fluid in ear canals; those animals that have an absence of ear canal fluid can suffer from collapsing ear drums, which is associated with degenerating auditory nerves. This degeneration then leads to deafness (Sunquist, 2006).

Two other well-known harmful mutations include the chickens' "frizzle" trait and the fruit flies' *vestigial* gene. The "frizzle" trait was discovered among chickens in 1936. Researchers Elizabeth Upham and Walter Landauer found that chickens that expressed the dominant "frizzle" gene had defective, outwardly curling feathers and higher metabolisms and blood flow rates, atypical body temperatures, increased digestive capacities, and lower egg production rates than chickens that did not have this allele. Fruit flies (*Drosophila melanogaster*) can have a defect in their *vestigial* gene that codes for wing formation as well as a number of other developmental processes (Zider *et al.*, 1996), such as the position of the bristles on a fly's scutellum (their third dorsal section in the thoracic segment), the number of egg strings in a fly's ovaries, and decreased life expectancy (Miglani, 2002). Flies that receive a defective *vestigial* gene from each parent have vestigial wings and are unable to fly, while flies receiving only one mutated *vestigial* gene will have normally shaped wings, as the mutation is recessive. For fruit flies, the dominant V allele codes for long wings, while the recessive v allele codes for vestigial wings; flies with the VV or Vv genotypes will have long wings, while flies with the vv genotype will have vestigial ones. Most traits are a result of two or more genes that result in a single phenotype (this is referred to as polygenic).

Epic Epistasis

We've discussed the effects of single genes, but what about the genetic variability that occurs with *multiple gene effects*? Multiple gene interactions are deviations from the expected phenotype when multiple genetic mutations are combined. For example, one might expect that a plant will have purple flowers or a human might have brown hair, based upon the visible contributions of the parents, but there's another gene "in the background" working with the gene for flower or hair color. This is called epistasis, and it's when the effect of one gene is dependent upon the presence of another gene that acts upon or "modifies" it. There are three different versions of epistasis: dominant masking epistasis, which occurs in the coat color of horses; recessive masking epistasis, which occurs in the coat color of Labrador Retrievers; and modifying epistasis, which can be observed in the coat color of Doberman Pinschers.

Before we discuss the specifics of dominant masking, recessive masking, and modifying epistasis, let's examine the basic principle of epistasis. A classic example is the Blue-eyed Mary (*Omphalodes verna*), a herbaceous perennial plant that sometimes yields unexpected petal colors. This plant has a gene with alleles that control the color of the flowers for either magenta or blue, but there's another gene that controls whether the flowers have color or are all white instead. Think of this as a decision matrix: at the top, you'll have options for color or no color. If "no color" is selected, all other directions for color specifics will be rendered moot. But if "color" is selected, then there's a range of options. This is similar to a version of one of those tests that a grade school teacher liked to give to measure your ability to pay attention: at the top of the sheet it says to simply "write your name and then turn the paper over if this request is listed twice;" think of this as the gene for "color" or "white." The rest of the sheet has a series of questions that must be answered. Think of these as the gene for magenta or blue. If the plant has two copies of the white allele, then the flowers are neither magenta nor blue—when presented in duplicate, the directions at the top of the page override the rest of the information.

First let's examine dominant masking epistasis, which occurs in horse coat color. With horse coat color, there are three possible genotypes: G--- for a horse that is gray at maturity, gg-- for a horse with no gray in their coat, ggE- for a black coat, and ggee for a chestnut coat. (In genetics, the "-" is essentially a placeholder denoting that any of the alleles would be a possibility. For example, the G would determine the coat color, as it is the dominant allele, and the alleles that followed would not have an impact on that trait). When the gene for a gray coat is dominant (G) the horse will be gray at maturity, even if they were born a different color, because G masks the expression of other coat color genes. When the gene for gray coat is recessive (gg), the genotype for other coat colors isn't masked, and they will not have a gray coat. Horses that have a dominant E genotype will have a black coat, while horses that have a recessive ee genotype will have a chestnut coat.

Next, let's look at recessive masking epistasis. According to the official AKC breed standard, all Labrador Retrievers should have a double coat with an insulating, downy undercoat and a denser, thicker top coat to provide protection against water and inclement weather. But their coats vary in color, coming in three AKC-approved hues: yellow (which ranges from "fox red" to a light cream), chocolate (which ranges from light to dark), and black. These coat colors occur due to genes influencing the expression of two pigments, pheomelanin and eumelanin, that are in the dog's skin and fur. Pheomelanin produces hairs with a red hue, while eumelanin has two pigment types: black and brown. There are two eumelanin color phenotypes: black (BB, Bb) and chocolate (bb). When there is the presence of brown eumelanin in the absence of other pigments, the result is yellow hair. When there is the presence of black eumelanin in the absence of other pigments, the result is gray hair. Black eumelanin continues to be produced throughout the lifetime, resulting in the gray hair that many of us experience at some point in our lives.

None of these genes act independently, but act together to affect this one trait. Masking epistasis, such as what occurs with the Labrador's coat color, happens when one gene "masks" or prevents the expression of another, so that the phenotype of the former, but not the latter, will be expressed. Among Labrador Retrievers, there are three possible genotypes for coat color: chocolate, black, or yellow, designated by the genotypes bbE- for chocolate, B-E- for black, and --ee for yellow (Fig. 3.4). If a puppy inherits the homozygous recessive ee genotype of the E gene, he will have a yellow coat. Even if the puppy also has the B gene, the ee gene will mask it. The Yellow coat color is a little bit more complicated than that, though. There's a mutation in the Extension (or E) trait that's unique to yellow Labrador Retrievers and Golden Retrievers. This mutation, which is directed by the melanocyte-stimulating hormone receptor gene (*MC1R*), shortens the protein and disallows eumelanin deposition (Everts *et al.*, 2000). Perhaps because of domestication in dogs, there is also a mutation in this gene that results in melanism variation (Protas and Patel, 2008).

But what about "white" Labrador Retrievers? According to the AKC, "The Labrador Retriever coat colors are black, yellow, and chocolate. Any other color or a combination of colors is a disqualification. A small white spot on the chest is permissible, but not desirable. White hairs from aging or scarring are not to be misinterpreted as brindling."

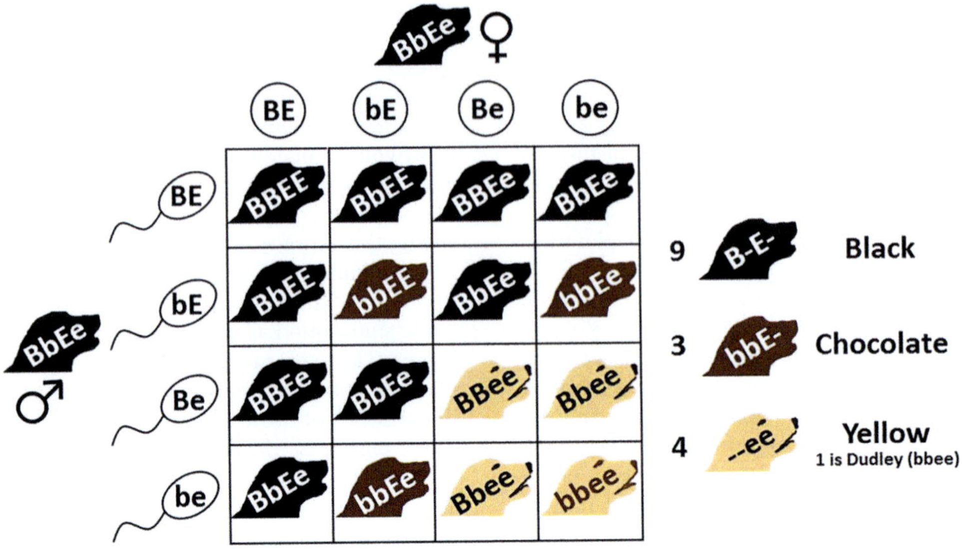

Fig. 3.4. Labrador Retriever coat color genetics shown as Punnett square. Permission to use based on Creative Commons Attribution-Share Alike 4.0 International license. Available at: https://commons.wikimedia.org/wiki/File:Labrador_punnett.png.

(Recall that black eumelanin continues to be produced throughout one's lifetime, thus resulting in gray hairs.) According to the AKC, the acceptable range of pigmentation for yellow Labradors can vary from a very light cream to a ruddy hue, thus giving those dogs on the paler end of the spectrum a "white-coated" appearance. The variability occurs due to differences in the expression of pheomelanin, but the gene responsible for these differences is unidentified (Schmutz and Berryere, 2007).

Now "white" Labradors either have the *appearance* of white hair due to this very light shade of yellow, or they are a*ctually* white, but only due to the rare genetic disorder of albinism. So what causes the variation that ranges from ruddy to light cream? The E locus affects the expression of pheomelanin, which is only seen if the fur does not have eumelanin. An enzyme called tyrosinase makes pheomelanin and eumelanin, and certain mutations can lead to albinism or dilution of color (as evidenced in the Labrador's yellow coat). So, while the AKC and other kennel clubs only recognize the Labrador Retriever's three official coat colors, "designer" variants including "charcoal," "silver," and "champagne" are emerging. While aesthetically pleasing, these colors unfortunately result from a skin disease called color dilution alopecia. A "Dudley" is the term given to a type of Labrador with a unique pale coat color. Unlike traditional black, yellow, or chocolate Labradors, Dudleys also have a pinkish or flesh-colored nose and light-colored eyes.

Since we're talking about the genetics of coat color, let's revisit our discussion of the black-, gray-, and white-coated wolves from Chapter 1. Wolf coat color is influenced by three genes, including one that codes for a black coat or a gray coat. To explain these selective processes, we'll provide a simplified version of coat color genetics. In a real-world scenario, wolves' black coat color comes from the KB allele (Anderson *et al.*, 2009), which we'll designate as K for black (as it is dominant) and k for gray, with possible variants being KK (these wolves typically die *in utero*), Kk, and kk. Wolves with the Kk and KK variants will have black coats, while wolves with the kk variant will be gray. (With the KB allele, white is actually considered to be a variant of gray). With disruptive selection, the extremes are selected for, and not the middle, so you would see equal numbers of black K wolves and k wolves that had a white variant of the gray, but no k wolves with the true gray coat. With directional selection, only one extreme would be selected for, thus, there would be only black K wolves in the population and the chromosomes with the k allele would decrease from the population as fewer and fewer wolves would carry this allele. And with stabilizing selection, true gray kk wolves would be the most common in the population, with the black K gene gradually disappearing from the population. White-variant kk wolves would have lower rates of reproductive success and thus while the genetics for white coat color would technically still be present, they would be seen far less frequently.

Epi-Static Cling

A hypostatic gene is one whose phenotype becomes altered by the expression of an allele at a separate locus in an epistasis event. For example, with Labrador Retrievers, the chocolate coat color is the result of homozygosity for a gene that's epistatic to the black versus brown gene. Among humans, hair color genes are hypostatic to the baldness gene, meaning their phenotype is modified by the expression of an allele at a separate locus, with the gene for total baldness being epistatic to the genes for brown, red, or blond hair.

Lastly, let's examine modifying epistasis, which occurs when one gene modifies the other, creating new phenotypes. This can be observed with Dobermans' coat colors, which have four possible genotypes: B-D- has a black coat, bbD- has a red coat, B-dd has a "blueish" coat and bbdd has a fawn coat. When the D genotype is recessive (dd), it modifies B's phenotypic expression, creating a coat with a "faded" appearance. Blue is a faded or "diluted" version of black, while fawn is a diluted version of red. Epistatic mutations arise due to interactions between or within effects, yielding non-linear effects, wherein the phenotypic result (e.g. coat color), does not correspond with the phenotypic inputs—certain genotypes are working "behind the scenes" and masking or modifying other genotypes. Epistasis can exert a strong influence on the evolution of a species, as phenotypic traits within a population can change rapidly from the founders of the population.

Genes That Hold "Hands"

Linked genes are physically close to one another on the same chromosome, and thus more likely to be inherited together, whereas genes on separate chromosomes are never linked (Fig. 3.5). To simplify this, consider a bowl of alphabet soup. If your spoon dips in where an A and a T are adjacent, and a G is on the opposite side of the bowl, the A and T are more likely to be scooped up together than the A and the G or the T and the G. With humans, genes for eye and hair color are linked, resulting in certain eye/hair color combinations being frequently inherited, such as blue eyes and blonde hair and brown eyes and brown hair. It gets a little bit more complicated than this, however. In the 1900s, American evolutionary biologist and geneticist Thomas Hunt Morgan developed "linkage maps"—tables that display a species' genes relative to other genes. Morgan's first genetic linkage map involved fruit flies. With fruit flies, the genes for body and eye color are located closely together on the X chromosome, so they are transferred to offspring more frequently (but not always) than they would be if they were located farther apart. Females have two X chromosomes, and will have two copies of their X-linked genes. In fruit flies, there are no alleles for eye and body color on the Y chromosome, and red eye color (designated as w^+) is dominant to white eye color (w). Tan body color (y^+) is dominant to yellow body color (y). So for example, a red-eyed, yellow-bodied female fruit fly will be dominant for her red eye color and recessive for her yellow body color. Let's say she mates with a white-eyed, tan-bodied male. The female fruit fly's sons will inherit their eye and body color genes from their mother only, so they will have the same phenotype as their mother. Daughters, however, will inherit their mother's red eyes and their father's tan bodies.

Genetics plays a big role in behavior and personality, too; remember how quickly Belyaev and Sorokina's silver foxes (*Vulpes vulpes*) became tame and friendly? By selecting foxes with the fewest aggressive behaviors toward humans, subsequent generations weren't just friendlier; they resembled domesticated dogs, genetically and morphologically, as well. These Cold War Canines had decreased activity in their hypothalamus–pituitary–adrenal (HPA) axis and two neuropeptides, neuropeptide Y (NPY) and calcitonin gene-related peptide (CALCB), also had altered expression, indicating rapid changes in brain gene expression (Saetre *et al.*, 2004). Both their personalities and their appearances changed, revealing that the genes for a more "dog-like" appearance (coat color and pattern, droopy ears, upwardly curving tail) were likely linked to the genes for more "dog-like" behavior (wagging tails and licking humans). All behaviors are influenced by genes. (We'll examine this more later.)

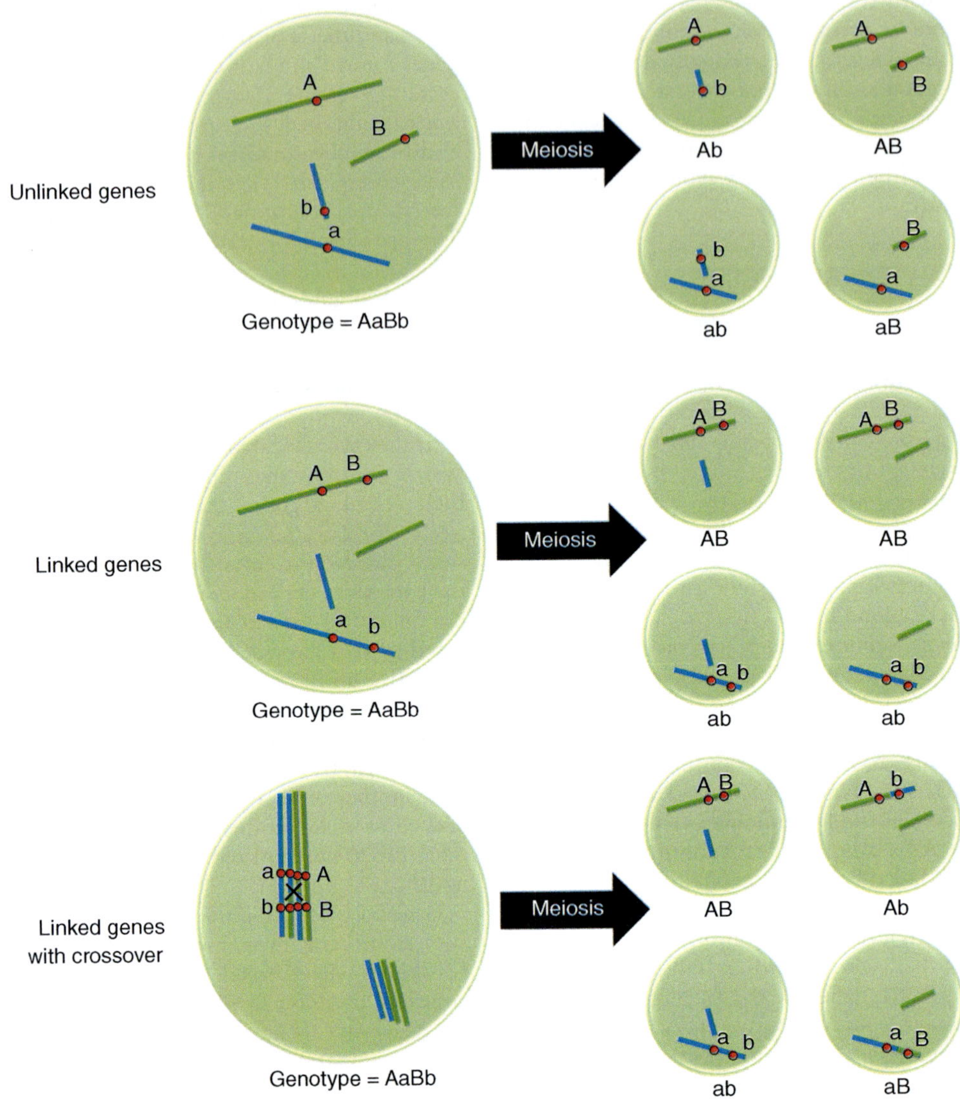

Fig. 3.5. Linked genes versus unlinked genes (with crossover). Permission to use based on Creative Commons CC0 1.0 Universal Public Domain Dedication. Available at: https://commons.wikimedia.org/wiki/File:Linked_Genes_vs._Unlinked_Genes_(with_Crossover).png.

More Than the Sum of Its Parts

In addition to linked genes, there are also "multiplicative effects" with genes. This occurs when the alleles at more than one gene locus have a higher contribution to the phenotype than the sum of their probabilities. With humans, in diseases with a genetic cause, the risk of disease from having two alleles is higher than having each one individually. For example, in genetically caused diseases such as ovarian cancer (Wacholder *et al.*, 2011) and panic disorder (Keck *et al.*, 2008), the risk of expressing the disease when you have two alleles is higher than each one individually. Genetic influences on attention deficit hyperactivity disorder (ADHD) may be multiplicative, as well (Nikolas and Burt, 2010). Multiplicative effects create even more variation—when effects of two different alleles are not just additive but multiplicative, they can create a much larger effect. On top of that, you can get a multiplicative effect from an interaction between the environment and genetics, so that when both interact, they create more of an effect than either one would individually. This contributes to the vast variation that we see in populations and the difficulty in predicting behavior. But this doesn't mean that genetics aren't important; it's a complex relationship.

So why is genetic variability so important? Because it's the foundation of any healthy population. There are three sources of genetic variation in a population: mutations (we discussed these earlier), sexual reproduction, and gene flow. Sexual reproduction provides genetic diversity from the genes in the mother's egg and the father's sperm. During sexual reproduction, cells double their DNA in meiosis and then are mixed in genetic recombination. Genetic recombination (also referred to as "genetic shuffling") occurs when genetic material is exchanged between organisms and leads to offspring that have combinations of traits that are not seen in their parents. Gene flow, also known as allele flow or gene migration, is the movement and transfer of genes or alleles from one population of a species to another through interbreeding populations.

The most common type of genetic variation are single nucleotide polymorphisms (SNPs; or "SNIPs" to their friends). These are variations in a single nucleotide occurring at specific positions in the genome. For example, replacing the nucleotide thymine (T) with the nucleotide cytosine (C) in a segment of DNA. SNPs are naturally occurring in a person's DNA.

Mitochondria are cellular organelles with double membranes that convert chemical energy into adenosine triphosphate, a food source that cells can use. Mitochondrial DNA (mtDNA) is the DNA that is located inside the mitochondria. mtDNA (also known as maternal DNA) is only passed on from mother to offspring and these organelles are often referred to as the "powerhouse" of the cell. Humans have 16,500 base pairs of mtDNA out of a total of approximately 3 billion total base pairs. All humans carry mtDNA from 200,000 years ago, and the source of this genetic material is referred to as "Mitochondrial Eve."

But how are these genes expressed? Gene expression is the process of transcribing the heritable information in a gene, and making them into a functional gene product, either RNA (ribonucleic acid, another type of nucleic acid) or protein. A haplotype, or haploid genotype, is a group of alleles on a chromosome that one inherits together from one parent. Haplotypic homozygosity is the probability of randomly selecting two identical haplotypes from a population. It is a measure of linkage disequilibrium (the non-random relationship of alleles at different loci in a population), and can provide insight about the subgroups of haplotypes.

Why Inbreeding Is Out

Recall the plight of the Bulldog—their genetic variability has been greatly reduced due to inbreeding, and the rate of harmful traits within its limited genome have potentially put it at the "point of no return." English and French Bulldogs are "fashionable" pets, but popularity comes at a cost. The results of an 80-year study on the cultural evolution of dog breeds suggest that fashion, rather than function, was a stronger determining factor in dog breed popularity (Ghirlanda *et al.*, 2013). And the best way to follow these fashionable phenotypic trends? Through mating those individuals that exhibited the strongest versions of those traits with one another—but at the cost of genetic variability.

Reducing genetic variability can lead to a rapid increase in certain physical characteristics, such as a shortened face and legs, but it can also lead to a rapid increase in genetic issues. And we don't mean to pick on Bulldogs here; many other breeds suffer from similar genetic issues because phenotypic traits have been favored without considering genotypic ones. That's one of many reasons why animal advocates are so staunchly anti-backyard breeder and anti-puppy mill: these breeding operations are less likely to pay attention to genetics and churn out puppy after puppy with increasingly harmful genetic issues. Many of the most popular breeds also have the highest rates of inherited ailments, with 25% of dog breeds suffering from 45 or more disorders (Summers *et al.*, 2010). Many of these popular dogs hide a lifetime of suffering behind their fashionable faces. By the age of six, Cavalier King Charles Spaniels, for example, have a 70% rate for canine syringomyelia, a neurological disorder where the brain exceeds the skull's capacity, creating fluid-filled cavities in the neck and spinal cord, causing extreme pain and spinal cord damage (Parker *et al.*, 2011).

Breed standards, based upon phenotypic traits, have been the driving force behind line breeding, a variation of inbreeding where direct relatives are mated with one another. Discrete breeds arose after selecting dogs with particular (and often pathologic, unfortunately) genotypes (Freedman *et al.*, 2016). Recall the rate of genetic issues that Dalmatians have, such as HUU. In 1973, geneticist Robert Schaible created the Dalmatian-English Pointer Backcross Project, where he mated an English pointer with an AKC champion Dalmatian, and then continued crossing dogs from subsequent litters (Safra *et al.*, 2006; Bannasch *et al.*, 2008). English Pointers have normal uric acid levels, and after 15 generations, Schaible was allowed to register his "new" line of Dalmatians, which were healthier than their non-outcrossed peers, and still had their distinguishable spots. Schaible was a geneticist using 19th century dog breeding practices to solve a problem that continues to plague us in the 21st century: cross-breeding introduced desirable physical traits and genetic variation and, historically, offspring from these crosses could be registered as "purebred" by the third generation. But over time, this practice became unpopular, much to the detriment of many dog breeds. Humans were creating population bottlenecks with dog breeds, over and over, but as Schaible's study showed, there was a much healthier way to go about dog breeding.

A population bottleneck is the reduction in size of a population due to any of a variety of factors, including illness or environmental disasters, that reduces the genetic variation in the population. Introducing "new" genes (or "new to them," in the case of re-introducing the genes of extinct animals to the gene pool of extant animals) has been an increasingly popular approach for reanimating extinct species and adding genetic diversity to extant ones. The bottleneck that occurred during dog domestication, and during the separation of different dog populations to create breeds, is expected to shape patterns of deleterious variation across the dog genome (Freedman *et al.*, 2016).

In 2005, the dog genome was sequenced, and with rather surprising results (Wayne and Ostrander, 2007). Using a purebred female boxer named Tasha, researchers discovered that dogs have approximately 19,500 genes (Lindblad-Toh *et al.*, 2005), slightly less than the 22,000 that humans have (Kirkness *et al.*, 2003). Of these 19,500 genes, at least 70% of them have counterparts in the human genome, with 5% of them being identical to human genes (Lindblad-Toh *et al.*, 2005). The researchers used comparative genomics to identify rapidly evolving nuclear genes that revealed the relationships among the bush dog and the maned wolf, the kit fox and the Arctic fox, and the domesticated dog and the gray wolf (Lindblad-Toh *et al.*, 2005). Some of the genes that had undergone rapid evolution in dogs also had accelerated evolution with humans, including some related to neurological functions.

Dr Elaine Ostrander runs the National Human Genome Research Institute's (NHGRI) Dog Genome Project, which has a particular interest in the variation in individual dog breeds. Some of the research from Dr Ostrander's laboratory has included an examination of genetic selection of athletic success in sport-hunting dogs, genomic analyses on the geographic origin, migration, and hybridization of modern dog breed development, and building demographic profiles in domestic dog breeds to optimize genetic trait mapping (National Human Genome Research Institute, 2021).

So what can research involving gene sequencing tell us? For some scientists, it helps us move forward, while for others, it gives us a window into the past—and perhaps recreating it, too. Dr George Church, of Harvard University, is attempting to recreate the woolly mammoth through editing 44 woolly mammoth genes. Mammoths aren't the only species that scientists are trying to reanimate, though. Some species have already been brought back, such as the 1918 influenza virus, the human retrovirus HERV-K, and the Pyrenean ibex, an ungulate that was abundant on the Iberian Peninsula during Medieval times, but was left with only one living individual, a female named Celia, by 1999. Scientists captured Celia and obtained tissue samples from her; after her death in 2000, attempts were made to clone her by transferring nuclei from her cells into the eggs of domesticated goats. While 208 goats were impregnated, there was only one live birth, and she died after 7 minutes because of a lung defect (Choi, 2009). The same process has been tried with animals that are nearing extinction. Scientists successfully cloned the endangered banteng, a southeastern Asian species of wild cattle (Holden, 2003). Similar attempts could be made to help other endangered species with low genetic diversity, such as the Tasmanian devil (*Sarchophilus harrisii*), koala (*Phascolarctos cinereus*), and cheetah (*Acinonyx jubatus*).

Continuing to reanimate extinct species such as the woolly mammoth could provide a healthy advantage to many ecosystems, such as the tundra that mammoths once roamed. In their time, woolly mammoths were a keystone species—animals that the ecosystem largely depended upon. In their absence, the ecosystem changed dramatically. Canada and Russia's tundras were a healthy grass and ice-based ecosystem 4000 years ago, but today, they are melting, which could lead to increased greenhouse gases. Reanimating woolly mammoths, which were an integral part of this ecosystem for millennia, could help diminish these effects by helping the region stay at a colder temperature. The mammoths would do so by eating dead grass, which would clear the way for sunlight to reach new grass with root systems that would keep the ground intact, decreasing the likelihood of erosion, by clearing out snow to allow the air to penetrate the soil, and by felling trees to increase reflected light (Church, 2013).

Koalas (*Phascolarctos cinereus*) have had low genetic diversity for many generations and likely suffered a population bottleneck during the Late Pleistocene, around the same time that a larger species of koala, *Phascolarctos stirtoni*, became extinct. The California condor (*Gymnogyps californianus*) experienced a dramatic population bottleneck: all modern members of the species descended from only 14 genetic founders. As a result, more than 80% of the unique haplotypes (a group of alleles on a chromosome that one inherits together from one parent) in this species have disappeared from the gene pool, and the genetic bottleneck has led to increased inbreeding in an attempt to prevent this species' extinction (D'Elia *et al.*, 2016).

Perhaps the most famous bottleneck occurred during the transition from the Pleistocene to the Holocene epochs, 10,000–12,000 years ago, when many of the earth's larger animal species went extinct. During this mass extinction, commonly referred to as the Quaternary Extinction Event, the continent of South America lost 82% of its megafauna, Australasia lost 71%, North America lost 70%, Europe lost 59%, Asia lost 52%, and Subsaharan Africa lost 16%. Of the animals that survived this mass extinction, including cheetahs, populations were small and isolated, thus lacking genetic diversity. Rather than being genetically heterozygous, they were homozygous; today's cheetahs suffer from the effects of long-term inbreeding. Each cheetah shares 99% of its DNA in common with others of its species. Many cheetahs have kinked tails, immune disorders, and susceptibility to infectious diseases. Cheetahs are particularly vulnerable to catastrophic population losses if a novel disease is introduced to the population, as they lack the genetic diversity to withstand such an onslaught.

To get a better idea of what this means, let's examine the coefficient of relatedness (COR) and the coefficient of inbreeding (COI). The COR, abbreviated as r, measures the degree of consanguinity between two individuals. One's r ranges from "0" (no recent common relatives) to "1" (for identical twins). Recall from the Introduction that among humans, as many as 250,000 people have a COR to their partner that is 0.0625 or higher. This would be the r for second cousins. r is calculated in diploid organisms (those of us who have two complete sets of chromosomes, one received from each parent—this includes you, and your dog, and thousands of other species). With each parent contributing half of its genes to its child, the COR between parent and child would be 0.5. The closer that individuals get to shared genes, the higher their r gets; for genetic diversity and perpetuation of a species, having a low r is an attractive quality to have. As first cousins, Charles Darwin was concerned about having an r of 0.125 with his wife, so imagine how concerned he would have been for the cheetah, whose r is approximately 0.99, which is about as close to 1 as a species can get, unless you're talking about identical twins.

The COI, represented by F, is the probability that an individual would have two alleles at any locus that are identical by descent. When offspring are inbred through more than one line of descent, their total COI is the sum of the separate coefficients. First cousins, like Charles and Emma Darwin, share two grandparents, so their $F = (½)^5 + (½)^5 = (½)^4 = 1/16$, or 0.0625. Across multiple species, high COIs have been correlated with mental and physical health issues. Among gazelles, F correlated with a higher parasite load (Cassinello *et al.*, 2001). Among humans, F has been correlated with higher rates of schizophrenia (Mansour *et al.*, 2010), bipolar I disorder (Mansour *et al.*, 2009), and TB (Stagas *et al.*, 2011). Among Mexican gray wolves, F has been correlated with reduced sperm quality, and thus reduced reproductive success (Asa *et al.*, 2007). And among Icelandic Sheepdogs, there was a significant relationship between F and incidents of hip dysplasia (Ólafsdóttir and Kristjánsson, 2008). We'll revisit the importance of F and r in

Chapter 5 when we take a closer look at how population bottlenecks have impacted specific dog breeds.

A selective sweep is a decrease in the genomic variation surrounding a mutation due to positive selection for the mutation. There are three kinds of sweeps: a hard selective sweep, which occurs when beneficial mutations are uncommon, but once they do occur, increase rapidly and reduce genetic variation within a population; a soft sweep, where a previously neutral mutation becomes beneficial due to environmental changes; and a "multiple origin soft sweep," which occurs when mutations are common and no genomic background reaches high frequency relative to any other genomic background. Humans have experienced selective sweeps across multiple characteristics, including skin color, lactose tolerance, and high-altitude adaptation. Tens of thousands of years ago, as humans emigrated from Africa to other parts of the world, their skin began getting lighter across generations. In Africa, darker pigmentation provides protection from the sun's direct ultraviolet rays, but in areas with less direct sunlight, this pigmentation was no longer selected for—lighter skin was more favorable in these less-sunny areas. Humans' recent geographic distribution isn't the only dramatic change they've undergone. Approximately 10,000 years ago, humans moved toward an agricultural lifestyle, domesticating livestock and using milk to supplement their diets. Humans, like all mammals, drink their mother's milk when they're born, but as they age, many typically suffer from lactose intolerance because they no longer produce the enzyme lactase, which digests dairy products. Our ancestors who possessed the allele to create lactase into adulthood had a favorable advantage over those that did not, and there was a selective sweep throughout Europe to possess this ability.

The ancestors of modern dogs show evidence of shifting their diets once they began to eat alongside hunter-gatherers, rather than subsisting on food that they hunted or happened upon. One of the strongest indicators of selection between dogs and wolves is *CCRN4L* (nocturnin), which is a gene that interacts with peroxisome proliferator-activated receptor (PPAR)-γ and mediates metabolism (Freedman *et al.*, 2016). Among non-human animals—especially domesticated ones—selective sweeps have shaped evolution. All gray horses share a single causative mutation for their coat pigmentation. At some point during their domestication, humans selected a few horses whose coats turned "white" with age and selectively bred them. Up to 10% of the worldwide horse population today is gray. A locus on chromosome four in pigs has undergone selective sweeps, resulting in increased muscle growth and decreased fat production in pigs raised for meat. During the 1980s, domesticated pigs were cross-bred with European wild boar. The resultant phenotypic variation included multifactorial traits such as body composition and growth (Andersson *et al.*, 1994). Most pigs raised in the west for meat today carry an allele that favors muscle growth, increased heart size, and reduced subcutaneous fat depth (Andersson, 2012). Selective sweeps also occurred with chickens that were raised for meat (broiler chickens) or for eggs (layer chickens), selecting for meat and egg production, respectively.

While selective sweeps can rapidly change the complexion of a population, genetic admixture can increase a population's variability or, paradoxically, wipe it out completely. Genetic admixture is the process by which isolated populations initiate previously non-existent gene flow. For example, an individual will have DNA from a distantly related population or species because their parents were from disparate populations or species that interbred. While this introduces new genetic material into a population, these genotypes can be ill-adapted to the environment (consider a population being introduced to pathogens to which their immune systems were completely naïve). There are few modern human populations today that don't possess genetic admixture; a likely candidate for this would be the geographically isolated peoples of India's North Sentinel Island, which is in the Andaman Islands archipelago in the Bay of Bengal. The Sentinelese defend their home to the death, disallowing the presence of *any* outsiders. This vigilance has probably preserved their endangered population.

Admixture has occurred across multiple canine species. A 2014 study (Monzón *et al.*, 2014) used SNPs to examine admixture between coyotes, wolves, and domesticated dogs. The study examined 427 canids, finding that coyotes that lived in areas of high deer density had more wolf-like traits, coyotes in Ohio were highly admixed with dogs and wolves, and eastern coyotes had varying levels of admixture. A later study (Pilot *et al.*, 2018) found that admixture between wolves and domesticated dogs had occurred across Eurasia for thousands of years.

So, what do genes and trends have in common? When a new book in a best-selling series comes out, everyone wants one. The same goes with a bold new fashion—it often sells out within minutes. While this is great news for writers and fashion designers, that same trend can be deleterious for genetic variability among domesticated animals. When an animal achieves high status, everyone wants to breed their female animal to him. This is called the "popular sire effect", wherein there's a reduction in genetic diversity in a population due to excessively mating with a sire that has desirable traits.

Carol Beuchat, PhD, of the Institute of Canine Biology, notes that geneticists warn breeders to avoid the Popular Sire Syndrome (Beuchat, 2013), but paradoxically, the most common advice that breeders receive is to "breed the best to best." Well, how do you operationally define "best"? Is it the best physical specimen of the animal (focusing on phenotype, but ignoring the genotype behind these physical traits)? Is it the best genetic match with another animal (e.g. two individuals who aren't closely related and don't both have genes for a rare genetic condition)?

Breeding closely related individuals, even if they're both the cream of their respective crops, is a potential recipe for disaster. More than 100 years ago, William Haynes wrote about the "Effect of the popular sire," when 20% of the sires of Terriers produced 40% of all puppies (Haynes, 1915). This occurs today, but at a much larger scale: when a male dog wins big at a dog show or other competition, he's the next big thing for his breed, and people will clamor to breed their female dogs to him. Mutations are common; those that are deleterious are typically removed from the gene pool through natural selection. Dr Beuchat writes: "... the recessive, silent [mutations] remain in the genome as the 'genetic load'..."

The longer a population has a small size, and thus a smaller pool of potential mates to spread different genes, the higher one's genetic load is likely to be. Dogs aren't the only species that has seen deleterious effects with restricted population sizes. Domestic horses have a genetic load that's 1.5–11% higher than the genetic load seen in ancestral wild horses (Schubert *et al.*, 2014). As the population of Neanderthals continued to dwindle, their genetic load increased in comparison to their *Homo sapiens* cousins (Do *et al.*, 2015).

According to Dr Beuchat, "Every dog—in fact, every organism—has its own unique collection of damaged alleles that causes no harm as long as there is also a copy of a normal allele of each that can do the job it is supposed to ..." In ordinary circumstances, with random breeding at normal rates, these silent, recessive mutations won't have an impact on the overall population, breed, or species, but when a male dog wins big and breeders are clamoring to breed "the best to the best," that dog can potentially sire a dozen or more litters in 1 year alone. While the first generation of puppies appear to have no ill effects, the second generation might. Beuchat writes: "Perhaps there were a few half-sibling matings, or father-to-daughter, and some puppies are produced that are homozygous for mutation. Perhaps the mutation is lethal and these are stillborn pups, or maybe the puppies are born with a disease. But the breeders will be mystified—they have never had this problem in their line, or even in the breed, so maybe it's just bad luck? Nobody can see yet that this is just the tip of the iceberg."

And this iceberg has the potential to sink the entire ship. In subsequent generations, more ill effects appear, and more and more of his descendants will carry his once-rare mutation. This happened with an American Quarter Horse named Impressive, who was born in 1968 and lived up to his name in the show ring. He was one of the best of his generation and sired more than 2250 foals. His foals excelled in the show ring; they were among the best of their generation, as well. And then, just like Beuchat warned, many owners bred back closely related individuals to one another. Impressive ended up with tens of thousands of descendants in a very short amount of time. Not all of these descendants were healthy, though; some suffered from muscular twitches and temporary immobility. This condition was identified as hyperkalemic periodic paralysis (HYPP) and it was traced back to Impressive, who likely had a silent, recessive mutation that was amplified by the breeding of so many close relatives. The disease earned the moniker "Impressive Syndrome," and breeders soon asked for genetic tests to ensure that their horses didn't carry the disease.

Let's revisit the term "genetic load," which was referred to in Dr Beuchat's piece on the "popular sire effect". A genetic load is a reduction in the mean individual fitness of a population due to the presence of deleterious alleles or allelic combinations in comparison to a genotypically ideal (genetically diverse, heterozygous) population. When there's a high genetic load in a population, the chance of extinction increases. While a high genetic load can happen with either natural or artificial selection, one of the quickest ways to increase it is with the "popular sire effect" and inbreeding.

Gene On, Gene Off

The study of genetics is also the study of the environment and behavior. Genotypes, phenotypes, and behaviors are all linked. People used to argue over whether it was nature or nurture that contributed to an individual's personality and behavior when the answer has always been, "It's both." But it's also a bit more complicated than that. The body has trillions of cells and each contains a nucleus with our DNA, which is composed of gene sequences that provide the directions for making the proteins that program each cell. Think of genes as being "off" or "on"—this is how cells are differentiated from one another. Some genes are active, while others aren't; some cells will become liver or lung cells, while others will be bone or blood cells. During development, the genes within each cell are turned off and on as they respond to environmental factors, including temperature, the amount of light, and available nutrients. In humans, each cell has numerous genes, with somatic cells (any cells not differentiated as sex cells) typically having 46: one copy of chromosomes 1–22 from each parent, an X chromosome from the mother, and an X or Y from the father. Each person has a total of 20,000–25,000 protein-coding genes, with each cell turning on only a fraction of these genes. The remainder of the genes are turned off. So only some genes are active—the switch is turned to "on"—in some tissues and organs, but not in others. Genes are turned on and off during development and in response to environmental changes, such as infection and metabolism.

What happens when the wrong gene gets set to "on?" Each cell in the body is programmed for a certain job. If the wrong genes become turned on within a cell, the results can be catastrophic, and may include susceptibility to diseases such as cancer.

Man's Best GMO

There's a lot of buzz about genetically modified organisms (GMOs), and most of it is bad. It's really a case of semantics, and not science, that leads people to hate GMOs. Without genetic modification, we wouldn't have most of our favorite familiar foods: corn, watermelon, cotton, bananas, alfalfa, sugar beets, squash, zucchini, papaya, and scores of other flowers, fruits, and vegetables—at least not the way we know them. North American sweetcorn originated from the teosinte plant, a Mexican grass that was almost inedible. Before we modified watermelons, the meaty portion of the fruit was a pale green with swirls of red and large seeds. Modern watermelons have a larger red interior (this is the placenta of the fruit). Thousands of years ago, people used wild cotton to spin into yarn to make fabric, selecting those plants that yielded more, or better, yarn to breed with other high-yielding plants, resulting in the domesticated cotton that we're familiar with today. Cotton was further genetically modified with a bacterium called *Bacillus thuringiensis*; plants with this trait have a selective advantage over those that don't as *Bacillus thuringiensis* produces a protein that's toxic to many insect larvae. Today's bananas are unrecognizable in comparison to the bananas that were first discovered by humans as long as 10,000 years ago in modern-day Papua New Guinea and Southeast Asia. Domesticated bananas came from two wild banana varieties, *Musa balbisiana* and *Musa acuminata*, which had larger seeds and a rounder shape than today's smaller-seeded, crescent-shaped fruit.

GMO crops have been altered for taste, appearance, and size, to resist destructive pests and viruses and, in some instances, to resist bruising and prematurely going bad. And this isn't a new practice: humans have cultivated and engineered fruits and vegetables for 9000 years. In his book, *The Omnivore's Dilemma: A Natural History of Four Meals*, author Michael Pollan notes that corn is in—or has a relationship with—almost everything that we eat (Pollan, 2006). Corn syrup, which is made from the starch of corn, is used to sweeten or enhance the flavors of many foods, increase volume, and deter crystallization. High-fructose corn syrup is found in dozens of familiar foods, including candy, popsicles, soda, juice, coffee creamer, yogurt, bread, cereal, granola bars, salad dressing and other sauces and condiments, jam, and jelly. But even more impressive than the ubiquitous corn plant is man's relationship with his first GMO—the domesticated dog. Yes, man's best friend was also man's first GMO.

A GMO has been modified through selection, induced mutations, and hybridization to achieve a specific result. Artificial selection has been a part of humanity's relationship with dogs since the beginning. Humans chose the specific ancestors of gray wolves with suitable traits to mate with other suitable dogs. For tens of thousands of years, we have genetically modified dogs, with the end result being *Canis familiaris*. Genetic mutations are the movers and the shakers of evolution. They occur when a gene includes a change that creates a disruption in the encoded message. Sometimes, this message is then passed on to offspring, and humans have greatly accelerated this process with dogs by selectively breeding them. This is descent with modification: the chance arrival of new forms through DNA mutation.

So if genes are the blueprint, how do they work? For dogs, body size differences are largely explained by differences in a single gene, *IGF1*, among dog breeds. *IGF1* codes for a protein hormone called insulin-like growth factor 1, also referred to as somatomedin C. Growth hormone, originating in the pituitary gland, stimulates the liver to grow cells in the body and has a strong association with small stature across dog breeds. Humans have manipulated dogs' genes for size, temperament, coloration, and a slew of other characteristics, many of them simultaneously. And while selective breeding differs from the traditional view of GMOs, where scientists in white laboratory coats are doctoring the genes of an organism, it is, nonetheless, genetic modification.

Search and Replace

With all of the discussion about artificial selection and GMOs, would it be possible to "repair" some of the less desirous mutations that lead to diseases and other unhealthy conditions? Genome editing has jumped from the realm of sci-fi into the field of modern science. A process called "prime editing" rewrites genetic information, allowing scientists to target diseases such as Tay-Sachs and sickle cell (Anzalone *et al.*, 2019) in humans. Researchers believe that prime editing could have a dramatic impact on the known genetic variants related to human diseases, correcting up to 89% of them (Anzalone *et al.*, 2019). Even though humans and dogs last shared a common ancestor 95 mya, there are approximately 360 genetic disorders that dogs and humans share in common, including allergies, arthritis, cancer, dementia, dental disease, diabetes, diseases related to the heart, kidney, and liver, and physical issues such as hip dysplasia. And many of our cats can get these disorders, as well. So how many breeds of dog (and cat!) would benefit from similar genetic interventions with search and replace technology?

Conclusion

While it's clear that genes play a role in the behavior of species, breeds and individuals, predicting the exact role genes play in an individual is difficult due to the complexity of gene expression and the interaction of genes with an individual's environment. Here we have outlined the molecular roots of some of the complexity in behavioral genetics, including the role of mutations, admixture, evolutionary sweeps, and epistasis, as well as the ordinary independent assortment of alleles that sexual organisms inherit from each parent, in creating variability.

To step back and look at the big picture on the influence of genes on behavior, it's clear that genetics influences behavior in complex, but undeniable ways, and that selection acts on the genetic underpinnings of behavior. But, predicting the behavior of one individual is challenging because that individual could have inherited two uncommon alleles (recessive traits), or even more rarely, could have a new mutation, or could be expressing a genetic trait that must be triggered by the environment. However, it is *far more likely* that the individual demonstrates behavioral traits that are highly similar to the rest of the population (at the species level), as well as the breed level (e.g. the friendly Golden Retriever), as well as one or both parents. Overall, it's better to think about the likelihood of genetic predispositions or tendencies given a species, breed, and an individuals' parents. This information, while imperfect due to the complexities and exceptions discussed here, is generally useful and accurate, and can be a guide in selecting a puppy or helping with a behavioral issue.

References

Anderson, T.M., vonHoldt, B.M., Candille, S.I., Musiani, M., Greco, C. *et al.* (2009) Molecular and evolutionary history of melanism in North American gray wolves. *Science* 323(5919), 1339–1343.

Andersson, L. (2012) How selective sweeps in domestic animals provide new insight into biological mechanisms. *Journal of Internal Medicine* 271(1), 1–14.

Andersson, L., Haley, C.S., Ellegren, H., Knott, S.A., Johansson, M. *et al.* (1994) Genetic mapping of quantitative trait loci for growth and fatness in pigs. *Science* 263(5154), 1771–1774.

Anzalone, A.V., Randolph, P.B., Davis, J.R., Sousa, A.A., Koblan, L.W. *et al.* (2019) Search-and-replace genome editing without double-strand breaks or donor DNA. *Nature* 576(7785), 149–157.

Asa, C., Miller, P., Agnew, M., Rebolledo, J.A.R., Lindsey, S.L. *et al.* (2007) Relationship of inbreeding with sperm quality and reproductive success in Mexican gray wolves. *Animal Conservation* 10(3), 326–331.

Bannasch, D., Safra, N., Young, A., Karmi, N., Schaible, R.S. *et al.* (2008) Mutations in the *SLC2A9* gene cause hyperuricosuria and hyperuricemia in the dog. *PLoS Genetics* 4(11), e1000246.

Belkin, D. (2006) Don't it make my blue eyes brown: Americans are seeing a dramatic color change. *The Boston Globe*. Available at: web.archive.org/web/20141022125010/http:/www.boston.com/yourlife/articles/2006/10/17/dont_it_make_my_blue_eyes_brown/?page=2 (accessed 28 December 2023).

Benson, K.F., Li, F.Q., Person, R.E., Albani, D., Duan, Z. *et al.* (2003) Mutations associated with neutropenia in dogs and humans disrupt intracellular transport of neutrophil elastase. *Nature Genetics* 35(1), 90–96.

Beuchat, C. (2013) The pox of the popular sire. The Institute of Canine Biology. Available at: www.instituteofcaninebiology.org/blog/the-pox-of-popular-sires (accessed 28 October 2023).

Carey, A.H., Chan, K.L., Short, F., White, D., Williamson, R. *et al.* (1993) Evidence for a single gene effect causing polycystic ovaries and male pattern baldness. *Clinical Endocrinology* 38(6), 653–658.

Cassinello, J., Gomendio, M. and Roldan, E.R.S. (2001) Relationship between coefficient of inbreeding and parasite burden in endangered gazelles. *Conservation Biology* 15(4), 1171–1174.

Choi, C.Q. (2009) First extinct-animal clone created. *National Geographic Science*. Available at: www.nationalgeographic.com/science/2009/02/news-bucardo-pyrenean-ibex-deextinction-cloning (accessed 28 October 2023).

Church, G. (2013) George church: de-extinction is a good idea. *Scientific American*. Available at: www.scientificamerican.com/article/george-church-de-extinction-is-a-good-idea (accessed 28 October 2023).

D'Elia, J., Haig, S.M., Mullins, T.D. and Miller, M.P. (2016) Ancient DNA reveals substantial genetic diversity in the California Condor (*Gymnogyps californianus*) prior to a population bottleneck. *The Condor* 118(4), 703–714.

Do, R., Balick, D., Li, H., Adzhubei, I., Sunyaev, S. *et al.* (2015) No evidence that selection has been less effective at removing deleterious mutations in Europeans than in Africans. *Nature Genetics* 47(2), 126–131.

Dodgson, S.E., Day, R. and Fyfe, J.C. (2012) Congenital hypothyroidism with goiter in Tenterfield terriers. *Journal of Veterinary Internal Medicine* 26(6), 1350–1357.

Everts, R., Rothuizen, J. and van Oost, B.A. (2000) Identification of a premature stop codon in the melanocyte-stimulating hormone receptor gene (MC1R) in labrador and golden retrievers with yellow coat colour. *Animal Genetics* 31(3), 194–199.

Freedman, A.H., Lohmueller, K.E. and Wayne, R.K. (2016) Evolutionary history, selective sweeps, and deleterious variation in the dog. *Annual Review of Ecology, Evolution, and Systematics* 47(1), 73–96.

Frischknecht, M., Niehof-Oellers, H., Jagannathan, V., Owczarek-Lipska, M., Drögemüller, C. *et al.* (2013) A *COL11A2* mutation in labrador retrievers with mild disproportionate dwarfism. *PLoS One* 8(3), e60149.

Fyfe, J.C., Kampschmidt, K., Dang, V., Poteet, B.A., He, Q. *et al.* (2003) Congenital hypothyroidism with goiter in toy fox terriers. *Journal of Veterinary Internal Medicine* 17(1), 50–57.

Fyfe, J.C., Lynch, M., Olsen, J. and Louër, E. (2013) A thyroid peroxidase (*TPO*) mutation in dogs reveals a canid-specific gene structure. *Mammalian Genome* 24(3–4), 127–133.

Galton, D. (2009) Did Darwin read Mendel? *QJM: An International Journal of Medicine* 102(8), 587–589.

Generatio: Center for Animal Genetics (2023) Inherited diseases in dogs. Available at: www.centerforanimalgenetics.com/services/dog-genetic-testing/hereditary-disease-testing-for-dogs/muscular-hypertrophy (accessed 28 October 2023).

Ghirlanda, S., Acerbi, A., Herzog, H. and Serpell, J.A. (2013) Fashion vs. function in cultural evolution: the case of dog breed popularity. *PLoS One* 8(9), e74770.

Grant, M.D. and Lauderdale, D.S. (2002) Cohort effects in a genetically determined trait: eye colour among US whites. *Annals of Human Biology* 29(6), 657–666.

Habib, A.M., Okorokov, A.L., Hill, M.N., Bras, J.T., Lee, M.-C. *et al.* (2019) Microdeletion in a *FAAH* pseudogene identified in a patient with high anandamide concentrations and pain insensitivity. *British Journal of Anaesthesia* 123(2), e249–e253.

Hartl, D.L. and Jones, E.W. (2005) *Genetics: Analysis of Genes and Genomes*. Jones & Bartlett Learning, Burlington, MA.

Haynes, W. (1915) Effect of the popular sire. *Journal of Heredity* 6, 494–496.

Holden, C. (2003) Banteng cloned: newborn calf appears to be the first healthy clone of an endangered species. *Science*. Available at: www.science.org/content/article/banteng-cloned (accessed 28 October 2023).

Kambadur, R., Sharma, M., Smith, T.P.L. and Bass, J.J. (1997) Mutations in myostatin (*GDF8*) in double-muscled Belgian Blue and Piedmontese cattle. *Genome Research* 7(9), 910–916.

Keck, M.E., Kern, N., Erhardt, A., Unschuld, P.G., Ising, M. *et al.* (2008) Combined effects of exonic polymorphisms in *CRHR1* and *AVPR1B* genes in a case/control study for panic disorder. *American Journal of Medical Genetics Part B* 147B(7), 1196–1204.

Kirkness, E.F., Bafna, V., Halpern, A.L., Levy, S., Remington, K. *et al.* (2003) The dog genome: survey sequencing and comparative analysis. *Science* 301(5641), 1898–1903.

Lindblad-Toh, K., Wade, C.M., Mikkelsen, T.S., Karlsson, E.K., Jaffe, D.B. *et al.* (2005) Genome sequence, comparative analysis and haplotype structure of the domestic dog. *Nature* 438(7069), 803–819.

Mansour, H., Klei, L., Wood, J., Talkowski, M., Chowdari, K. *et al.* (2009) Consanguinity associated with increased risk for bipolar I disorder in Egypt. *American Journal of Medical Genetics Part B* 150B(6), 879–885.

Mansour, H., Fathi, W., Klei, L., Wood, J., Chowdari, K. *et al.* (2010) Consanguinity and increased risk for schizophrenia in Egypt. *Schizophrenia Research* 120(1–3), 108–112.

Maron, D.F. (2018) Under poaching pressure: elephants are evolving to lose their tusks. *National Geographic*. Available at: www.nationalgeographic.com/animals/2018/11/wildlife-watch-news-tuskless-elephants-behavior-change (accessed 28 October 2023).

Miglani, G.S. (2002) *Advanced Genetics*. Alpha Science, Pangbourne, UK.

Monzón, J., Kays, R. and Dykhuizen, D.E. (2014) Assessment of coyote–wolf–dog admixture using ancestry-informative diagnostic SNPs. *Molecular Ecology* 23(1), 182–197.

Mosher, D.S., Quignon, P., Bustamante, C.D., Sutter, N.B., Mellersh, C.S. *et al.* (2007) A mutation in the myostatin gene increases muscle mass and enhances racing performance in heterozygote dogs. *PLoS Genetics* 3(5), e79.

National Human Genome Research Institute (2021) Dog Genome Project: Publications. Available at: https://research.nhgri.nih.gov/dog_genome/publications/index.shtml (accessed 28 October 2023).

National Library of Medicine (2022) OCA2 gene. Available at: https://ghr.nlm.nih.gov/gene/*OCA2* (accessed 28 October 2023).

National Library of Medicine (2023) HERC2. Available at: www.ncbi.nlm.nih.gov/gene/8924 (accessed 28 October 2023).

Nikolas, M.A. and Burt, S.A. (2010) Genetic and environmental influences on ADHD symptom dimensions of inattention and hyperactivity: a meta-analysis. *Journal of Abnormal Psychology* 119(1), 1–17.

Okane, C. (2019) African elephants are evolving without tusks because of poaching. *CBS News*. Available at: www.cbsnews.com/news/african-elephants-are-evolving-to-not-grow-tusks-because-of-poaching (accessed 28 October 2023).

Ólafsdóttir, G.Á. and Kristjánsson, T. (2008) Correlated pedigree and molecular estimates of inbreeding and their ability to detect inbreeding depression in the Icelandic sheepdog, a recently bottlenecked population of domestic dogs. *Conservation Genetics* 9(6), 1639–1641.

Parker, J.E., Knowler, S.P., Rusbridge, C., Noorman, E. and Jeffery, N.D. (2011) Prevalence of asymptomatic syringomyelia in Cavalier King Charles spaniels. *Veterinary Record* 168(25), 667.

Pedersen, N.C., Pooch, A.S. and Liu, H. (2016) A genetic assessment of the English bulldog. *Canine Genetics and Epidemiology* 3, 6.

Pettigrew, R., Fyfe, J.C., Gregory, B.L., Lipsitz, D., Delahunta, A. *et al.* (2007) CNS hypomyelination in Rat Terrier dogs with congenital goiter and a mutation in the thyroid peroxidase gene. *Veterinary Pathology* 44(1), 50–56.

Pilot, M., Greco, C., vonHoldt, B.M., Randi, E., Jędrzejewski, W. *et al.* (2018) Widespread, long-term admixture between grey wolves and domestic dogs across Eurasia and its implications for the conservation status of hybrids. *Evolutionary Applications* 11(5), 662–680.

Platt, O.S., Brambilla, D.J., Rosse, W.F., Milner, P.F., Castro, O. *et al.* (1994) Mortality in sickle cell disease. Life expectancy and risk factors for early death. *New England Journal of Medicine* 330(23), 1639–1644.

Płonek, M., Giza, E., Niedźwiedź, A., Kubiak, K., Nicpoń, J. *et al.* (2016) Evaluation of the occurrence of canine congenital sensorineural deafness in puppies of predisposed dog breeds using the brainstem auditory evoked response. *Acta Veterinaria Hungarica* 64(4), 425–435.

Pollan, M. (2006) *The Omnivore's Dilemma: A Natural History of Four Meals*. Penguin, London.

Protas, M.E. and Patel, N.H. (2008) Evolution of coloration patterns. *Annual Review of Cell and Developmental Biology* 24, 425–446.

Saetre, P., Lindberg, J., Leonard, J.A., Olsson, K., Pettersson, U. *et al.* (2004) From wild wolf to domestic dog: gene expression changes in the brain. *Molecular Brain Research* 126(2), 198–206.

Safra, N., Schaible, R.H. and Bannasch, D.L. (2006) Linkage analysis with an interbreed backcross maps Dalmatian hyperuricosuria to CFA03. *Mammalian Genome* 17(4), 340–345.

Schmutz, S.M. and Berryere, T.G. (2007) The genetics of cream coat color in dogs. *Journal of Heredity* 98(5), 544–548.

Schubert, M., Jónsson, H., Chang, D., Sarkissian, C., Ermini, L. *et al.* (2014) Prehistoric genomes reveal the genetic foundation and cost of horse domestication. *Proceedings of the National Academy of Sciences of the United States of America* 111(52), 5661–5669.

Shaffer, J.R., Li, J., Lee, M.K., Roosenboom, J., Orlova, E. *et al.* (2017) Multiethnic GWAS reveals polygenic architecture of earlobe attachment. *American Journal of Human Genetics* 101(6), 913–924.

Stagas, M.K., Papaetis, G.S., Orphanidou, D., Kostopoulos, C., Syriou, S. *et al.* (2011) Polymorphisms of the *NRAMP1* gene: distribution and susceptibility to the development of pulmonary tuberculosis in the Greek population. *Medical Science Monitor* 17(1), 1–6.

Strain, G.M. (2004) Deafness prevalence and pigmentation and gender associations in dog breeds at risk. *Veterinary Journal* 167(1), 23–32.

Summers, J.F., Diesel, G., Asher, L., McGreevy, P.D. and Collins, L.M. (2010) Inherited defects in pedigree dogs. Part 2: disorders that are not related to breed standards. *The Veterinary Journal* 183(1), 39–45.

Sunquist, F. (2006) Malaysian mystery leopards. The National Wildlife Federation. Available at: www.nwf.org/en/Magazines/National-Wildlife/2007/Malasian-Mystery (accessed 28 October 2023).

van Hagen, M.A., van der Kolk, J., Barendse, M.A., Imholz, S., Leegwater, P.A. *et al.* (2004) Analysis of the inheritance of white spotting and the evaluation of KIT and EDNRB as spotting loci in Dutch boxer dogs. *Journal of Heredity* 95(6), 526–531.

Wacholder, S., Han, S.S. and Weinberg, C.R. (2011) Inference from a multiplicative model of joint genetic effects or ovarian cancer risk. *Journal of the National Cancer Institute* 103(2), 82–83.

Wayne, R.K. and Ostrander, E.A. (2007) Lessons learned from the dog genome. *Trends in Genetics* 23(11), 557–567.

Zider, A., Flagiello, D., Frouin, I. and Silber, J. (1996) Vestigial gene expression in *Drosophila melanogaster* is modulated by the dTMP pool. *Molecular & General Genetics* 251(1), 91–98.

Zierath, S., Hughes, A.M., Fretwell, N., Dibley, M. and Ekenstedt, K.J. (2017) Frequency of five disease-causing genetic mutations in a large mixed-breed dog population (2011–2012). *PLoS One* 12(11), e0188543.

Zou, Q., Wang, X., Liu, Y., Ouyang, Z., Long, H. *et al.* (2015) Generation of gene-target dogs using CRISPR/Cas9 system. *Journal of Molecular Cell Biology* 7(6), 580–583.

4 Individuals Vary

So it was that a lone black wolf came to live among us, became woven into Juneau's story, part of who we are. We knew him as neighbor and friend, as he knew us. But this is what matters: years from now, parents will tell their kids, *Once upon a time there was a wolf called Romeo.* And they'll remember.

Romeo the Friendly Wolf (Jans, 2023)

Abstract

Chapter 4 illustrates how individuals can have very similar genetic backgrounds, such as belonging to the same breed or even the same family, but have different temperaments, behaviors, and sensitivities. In biology, "variation" indicates any differences in cells, individual organisms, or groups of organisms within a species caused by genetic differences (genotypic variation) or by the effect of environmental factors in the expression of genetic potentials (phenotypic variation). There is an examination of genetic recombination, variation within a species, and variation within a breed where different lines have been bred for different purposes, such as field and show Golden Retrievers. There is a discussion of epigenetics, looking at the environmental factors that can lead to changes in an individual's genes.

Romeo

What is it that makes some individuals stand out, their actions and temperaments impossible to forget? Individuals vary—and how and why they do is just as important as what happens with that variability in subsequent generations. Genetics, environmental influences, including epigenetics, temperament, and early life experiences, work together to create the variation that we see within individuals. These disparate sources of variability influence our ability to interpret species and breed differences, as well. Individual variation is particularly important to the story of the domesticated dog—how and why they entered our lives, and the perfect storm of factors leading up to their entrance. While rare, wolves like Romeo (Fig. 4.1) provide a hypothetical window into the past, a vision of individuals imbued with traits that made them attractive to humans.

Romeo's story trots on across the skyline. Late at night, his memory fills the spaces between Nick Jans' heartbeats. It nudges him awake. And it tints his voice with sadness. Years after he first witnessed a surreal meeting between canines, Jans relives it, a vision as crisp as that dark silhouette against the snow. Jans, who is a best-selling author (Jans, 2014) and photographer, has lived in northwest Arctic Alaska for more than three decades. He counts bears and wolves as his "first loves" and specializes in capturing Alaska's wild places and wild beings on film and paper. It's poetic justice that it was one of his first loves that captured him in return.

Jans recounted the day that Romeo, tall, dark, and very handsome, stood on the snowy landscape, his head tilted forward, ears erect, tail straight out as he met his "Juliet." Known as Dakotah to her friends, the pretty, almost-white Labrador Retriever was irresistible to the lone wolf. For 7 years, Romeo lived among humans, joining them on walks and runs, waiting patiently for their dogs to join him, and playing with them, much like another dog would. And years after this friendly wolf passed away, his unique story is still very much a part of the Alaskan landscape. A wolf is a wolf, but individuals vary. It was Romeo's departure from the expected wolf "norm" that touched and taught an entire community. A plaque dedicated to him carries the inscription, "The spirit of Juneau's friendly black wolf lives on in this wild place."

The sooty-coated wolf first made an appearance in 2003, shortly after a pregnant female wolf was struck and killed by a taxi. While DNA tests were never done on Romeo, he was classified as a rare Alexander

DOI: 10.1079/9781800624566.0004

Fig. 4.1. Romeo. Photograph courtesy of Nick Jans.

Archipelago wolf (*Canis lupus ligoni*), despite being twice the size of the typical *Canis lupus ligoni*. The Alexander Archipelago wolf is endemic to Southeast Alaska, representing the remnants of a population from the last glacial period. Fewer than 1000 of them exist today, but none of them—to date, at least—have shown the same traits that Romeo did. Those who knew him might characterize the unusual wolf as wise, curious, cautious, and playful; he came to recognize specific people and dogs and preferred to spend his time with them, rather than other wolves. Romeo's disappearance in 2009 led to the discovery that he had been illegally gunned down by two "hunters" who aren't worth naming here. Romeo's tragedy demonstrated what it may have been like for the ancestors of dogs to first join man's fire.

"Romeo is still very much a part of who we are, and not in an abstract way," Jans said in 2019. Eight years after Romeo's death, the community wanted to create something to remember him by. "We built an exhibit to honor his memory. It was finished during January of 2017; the actual building only took 4 days. And people would show up even during the construction, because they had known him."

The One Who Stood Alone

On a particularly stormy evening, Romeo's exhibit was finally completed. "The weather was about as bad as it gets in Juneau," Jans recalled, "but people still showed up." In total, 400 people stood up to honor the memory of this wolf who changed their lives forever—a wolf whose individuality made him stand apart from every other wolf they'd "met" or heard about. "The emotion was palpable," Jans said. "People took one look at him and burst into tears ... We all have different emotional thresholds." And we all have different temperaments, personalities, and sensitivities. Romeo, with his unusual personality, struck a chord with so many; his tragic demise and the lessons he taught a community brought most of those spectators to their limits. *But why?* Because Romeo was the wolf who stood alone—the wolf who chose us over his own species (Fig. 4.2).

Wolves, like dogs and humans, will vary widely, even within one family. "There are certainly very different wolves in one pack," Jans explained. "If you're watching a pack of wolves, some will disappear beyond the skyline, even if there's a den or a rendezvous area. Some will be indifferent or neutral: they'll be cautious about your presence, but it's not a big deal. And then there are some that want to approach. Sometimes this will happen, but not always. I've had some wolves come right at me, and from a very long distance away. I interrupted a moose hunt up a snowy creek. I was following fresh wolf tracks. I didn't mean to disturb them.

Fig. 4.2. Romeo and dog. Photograph courtesy of Nick Jans.

Most of the wolves, and the moose, went up over the skyline and were gone. And one wolf, who happened to be black, like Romeo, ran right at me, and stopped about 20 or 30 feet away. She was a young female. She was excited, but also curious. My snowmobile was stuck at the time; I was up to my armpits in snow. I wanted to get up higher so I could see her and not be below her. I remember the flashing of her eyes ... she was just looking at me."

Alone But Not Unique

While the black coat color is a dominant trait, black is a rare coat color variant in the northern open taiga or tundra wolf populations. Recall from Chapter 3 that there are three genes that determine wolf coat color. We'll focus on the gene that codes for the black coat color cline, or gradation: the KB allele. The KB allele is from southern forest populations, where black and gray are equally common colors (Anderson *et al.*, 2009). This allele is a β-defensin gene, and it's hypothesized that it improves disease resistance of heterozygotes. Those individuals that are heterozygous (having two alleles of a particular gene or genes; these would be the Kk wolves) for the black allele have enhanced resistance against disease, and thus higher lifetime survivorship. Genetically speaking, it's often a "safer bet" to be heterozygous: one is less likely to have a rare recessive (and potentially harmful) condition.

Black homozygous (KK) wolves like Romeo are thought to have reduced fitness; most die *in utero*. Surviving adult KK wolves are very rare (Freedman *et al.*, 2016). Research examining the population of Yellowstone wolves, half of which have gray coats and half of which have black, appears to support this. A 2014 study found that 55% of the Yellowstone wolves were homozygous gray (kk), 42% were heterozygous black (Kk), and only 3% were homozygous black (KK) (Hedrick *et al.*, 2014). According to Daniel R. Stahler, wildlife biologist at Yellowstone Park and one of the investigators on the study, "There seems to be strong selection for the heterozygote black and selection against the homozygote black." While heterozygous black wolves had strong levels of reproductive success, such was not the case for their homozygous black peers.

After our crash course in genetics, you'll hopefully remember that a single gene can have multiple effects. One that effects color variation could also have other effects, including cellular function or temperament. Humans with red hair have a melanocortin 1 receptor (*MC1R*) polymorphism and also have lower pain tolerance and increased heat sensitivity. Rats, deer mice, and foxes with a black coat caused by an agouti signaling protein (*ASIP*) polymorphism are less active and more docile. And a study with horses found that black mares are more independent than bay ones (Jacobs *et al.*, 2016). Perhaps, in addition to having lower disease resistance, characteristics like curiosity and boldness are also associated with black homozygous wolves.

Traits such as these—curiosity and boldness—likely characterized the ancestors of the canines who first came to share their lives with us. "Romeo was the wolf who came to live by our fire, and there are a few of those," Jans said. "With someone else, that wolf who approached during the moose hunt would have been dead. It's difficult to know how many times, over the course of our history, humans missed the opportunity to know a wolf."

While "friendly" wolves are rare, another wolf in Dillingham, Alaska, was also receptive to people. Fish and game biologist Neil Barton, who knew Romeo, said that the Dillingham wolf was also friendly and hospitable toward people. But unlike Romeo, this wolf lasted only a few weeks before she was shot and killed by a local.

"Dillingham is a typical Alaska town where a seen wolf is a dead wolf," Nick said.

Jans' friend, Dwight Arnold, bred sled dogs, much like the Inupiaq Native Americans had, for centuries. Arnold bred his dogs with wolf hybrids and was familiar with the genetic and behavioral variation these crosses yielded. "Maybe out of several litters, he'd get one good sled dog you could handle," Jans recalled. Another of Jans' acquaintances, a native who bred sled dogs with wolves, raced in the Iditarod several times. "It's a roll of the dice. His dogs would be barely controllable at the starting line, jumping straight up in the air." Even with individual variation, breeding dogs with "friendly" wolves doesn't guarantee friendly offspring, nor is it a recommended practice. Wolves like Romeo are unique, but they're still wolves.

Wendy Spencer has been with Wolf Haven International since 1998. She has found that the similarities between wolves and dogs have been problematic for decades. "Wolves are hard-wired differently than dogs," she explained. "All wolves belong in the wild, not in captivity. They're evolutionarily designed for living in family groups, traveling long distances, maintaining large territories, and hunting cooperatively to take down large prey. Humans force wolves to conform to their misguided notions of what they want them to be, which is usually incongruent with a wolf's true nature, and the consequences are often tragic for the animals and humans involved."

For those first wolves who came in to live by our fire, the stakes were high, much as they are today. As Jans said, wolves would have varied reactions to the humans they met during their travels. There is also the individual variation among the humans they encountered. Some would be afraid and flee, some would be neutral, some would kill them, and a rare few would accept them, allowing them to cross over the periphery of their lives.

A Crisis of Variability?

Whether they're humans or domesticated dogs, individuals have remarkable variability. This is due to several factors, including genetics, epigenetics, differences during one's sensitive and critical periods of development, and different experiences throughout one's lifetime. But with growing knowledge about these factors comes an increased awareness that we can't use population insights to make inferences about individuals. A 2013 study on the science of the individual argued that individuals behave, learn, and develop in distinctive ways; so much so that it has created a scientific explanatory crisis that the authors refer to as a "Crisis of Variability" (Rose *et al.*, 2013).

This variability includes differences in behavior, neurology, physiology, and biology; differences have even been found between how identical twins age. Humans vary in many ways, including, but not limited to, their body temperature (98.6°F is only an average), blood pressure, immune response, vision, hearing, senses of smell and taste, cognitive capacities, including innate intelligence, memory, and capacity to learn, athleticism, metabolism, and aging. While the latter has strong heritable factors, lifespans and longevity don't appear to be heritable. Twins had an increased risk with certain kinds of cancer, including prostate, colorectal, stomach, lung, and breast cancers. But even identical twins don't get the most heritable forms of cancer at the same rate: colorectal, prostate, and breast cancer have a risk of only 9–21% before the age of 75 years when a monozygotic twin had that form of cancer

(Lichtenstein *et al.*, 2000). In comparison, the rates for occurrence of certain cancers with non-twin siblings is 2.31% for breast cancer, 2.41% for colorectal cancer, and 3.69% for prostate cancer (Lee *et al.*, 2015).

Even animals that have been artificially selected for generations, including dogs and horses, will vary in their appearance, health, cognitive capacities, and athleticism; you can plan for a certain outcome, but it is far from assured. Take, for example, the case of Barbaro the Thoroughbred racehorse and his full siblings, Nicanor, Lentenor, Margano, and Pennmarydel. The sire (father) of all five of these brothers was Dynaformer and the dam (mother) was La Ville Rouge, but the success of the foals varied widely. In the six races that Barbaro finished, he was never bested, including his triumph in the 2006 Kentucky Derby. The only time he didn't cross the finish line first was in his seventh and last race, the Preakness Stakes. He shattered a hind leg and was humanely euthanized the following year. His full brothers never reached this level; the most successful was Margano, born in 2009. He was stakes-placed and raced 30 times, with five wins, five seconds, and six thirds, for earnings of US$373,969. Five full brothers. Five very carefully planned breedings between the same parents. And five very different results.

The Milkman's Baby

Two full human siblings submitted their DNA to a genetics testing company. Their tests came back with slightly dissimilar results—but how? Were they not actually full siblings? Children inherit half of their DNA from their mother's egg and half from their father's sperm, but it's more complicated than that. During the creation of an egg or sperm, there's a process called genetic recombination wherein the number of chromosome pairs in a cell is halved from 46 to 23. During recombination, the chromosomes align themselves in pairs. Genetic material is exchanged before creating an egg or sperm cell, allowing for variation between each gamete. Each parent thus contributes 23 pairs of chromosomes to their offspring, for a total of 46 and, because of recombination, children don't receive identical DNA from their parents. So, it is entirely possible for full siblings to have slightly different results: they are approximately, but not *necessarily* 50% related to one another.

Imagine that your mother has two chromosomes in a pair. We'll designate the separate chromosomes "red" and "blue." The single chromosome you inherit from your mother's pair of chromosomes will contain information from both chromosomes in the pair, condensed through recombination, into one chromosome. This chromosome will then pair up with a corresponding chromosome from your father. Let's imagine that each of these two chromosomes is a stick comprising 10 blocks. If you broke up the chromosome sticks and threw the separate blocks into a bag, you'd have 20 total blocks to choose from, but only 10 would go into the offspring's chromosome. The resulting chromosome could have one of 60,466,176 possibilities, including red/blue, red/blue; red/red, blue/blue; blue/blue, blue/red, and any number of combinations of these permutations, as well. This new chromosome could be 9/1 red/blue, 5/5 red/blue, 1/9 red/blue, or anything in between, and in any pattern, as well. This, in a nutshell, is recombination. Except the "blocks" or "parts" can be broken up way more than ten ways. Let's consider this in terms of chromosome possibilities. Humans have 23 pairs of chromosomes, for a total of 46, while dogs have 39 pairs of chromosomes, for a total of 78. From two human parents combined, there are 8,234,608 possible combinations of 23 chromosome pairs, calculated at 2^{23}, with 23 being the haploid number. Additionally, there can be recombination of chromosome "parts," which brings us to an incalculable number, due to gene linkage and random recombination. With dogs, there are 39 pairs of chromosomes, calculated at 2^{39}, which would equal an astronomical 549,755,813,888 possibilities (almost 550 billion possibilities!). These are *minimum* numbers, too—partial recombination only increases these figures. Because there is some linkage between chromosome parts, we can't increase these recombinations as if they're independent of one another.

Let's break this down into a simpler image. Consider the bell curve you would have if you flipped a coin 1000 times. The probability of having heads or tails each time you flipped would be roughly 50%; the chances of landing heads-side up 100% of the time or 0% of the time would be highly improbable. The likelihood of flipping heads 30 times in a row, for example, would be 2^{30}, or one in 1,073,741,824.

So, someone would be likely to flip a head 30 times in a row once in more than 1 billion attempts ... which is quite unlikely, indeed. During recombination, there are many potential variations of "red" and "blue." Each of the chromosomes that a person inherits from their parent is made up of a mix of two of their chromosomes. The DNA could be mixed in an infinite number of ways, with each offspring being 50% related to their parents, because all of the DNA came from their parents, but they aren't necessarily 50% related to their siblings, because their siblings inherited different patterns of red and blue, and at different ratios. There will be some overlap, but also a lot of loci that are unique to the individual. The different amounts of recombination typically average out to 50%, though. Given the possibilities in our genetic shuffling, it's surprising that we're even as closely related as we are (Fig. 4.3).

It's a slightly different story for dogs. Domesticated dogs have been matched up for breeding by humans for generations, with individuals paired with other individuals who shared particular highly desirable physical or behavioral traits (and a lot of their genetic material, as well, as they were often closely related). This artificial selection has been a different driver of variation among dogs than it has with humans. Excepting arranged marriages, most humans date, wed, and have children because they share deep affection, attraction, emotional attachment, and proximity. There are certain characteristics, such as familiarity and symmetry, that will attract two individuals, but creating an offspring that's as close to their carbon copy—much like we do with purebred dog breeding—isn't part of that decision-making process. Arranged marriages, which were more common before the 20th century, were often endogamous, and this is our closest approximation to artificial selection with humans. Endogamy is the practice of marrying within a particular group, whether it's consanguineous (related by blood), social, or ethnic; there's a higher likelihood among these pairings that individuals will look similar or be more closely related than relationships outside of the arranged parameters.

Another driver of genetics is assortative mating. Also known as homogamy, assortative mating is a mating pattern that occurs when individuals with similar phenotypes mate with one another. Compared with random mating, you get less

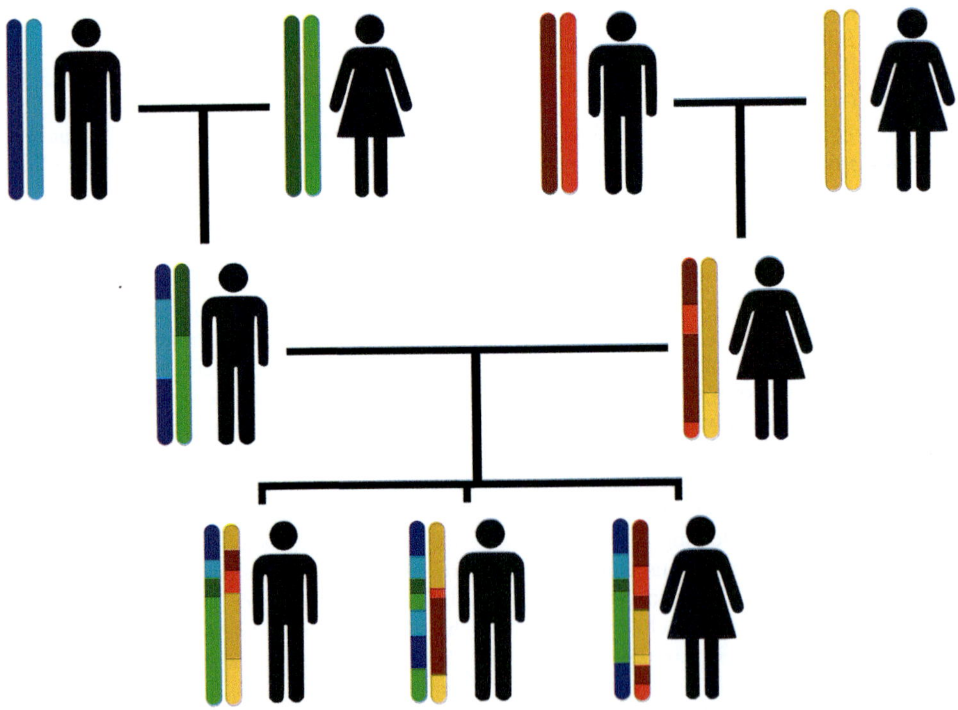

Fig. 4.3. Schematic representation of the autosomal transmission from grandparents to three siblings. This file is licensed under the Creative Commons Attribution-Share Alike 4.0 International license and was provided by Wikipedia. Available at: https://commons.wikimedia.org/wiki/File:Autosomal_inheritance_GPto3.svg.

variability in the offspring with assortative mating, because the parents are more similar than expected by chance. Assortative mating is a form of sexual selection. Under human control, the effects are amplified, as only individuals with those certain desired phenotypic traits will have the opportunity to mate. Assortative mating is particularly evident both with dog breeds and with different lines *within* dog breeds. While a breed has often been considered to be a uniform group, set by breed standards, there are also often subgroups, or "lines," within one breed that differ in coat color, temperament, and behavior. A 2007 study examined assortative mating and fragmentation within dog breeds, finding that genetic uniformity wasn't always the case within breeds, as some breeds are separated into different lines (e.g. show or field work) (Björnerfeldt *et al.*, 2007). The researchers examined genetic patterns with 164 Poodles, comparing them to 133 dogs from eight other breeds, including the Siberian Husky, Labrador Retriever, German Shepherd, Bull Terrier, Wire Fox Terrier, Smooth Fox Terrier, Giant Schnauzer, and Miniature Schnauzer. Poodles were selected for this study because the Fédération Cynologique Internationale (FCI) places them in four size groups: Toy, standing 28 cm at the withers, Miniature, standing at 28–35 cm, Medium, standing at 35–45 cm, and Standard, standing at 45–60 cm. In addition to being grouped by size, Poodles can only be registered if they have certain coat colors, and those colors are required to be uniform throughout. Coat color, paired with size restrictions, has fragmented the genetic diversity of Poodles into five distinct groups. The genetic patterns observed in the eight comparison breeds indicated that fragmentation, driven by breed standards, is likely common among many dog breeds.

Sister, Sister

On July 3, 2006, Amanda Wanklin gave birth to fraternal (two different ova, two different sperm) twin girls. Wanklin is fair-skinned and haired, while her husband, Michael Biggs, is of Jamaican descent. One of their twins, Marcia, was born with her mother's fair complexion, while their other twin, Millie, was born with her father's skin tone. On April 23, 2016, Whitney Meyer also gave birth to fraternal twin girls—with one of them, Kalani, being fair like her mother and the other, Jarani, having darker skin like her father, Tomas Dean. And in February of 2015, Libby Appleby gave birth to monozygotic (genetically identical, resulting from the division of a single fertilized ovum) twin girls. While their genotypes were identical, their phenotypes were not. Wait … what? Physically unidentical identical twins? Yes. Amelia was born with dark skin, eyes, and hair, like her father, Tafadzwa Madzimbamuto, while Jasmine was born with fair skin, blue eyes, and lighter brown hair. Instances of differently pigmented twins are highly unusual, but not unheard of, because skin color isn't a binary trait, but a quantitative one. For Amelia and Jasmine, a change occurred between them very early during their development, shortly after they separated as embryos. The two most likely causes of this phenotypic difference between the twins was either a somatic mutation or an epigenetic modification. Somatic mutations are alterations in the genes that could be passed on to the offspring of the mutated cell during cell division. Epigenetic modifications can affect whether a gene is expressed, but do not alter the DNA sequence. Li *et al.* (2014) noted how somatic driver mutations in single nucleotide polymorphisms (SNPs) could lead to differences between the genomes of twins who were otherwise identical. By some estimates, having fraternal twins with different pigmentation is a one-in-a-million occurrence, while having identical twins with different pigmentation is far less common—Amelia and Jasmine were the first known monozygotic twins to share all of their genetic material in common, but not their phenotypic traits.

So why does it matter if a trait is binary or quantitative? According to Dr Claire Steves at King's College London's Department of Twin Research, multiple genes control for skin pigmentation and identical twins are highly likely—but not *certain*—to share these genes completely (Brennan, 2016). This is important because it significantly contributes to the individual variation that we see within a population. Binary traits have only two distinct phenotypic values (e.g. present or absent) while quantitative traits depend upon the aggregate actions of numerous genes and the environment and have a continuous phenotypic distribution. Binary traits include albinism, Huntington's Disease, and cystic fibrosis (either you have this, or you don't), while skin color, height, fingerprint ridge counts, cholesterol levels, and intelligence levels are all quantitative traits, meaning that there's a range. Within any family, there will be a distribution among members for quantitative traits. Some might be very tall, while others are very short, and some might have olive skin tones and dark tresses, while others are peach-complected and blonde.

With identical twins, one would expect the same skin pigmentation and the same likelihood for genetic diseases, but neither is always the case. Recall that identical twins don't get even the most heritable forms of cancer at the same rate. But what about when one twin has two kinds of cancer? Joan Elvin and Jean Christ of Washington State are identical twins; while Joan had both ovarian and breast cancer before the age of 50 years, Jean has had neither. Both women are fit, active, and eat healthily. Their case is particularly interesting to scientists studying how identical twins can differ—and why.

The Long and Winding Road

When Charles Darwin traveled across the Galapagos islands, the endemic finches and mockingbirds, isolated from other breeding populations for thousands of years, revealed a remarkable level of radiation from a founding species, and as a result, a remarkable rate of *inter*-species variability. The domesticated dogs he met showed a remarkable range of *intra*-species diversity, and these animals helped shape his *On the Origin of Species* and his later works, as well. In biology, variation is any difference between cells, individual organisms, or groups of organisms, caused by genotypic variation or environmental influences on the expression of their genetic makeup (phenotypic variation). Variation in a species can occur in one of three ways: through sexual reproduction, gene flow, and mutation. Sexual reproduction introduces variation because the egg and sperm contain different gene combinations from the parent. Gene flow introduces variation because individuals from different populations (and most likely different founding gene pools) introduce new alleles into a population. This occurs when plants spread their pollen or when animals leave their natal group to produce offspring with unrelated individuals. The ultimate source of genetic variation, however, is through mutations, which are the permanent alteration of the nucleotide sequence of the genome. While most mutations will be neutral, some will be harmful and others helpful. Neither neutral nor harmful mutations would be strongly selected for in a free-living population. When a mutation provides a selective advantage, however, these new alleles can be selected for.

Populations with wider variability are more capable of adapting to environmental changes. With natural selection, it's more likely that genetic variation is maintained at some level. Recall from Chapter 1, the gray peppered moth (*Biston betularia*) during the Industrial Revolution. The population was capable of adapting to the environmental changes (light bark, and then bark darkened by pollution, followed by the return of light bark after introducing environmental protections) because there was enough variability in the population. Lighter or darker moths may have been more common, given the change in the environment and the selective advantage of each phenotype, but there was still a range of colors and a range of shades within both the "dark" and "light" phenotypes. Imagine if this was instead an artificial selection experiment? Would the population of moths still have maintained that light colored allele(s) to go backwards? Evolution isn't like a "Choose your own adventure" book; you can't turn back the page if you choose the wrong path or when a species goes extinct. Many scientists have tried to bring back extinct species, using some remnant DNA and the DNA of closely related species. It's a good start for thinking about re-introducing diversity within a breed suffering from low diversity rates, but hasn't been a successful strategy, at least thus far, for bringing extinct species back.

The environment adds variation to a population and it can be just as complex. Remember how genes can be "on" or "off?" Environmental factors can "turn on" some genes that would otherwise lie dormant, and it can interact with genetic predispositions in complex ways. We're a long way from understanding the complex interactions of genes and the environment in producing human behavior, much less dog behavior. But with our history of co-evolution, a closer examination of canines is as much a story about them as it is about us. For example, heritability estimates explain approximately 50% of the variation in human intelligence and approximately 40–50% of the variation in human personality measures such as sociability, anxiety, agreeableness, and activity (Berk, 2006). Both twin studies and adoption studies in humans suggest that schizophrenia, bipolar disorder, and autism are highly heritable (with estimates above 70%), and antisocial behavior and depression is only moderately heritable at 30–40% (Berk, 2006). The problem with heritability calculations is that they are specific to the population studied and don't necessarily transfer to other populations. Nevertheless, they do suggest that genetic effects on behavior are significant, and can themselves vary between individuals and the environment. We need to keep in mind the complex interactions of genes and environment in trying to understand behavior. Basically, our genes can influence our environment and our environment can influence our genes. This concept is called epigenesis (Gottlieb, 1998, 2007), and it's the next big thing in genetics research.

Like all "big things," epigenesis isn't without controversy: picture how the actions of your direct ancestors (parents and grandparents) impacted their genetics and, subsequently, yours. And imagine how your actions are impacting your genetics and have—*or will*—impact those of your children. If you're saying, "Wait, how can parents' actions, during their lifetimes, affect their children? Isn't this Lamarckism?" you're not alone.

Lamarckism was the brainchild of French biologist Jean-Baptiste Lamarck (1744–1829), who hypothesized that organisms could pass on characteristics that they had acquired during their lifetimes (see Introduction). While the hypothesis, known as the "inheritance of acquired characteristics," was certainly provocative, it was a wrong turn in evolutionary science. Lamarckians believed, for example, that a giraffe's long neck came from its parents stretching repeatedly to reach the tallest branches of a tree; with each stretch, the offspring would have a slightly longer neck, and so on until we achieved the long-necked creature we know today. While tantalizing, this isn't how evolution works. If, for example, a mother dog's tail is docked and her ears are cropped, her puppies aren't born with these modifications. Even if ten generations of female dogs each have these procedures done, their puppies' ears and tails will still be the phenotypic reflection of their genotypic blueprints. To be born with a certain phenotype, you need to have the genotype for it, and research in genetics revealed how traits were passed on from one generation to the next. Research in epigenetics shows how the genes for these traits can be turned on or off.

In a 2002 Swedish study, scientists found a fascinating relationship between male juvenile nutrition and the health of that individual's children and grandchildren. In the study, when fathers had insufficient food during critical developmental periods shortly before the onset of puberty, their children were less likely to die from cardiovascular disease than the children of fathers who did have sufficient juvenile nutrition (Kaati *et al.*, 2002). This is particularly interesting because the number of adipose (fat) cells that one has is fixed during puberty, remaining constant irrespective of weight gain or loss (National Institutes of Health, 2008). Thus, one's nutritional intake during this time could have wide-reaching effects for the duration of their lifespans. Among children whose paternal grandfathers overate when food was typically scarce, and their bodies were accustomed to a meager diet, diabetes-related deaths increased: these grandsons died more than 30 years earlier than if their grandfathers had eaten steadily. Those same mortality rates decreased when their fathers had sufficient food (Kaati *et al.*, 2002). Unlike the Lamarckian belief that stretching giraffe necks could be passed on to subsequent generations (it didn't act upon the individuals' genotypes), the amount of food consumed did act upon how the fathers' and grandfathers' genes were expressed, and how they would be active or inactive in their descendants.

Carnival of Light

Humans aren't the only species that has exhibited epigenetic effects. Scientists also found an interesting relationship between light and temperature with rabbit epigenetics. In 1913, American geneticist Alfred Henry Sturtevant discovered that light and temperature had an impact on the development of fur pigmentation among Himalayan rabbits. These rabbits carry the C gene, which is responsible for developing pigmentation in the skin, eyes, and fur—and Dr Sturtevant discovered that the expression of the gene is temperature-dependent (Sturtevant, 1913). Sturtevant found that the C gene is maximally active from 15°C to 25°C and inactive above 35°C. Where the rabbit's body was warm (above 35°C), their coat wouldn't produce pigment, but where it was considerably lower than 35°C, the coat would be black, creating a "seal point" coloration pattern. In 1930s, scientists studied this phenomenon with cats, finding that their coloration patterns were also temperature-dependent. When Siamese cats were brought to Moscow and kept in rooms ranging from 3°C to 16°C, their typically cream-colored coats shed out to a much darker hue (Iljin and Iljin, 1930). While we have interesting examples of heat's role in the development of genes with rabbits and cats, we have yet to find studies about temperature affecting pigmentation in dogs.

Light, like temperature, can also influence the expression of genes. In 1917, American geneticist and evolutionary biologist, Thomas Hunt Morgan, discovered that the exposure of caterpillars to certain light colors affected the wing colors they would later develop. Working with *Vanessa urtica* and *Vanessa io* caterpillars, he placed them under blue, green, or red light. Another group of caterpillars were kept completely in the dark. When they reached maturity and became butterflies, those that had been exposed to red light had intensely colored wings. Those that had been exposed to blue light had pale wings, larger than the wings of the others in the experiment. Those that had been exposed to green light had darker wings with no size difference (Sturtevant, 1913). Research with zebrafish also illuminated how light can manipulate gene expression. Shinzi Ogasawara with Hokkaido University's Creative Research Institution manipulated the timing and duration of gene expression in zebrafish by controlling the process of translating messenger RNA (mRNA) to protein. mRNA carries the genetic information from the DNA to the ribosome in the form of three-base-code "words," with each specifying the amino acid sequence for the protein products of gene expression. Ogasawara injected fluorescent protein mRNAs into zebrafish embryos and then irradiated them with either ultraviolet or blue light. The embryos that were irradiated with blue light did not have the fluorescent protein, while those that received ultraviolet light produced more fluorescent protein. By controlling the expression duration of "squint," a gene that codes for body axis formation with mRNA, Ogasawara also unintentionally created a double-headed zebrafish (Ogasawara, 2017). While this is an extreme modification, we could also ask: what else are we unintentionally changing in our dogs when we breed for very short legs, numerous skin folds, or large, brachycephalic heads?

Because the Wind is High

Light isn't the only environmental factor that can impact gene expression. Recent research out of the Himalayas, the planet's tallest mountain range, provides evidence that altitude, too, influences gene expression. The animals that reside in the Himalayan mountains of China, India, and Nepal often differ physically and behaviorally from those that live at lower altitudes. Himalayan wolves that live at these heights have comparatively paler coat colors and longer snouts than their lowland gray wolf brethren. Their calls have comparatively lower frequency and shorter duration, and their diets differ, too: Himalayan wolves will eat Tibetan gazelles, while gray wolves typically eat rodents. But that isn't all—the Himalayan wolves—who live at heights of 13,000 feet or more—are genetically distinct from gray wolves. A 2020 study showed that they have genes that enable them to thrive in thin mountain air (Werhahn *et al.*, 2020). The adaptations are similar to other high-altitude inhabitants, including domesticated yaks, Tibetan people, and Tibetan dogs, which appear to have interbred with Himalayan wolves.

This study used DNA from 280 wolf scat samples from Tajikistan, Kyrgyzstan, and western China, examining two mitochondrial DNA (mtDNA) loci, 17 microsatellite loci, four non-synonymous SNPs in three nuclear genes related to the hypoxia pathway, and ZF ("zinc finger") genes on both sex chromosomes. The researchers found a correlation between the divergent Himalayan wolf mtDNA haplotype and a hypoxia adaptation. According to University of California, Davis, canine evolutionary biologist Ben Sacks, this high-altitude research is big news, because it shows that the Himalayan wolf is genetically distinct from gray wolves and should be recognized as a distinct species (Morell, 2020). It also exemplifies directional selection.

Beyond light, temperature, and altitude, stress can also profoundly affect one's epigenome. Researchers at the University of Linköping in Sweden created a stress-inducing henhouse where circadian rhythms were manipulated to be unpredictable. With no set schedule to follow, the hens became stressed. They didn't know when to roost or when to eat. The increased stress reduced their ability to solve spatial learning tasks (finding food in a maze). Their offspring, who were raised without contact with them, also had reduced ability in this task compared to a control group whose parents hadn't been raised in a stressful environment. When researchers examined the genomes of the stressed parents and their offspring, they found that at least 31 genes in the hypothalamus and pituitary had been up- or down-regulated. These same genes hadn't been affected in the offspring from the non-stressed parents. The offspring from the stressed parents were raised in a non-stressful henhouse, but their offspring still struggled in the spatial learning tasks in comparison to a control group (Lindqvist *et al.*, 2007). Thus, the chickens inherited the effects of stress from prior generations, and it impacted their gene expression. This brings "nature versus nurture" to a whole new level: the two are equally, and simultaneously, important.

Epigenesis has been important in the evolution of dogs, as well. Remember that with epigenetics, phenotypic changes correspond to changes in gene expression. During domestication, a species changes both genotypically and phenotypically, but at a much faster rate than would be expected in comparison to changes that occur during natural selection. Domesticated animals—and dogs in particular—have been raised in environments where human actions would have epigenetic effects. Just like the shadow of stress on the genes of the hen's offspring, for tens of thousands of years, dogs around the world have been raised in human homes, with those epigenetic marks that were most beneficial. Those markers translated into phenotypes and behaviors that were attractive to humans—and have been selected for by humans. While dogs and wolves share 99.6% of their DNA in common, that 0.4% difference is deceptively large in terms of gene expression—and epigenetics could be the answer to how these genes are expressed.

Pörtl and Jung (2017) examined whether dog domestication was due to epigenetic brain modulation. The researchers examined whether certain changes that have occurred in the canine brain, such as those in the hypothalamic–pituitary–adrenal (HPA) axis, were important in the continued process of domestication. The HPA axis is a set of interactions between the adrenal glands, located atop the kidneys, and two areas of the brain: the pituitary gland and hypothalamus. The HPA axis is the stress control response center, regulating how the body responds to stress. Thus, it's highly important to social behavior—and to areas of research concerning the development of social partnerships, such as domestication. The researchers used the Active Social Domestication (ASD) model, which states that dog domestication was a unique phenomenon with an active social process on both sides (Pörtl and Jung, 2017). Belyaev and Sorokina's longitudinal experiment with silver foxes (*Vulpes vulpes*) revealed the physical and behavioral changes that occurred to a canid during domestication, and a 2018 study revealed the genetic changes that occur, as well. The study investigated gene expression in the anterior pituitaries of Belyaev and Sorokina's foxes, finding differences between the pituitaries of foxes that had been bred to be aggressive and foxes that had been bred to be tame (Hekman *et al.*, 2018). Recall the research of Anna Kukekova, who sequenced the genes of ten silver foxes with temperaments ranging from aggressive to tame (Chapter 2). Focusing on the 103 genomic regions that differed between the two temperaments, they found that the tamest foxes had a version of a gene called *SORCS1* that was not present in aggressive or conventionally bred foxes. Again, this version is associated with hypersociality. While the hens raised in stressful situations went on to have descendants who also had the epigenome of high stress (paired with poor results in spatial learning tasks), domesticated animals such as the foxes typically have lower stress levels than their free-living peers when introduced to unfamiliar people and novel objects. The silver foxes had both lower (visual) stress levels as well as decreased cortisol levels, accompanied by increased serotonin levels (Trut *et al.*, 2009), which is colloquially referred to as the "happy chemical." During domestication, the HPA axis was epigenetically downregulated; this modification can explain prosocial behavior (Meaney and Szyf, 2005). The HPA axis, that connection between the endocrine system and the brain, is an axis that's finding continued support for its importance in domestication.

Among humans, smoking cigarettes creates epigenetic changes. According to Dr Stephen Baylin, Professor of Oncology at the John Hopkins Kimmel Cancer Center, studies have indicated that the smoke sensitizes airway cells to genetic mutations that are known to cause lung cancer (Johns Hopkins Medicine, 2017). Maternal smoking epigenetically harms the fetus, as well; the fetuses of mothers who smoked while pregnant show DNA patterns similar to those found in adults who smoke cigarettes. Consider, then, the potential for differences among two full siblings, where the mother didn't smoke during the first pregnancy, but did during the second. This variable alone would determine whether certain genes were turned "on" or "off," leading to increased differences among these full siblings. Let's extend that full siblings example. Even if we try to control for all variables, such as eating the same healthy foods, exercising the same amount, and refraining from activities that are harmful during a pregnancy (such as cigarette smoking and alcohol consumption), there will still be different environmental factors with each pregnancy that have the potential for epigenetic effects.

The science of behavioral genetics in humans is not qualitatively different from the behavioral genetics of other mammals. Scientists use data collected from humans to make hypotheses about animals, and data from animals to make hypotheses about humans. We have animal models of human diseases, and human medicine can affect animal medicine, as well. The biology of mammals is similar enough to make these leaps, and we have found many examples of behavioral similarities, as well, particularly in the recent popularity of studying personality or temperament in animals. So, you cannot say that because behavior is complex and varied, it must all be due to experience and learning. Yet this is something we hear from some dog professionals. The scientific field has accepted that both dog and human behavior is affected by genetics, and given our biology and shared ancestry as mammals, it makes sense.

Genetics isn't the only source of variability; it's also important to take into consideration events that occur across the lifespan, particularly those that occur early during life. The early life experiences that have the highest potential to have a strong impact later in life include traumatic events. In dogs, trauma can take on many appearances, none of which they can verbalize to us. Organisms can also have experiences throughout their lifespans that build upon their other experiences to have a cumulative effect (e.g. the reoccurrence of a traumatic event can increase in intensity each time it happens). Let's take a look at how these early life experiences also shape later behavior.

Early Life Experiences

A growing body of research shows that early life experiences, including socialization and maternal care, aversive experiences, and insufficient stimulation, shape an individual's physiological and behavioral development, across multiple species. Early experiences also have a profound influence on whether a dog will develop behavioral issues during its lifetime. A 2016 study found that home rearing environment had an impact later on in a dog's energy level, excitability, distractibility, and separation-related behavior (Harvey *et al.*, 2016). Researchers used a questionnaire called the Puppy Walker Questionnaire (PWQ), at 5, 8, and 12 months of age, for 276 puppies. The PWQ had 59 categories, 22 of which were taken from the Canine Behavioral Assessment & Research Questionnaire (C-BARQ). (It should be noted that the C-BARQ was not intended to be used in pieces, with validity found only when it is used in its entirety; we will return to the C-BARQ in later chapters.) Question categories for the PWQ included trainability, body sensitivity, distractibility, general anxiety, stair anxiety, and miscellaneous. The study found that there were increased scores for trainability, excitability, and energy level among the puppies that interacted with children more frequently, as well as energy level increases when there was another dog in the household.

"Behavioral disorders," which include pathological behaviors, such as those that are contextually inappropriate or excessive in duration, intensity, or frequency (Dietz *et al.*, 2018), are one of the main reasons dogs are euthanized or sent to the shelter in the US, Canada, Finland, and Australia. Whether a dog is eliminating inappropriately or biting the hand that feeds them, behavioral issues account for 10–16% of all euthanasia requests; they're the reason that 11–34% of dogs end up in the shelter (Lambert *et al.*, 2015). But where do these behaviors originate from, and why are some dogs more prone to behavioral problems than others? Sadly, many dogs begin their lives in puppy mills, where socialization plays second fiddle to monetary compensation. As a result, commercially bred dogs have a high incidence of behavioral issues, including dog- and human-directed aggression (McMillan, 2017).

Recall the importance of sensitive and critical periods of socialization. Most mammals, including humans, wolves, and domesticated dogs, establish a bond of attachment with their mothers before these periods of socialization, even with the comparatively early point of weaning for domesticated dogs. The central nervous system (CNS) is most malleable during the sensitive period for socialization. This malleability during the sensitive period is referred to as "neural plasticity," meaning that the CNS has a greater capacity to make experience-based adjustments at this time (Knudsen, 2004). Among humans, sensitive periods in development correspond to neural plasticity wherein learning cognitively complex tasks such as language and music happen more readily than during other points in an individual's development (White *et al.*, 2013). After the sensitive period has passed, learning these tasks is far more difficult; consider, for example, an adult trying to learn a new language in comparison to a second-grader. The rate of learning and language acquisition will be much faster for the child than for the adult. Scott and Fuller's 1965 study was the first to examine the sensitive period for socialization in dogs, and all work on dog socialization that followed built upon their findings (Scott and Fuller, 1965). Some of this work showed that there are age-related correlations between many of the skills that are most "dog-like," including the ability to follow human pointing. Bhattacharjee *et al.* (2017) found that puppies had decreased rates of human avoidance and could follow proximal pointing, but these skills decreased when they were introduced later on in their development. Humans' closest cousins, the chimpanzees and bonobos, aren't able to follow pointing as well as dogs do, but dogs still need to be exposed to humans from an early age to learn what pointing means.

Across all other categories, including breed group, breed, age, and gender, clinicians have consistently seen a correlation between behavioral issues (such as anxiety and aggression) and the sensitive period of development. Around the age of 1.5–2 years, most dog breeds experience the process of social awareness and the development of inhibition. They are similar to humans who are approximately 18–22 years of age: they're legally adults, and can vote, consume alcohol, join the armed forces, and get credit in their own names, but aren't yet neurologically mature. The brain isn't fully developed until age 25 years and, because of this, some legislators in the US are asking for the legal drinking age to be increased by 4 years. Studies have also shown that those who smoke marijuana before this time will have significant abnormalities in the brain regions related to motivation and emotion, with the density, shape, and volume of neuronal abnormalities increasing with the number of joints smoked per week (Gilman *et al.*, 2014). Marijuana smokers in this age group had more cognitive deficits and mental health issues than their non-smoking peers, but consumption after this time appears to have a minimal impact on neuronal changes.

Hopefully our dogs aren't consuming alcohol or marijuana, regardless of their age, so what does this mean for them? Dogs, like humans, have a juvenile period when they lack full and complete inhibition. As subadults, many of their choices are still marked by impulsivity and recklessness. With humans, this stage is estimated to end at age 22 years, with the brain fully developing at age 25 years. Substances taken before that time could lead to arrested development and altered brain structure. For dogs, that stage is estimated to end at the age of 1.5–2 years; social situations that haven't been introduced to them before that time can lead to issues later in life.

For example, Riley, a 3-year-old Australian Shepherd, had begun showing aggressive behavior toward children. The behavior began when Riley was aged 2 years. After further questioning, it was discovered that Riley was a replacement for the couple's children, who had recently moved out, and that they had no prior experience with dogs. Riley never had any training. When he began to misbehave, he was isolated rather than redirected and shown the appropriate way to act. The household expectations were also inconsistent; there were few rules, and they were only enforced by the husband, and not by the wife. Fortunately for Riley, he'd been exposed to enough positive social interactions before this time, with both dogs and with humans, which provided the neural foundation for him to learn how to interact with others. The case involved showing the couple how to reward appropriate behavior and provide appropriate boundaries and expectations for Riley. This often isn't the case for dogs who have lived without these early social interactions. While dogs have co-evolved with humans for millennia, a crucial component of the human–canine relationship is having early exposure and social interactions. It's more challenging for dogs who have lived without human socialization and are then adopted into homes. One of these dogs, named Gus, was a feral beach dog from the Dominican Republic. Feral dogs are a portion of the approximately 900 million free-roaming dogs, worldwide (Gompper, 2013) (Fig. 4.4). They can be found almost everywhere that humans live, including in all 50 US states (Bergman *et al.*, 2009), Mexico, Canada, India, Israel, the Caribbean, South America, and Africa. Feral dogs tolerate but avoid humans, rather than socializing with them. Their lack of human interaction classifies them as feral rather than a stray or a street dog. While feral dogs don't have food or shelter intentionally provided by humans, they indirectly depend upon humans for resources (Boitani, and Ciucci, 1995); their relationship is based upon commensalism and not a true partnership. In a commensal relationship, one organism benefits, while the other is neither harmed nor helped. Typically, it neither harms

Fig. 4.4. Feral dog with puppies: a Sri Lanka wild dog nursing her pups. The mother shows many typical characteristics of the indigenous Sinhala Hound. Photograph courtesy of the author, Stephan Gillmeier, PD-Self. Available at: https://commons.wikimedia.org/wiki/File:Wilde_huendin_am_stillen.jpg.

nor helps the people of these areas to have dogs living on the periphery of their land, eating scraps and garbage. Thousands of years ago, the ancestors of wolves and dogs may have done the same thing, until a few individuals who were bolder, friendlier, or both, decided to cross that invisible line from the periphery of our properties to the heart of our homes.

It is in the home environment where our dogs develop those critical early social skills with their humans, learning roles and rules, expectations, limitations, and subtleties, and human language, both verbal and non-verbal. While it's possible for feral and free-ranging dogs to join a household (Miklosi, 2015), it has long been known that dogs that haven't had interactions with humans before 14 weeks of age will have a tendency to withdraw from human interactions (Freedman *et al.*, 1961). Dogs like Gus, who have not had these experiences, don't know how to navigate the human landscape. Gus was adopted at age two, which is outside the critical and sensitive learning periods for a dog. His early environment proved to be problematic for his new owner, Sundae Garner. Where most dogs would be social, Gus was aloof; rather than engage, he would withdraw. Gus's case isn't unique; dogs who have missed out on social interactions before critical and sensitive points of development will withdraw from human interactions. Gretchen was another dog that was a feral beach dog before she was adopted. Gretchen met her owners, Henry and Jo, while the couple was vacationing in Costa Rica. They fell in love with her, but shortly after she was brought to the US, Gretchen was diagnosed with severe social anxiety because she'd never learned how to act appropriately with humans. Her treatment plan included slow, careful introductions. While both Gretchen and Gus were able to overcome their socially deprived beginnings, the outcome for most cases like these are typically mediocre. Fortunately for both Gus and Gretchen, their adopters were dedicated to their rehabilitation.

Dogs, like humans, are social beings. When they are deprived of early, appropriate social interactions, both species exhibit abnormal behaviors and cognitive deficits (Battaglia, 2009; Dietz *et al.*, 2018). Human infants who aren't handled often will have a difficult time bonding emotionally compared to those who are given appropriate care and attention. Socially deprived infants will also have shorter attention spans and smaller brain sizes than those who have received attention. Consider this in terms of dogs who have been raised in a loving home, with canine and human companions, versus those who were "raised" in a poorly run commercial breeding operation or feral situation. The dog who was raised in an enriched social environment will likely fare far better with human companions, and with cognitive tests, than one who was raised without love, care, and attention (Dietz *et al.*, 2018). Even if they were genetically identical, if two individuals were raised in environments with differing levels of early social interactions, they would also exhibit different behaviors.

Conclusion

It might seem duplicitous to say there are both "breed standards" and exceptions to them, but hear us out. If a person had Irish or Scottish heritage, they would have a higher propensity to have red hair and freckles, but this genetic heritage wouldn't *guarantee* these traits. Nor would the absence of these traits indicate that they *didn't* have this heritage. Similarly, the presence of these traits wouldn't *guarantee* that they were Irish or Scottish. Given what we know about a host of factors, including recombination, assortative mating, epigenetics, and life experience, we can make broad generalizations about breeds and breed groups, but individuals will always vary within these discrete categories.

References

Anderson, T.M., vonHoldt, B.M., Candille, S.I., Musiani, M., Greco, C. *et al.* (2009) Molecular and evolutionary history of melanism in North American gray wolves. *Science* 323(5919), 1339–1343.

Battaglia, C.L. (2009) Periods of early development and the effects of stimulation and social experiences in the canine. *Journal of Veterinary Behavior* 4(5), 203–210.

Bergman, D., Breck, S.W. and Bender, S. (2009) Dogs gone wild: feral dog damage in the United States. *USDA National Wildlife Research Center – Staff Publications* 862.

Berk, L.E. (2006) *Child Development*, 7th edn. Allyn & Bacon, Boston, MA.

Bhattacharjee, D.N., Gupta, S., Sau, S., Sarkar, R., Biswas, A. *et al.* (2017) Free-ranging dogs show age related plasticity in their ability to follow human pointing. *PLoS One* 12(7), e0180643.

Björnerfeldt, S., Hailer, F., Nord, M.N. and Vilà, C.C. (2007) Assortative mating and fragmentation within dog breeds. *BMC Evolutionary Biology* 8(1), 28.

Boitani, L. and Ciucci, P. (1995) Comparative social ecology of feral dogs and wolves. *Ethology Ecology & Evolution* 7(1), 49–72.

Brennan, S. (2016) The mixed race twins with different colour skin and eyes: Amelia and Jasmine become UK's first sisters to be genetically identical but don't look the same. *MailOnline*. Available at: www.dailymail.co.uk/femail/article-3461832/First-identical-twins-different-skin-colour-born-UK.html (accessed 23 December 2023).

Dietz, L., Arnold, A.-M.K., Goerlich-Jansson, V.C. and Vinke, C.M. (2018) The importance of early life experiences for the development of behavioural disorders in domestic dogs. *Behaviour* 155(2–3), 83–114.

Freedman, A.H., Lohmueller, K.E. and Wayne, R.K. (2016) Evolutionary history, selective sweeps, and deleterious variation in the dog. *Annual Review of Ecology, Evolution, and Systematics* 47(1), 73–96.

Freedman, D.G., King, J.A. and Elliot, O. (1961) Critical period in the social development of dogs. *Science* 133(3457), 1016–1017.

Gilman, J.M., Kuster, J.K., Lee, S., Lee, M.J., Kim, B.W. *et al.* (2014) Cannabis use is quantitatively associated with nucleus accumbens and amygdala abnormalities in young adult recreational users. *Journal of Neuroscience* 34(16), 5529–5538.

Gompper, M.E. (2013) The dog–human–wildlife interface: assessing the scope of the problem. In: Gompper, M.E. (ed.) *Free-Ranging Dogs and Wildlife Conservation*. Oxford University Press, New York, pp. 9–54.

Gottlieb, G. (1998) Normally occurring environmental and behavioral influences on gene activity: from central dogma to probabilistic epigenesis. *Psychological Review* 105(4), 792–802.

Gottlieb, G. (2007) Probabilistic epigenesis. *Developmental Science* 10(1), 1–11.

Harvey, N.D., Craigon, P.J., Blythe, S.A., England, G.C.W. and Asher, L. (2016) Social rearing environment influences dog behavioral development. *Journal of Veterinary Behavior* 16, 13–21.

Hedrick, P.W., Stahler, D.R. and Dekker, D. (2014) Heterozygote advantage in a finite population: black color in wolves. *Journal of Heredity* 105(4), 457–465.

Hekman, J.P., Johnson, J.L., Edwards, W., Vladimirova, A.V., Gulevich, R.G. *et al.* (2018) Anterior pituitary transcriptome suggests differences in ACTH release in tame and aggressive foxes. *G3: Genes, Genomes, Genetics* 8(3), 859–873.

Iljin, N.A. and Iljin, V.N. (1930) Temprature [sic] effects on the color of the Siamese cat. *Journal of Heredity* 21(7), 309–318.

Jacobs, L.N., Staiger, E.A., Albright, J.D. and Brooks, S.A. (2016) The *MC1R* and *ASIP* coat color loci may impact behavior in the horse. *Journal of Heredity* 107(3), 214–219.

Jans, N. (2014) *A Wolf Called Romeo*. Houghton Mifflin Harcourt, New York.

Jans, N. (2023) *Romeo the Friendly Wolf*. Available at: https://nickjans.com (accessed 23 December 2023).

Johns Hopkins Medicine (2017) 'Epigenetic' changes from cigarette smoke may be first step in lung cancer development. *Hopkins Medicine.org*. Available at: www.hopkinsmedicine.org/news/media/releases/epigenetic_changes_from_cigarette_smoke_may_be_first_step_in_lung_cancer_development (accessed 31 December 2023).

Kaati, G., Bygren, L.O. and Edvinsson, S. (2002) Cardiovascular and diabetes mortality determined by nutrition during parents' and grandparents' slow growth period. *European Journal of Human Genetics* 10(11), 682–688.

Knudsen, E.I. (2004) Sensitive periods in the development of the brain and behavior. *Journal of Cognitive Neuroscience* 16(8), 1412–1425.

Lambert, K., Coe, J., Niel, L., Dewey, C. and Sargeant, J.M. (2015) A systematic review and meta-analysis of the proportion of dogs surrendered for dog-related and owner-related reasons. *Preventive Veterinary Medicine* 118(1), 148–160.

Lee, M., Czene, K., Rebora, P. and Reilly, M. (2015) Patterns of changing cancer risks with time since diagnosis of a sibling. *International Journal of Cancer* 136(8), 1948–1956.

Li, R., Montpetit, A., Rousseau, M., Wu, S.Y., Greenwood, C.M. *et al.* (2014) Somatic point mutations occurring early in development: a monozygotic twin study. *Journal of Medical Genetics* 51(1), 28–34.

Lichtenstein, P., Holm, N.V., Verkasalo, P.K., Iliadou, A., Kaprio, J. *et al.* (2000) Environmental and heritable factors in the causation of cancer – analyses of cohorts of twins from Sweden, Denmark, and Finland. *New England Journal of Medicine* 343(2), 78–85.

Lindqvist, C., Janczak, A.M., Nätt, D., Baranowska, I., Lindqvist, N. *et al.* (2007) Transmission of stress-induced learning impairment and associated brain gene expression from parents to offspring in chickens. *PLoS One* 2(4), e364.

McMillan, F.D. (2017) Behavioral and psychological outcomes for dogs sold as puppies through pet stores and/or born in commercial breeding establishments: current knowledge and putative causes. *Journal of Veterinary Behavior* 19, 14–26.

Meaney, M.J. and Szyf, M. (2005) Environmental programming of stress responses through DNA methylation: life at the interface between a dynamic environment and a fixed genome. *Dialogues in Clinical Neuroscience* 7(2), 103–123.

Miklosi, A. (2015) *Dog Behaviour, Evolution, and Cognition*, 2nd edn. Oxford University Press, New York, pp. 172–173.

Morell, V. (2020) High-altitude genes could turn Himalayan wolves into a new species: genetic analysis suggests they differ significantly from gray wolves. *Science News*. Available at: www.science.org/content/article/high-altitude-genes-could-turn-himalayan-wolves-new-species (accessed 23 December 2023).

National Institutes of Health (2008) Fat cell numbers in teen years linger for a lifetime. Available at: www.nih.gov/news-events/nih-research-matters/fat-cell-numbers-teen-years-linger-lifetime (accessed 23 December 2023)

Ogasawara, S. (2017) Duration control of protein expression *in vivo* by light-mediated reversible activation of translation. *ACS Chemical Biology* 12(2), 351–356.

Pörtl, D. and Jung, C. (2017) Is dog domestication due to epigenetic modulation in brain? *Dog Behavior* 3(2), 21–32.

Rose, L.T., Rouhani, P. and Fischer, K.W. (2013) The science of the individual. *Mind, Brain, and Education* 7(3), 152–158.

Scott, J.P. and Fuller, J.L. (1965) *Genetics and the Social Behavior of the Dog*. The University of Chicago Press, Chicago, IL.

Sturtevant, A.H. (1913) The Himalayan rabbit case, with some considerations on multiple allelomorphs. *The American Naturalist* 47(556), 234–238.

Trut, L., Oskina, I. and Kharlamova, A. (2009) Animal evolution during domestication: the domesticated fox as a model. *BioEssays* 31(3), 349–360.

Werhahn, G., Liu, Y., Meng, Y., Cheng, C., Lu, Z. *et al.* (2020) Himalayan wolf distribution and admixture based on multiple genetic markers. *Journal of Biogeography* 47(6), 1272–1285.

White, E.J., Hutka, S.A., Williams, L.J. and Moreno, S. (2013) Learning, neural plasticity and sensitive periods: implications for language acquisition, music training and transfer across the lifespan. *Frontiers in Systems Neuroscience* 7, 90.

5 Deep Roots, Broad Branches: The Range of Dog Breeds

Abstract

Chapter 5 demonstrates that there is more variability between dog breeds than there is between some species. It usually takes thousands of years—or longer—for speciation to occur. A slow accumulation of mutations cause inheritable changes to the phenotype. Most dog breeds originated during the Victorian era, falling far short of that typical timeline, although humans were the catalyst behind this accelerated process. The vast majority of differences in dogs, even among the most dissimilar of the species, is driven by relatively few loci, or regions, in the genome. These loci have a large phenotypic effect, yielding strong variation among breeds. There is a review of where different breeds came from and the effect of inbreeding on population structure from the pedigree analysis of purebred dogs, as well as looking at genetically linked dog diseases.

The Lovely Bones

Arora walked gingerly on the crust of the snow, her leather laced boots staying atop this time. Her dog, Keroua, padded quietly beside her, pulling a travois (a traditional Native American A-frame drag sled). The hide hammock between the travois' two poles carried smoked salmon, frozen seal meat, and tightly wrapped furs, and the poles scraped gently across the snowscape. Keroua's head bobbed once, twice, and then her gait shortened. Arora glanced over, just in time to see her dog's hind legs give out. She gently patted Keroua and untied the travois, pulling it herself. Snow fell lightly on them, pushed by a driving wind, and she glanced back to see a dozen other pilgrims, half of which were accompanied by dogs. Keroua had been struggling for the tail end of this journey, and she knew that her friend would need to rest soon. Glancing ahead, she saw several darkened curves in the horizon. Caves. She changed her trajectory, encouraging Keroua quietly. The other people in her party followed, their silhouettes hazy in the growing snowfall. As they entered the cave, Keroua darted ahead of her person, her ears and tail aloft. After a thorough inspection, she returned to Arora and touched her hand gently (Fig. 5.1). Arora nodded and waved her arm to the group behind her, beckoning them in. She took a spot not far from the entrance, unfolding her furs and asking Keroua to rest upon them. Her old dog quietly complied, but never took her eyes off of Arora.

Arora watched as the other members of her family slowly made their way into the cave; they would stay here for the night, and then set out again in the morning. Keroua huffed softly at her canine friends as they found their places in the cave. Arora brought her dog several pieces of salmon and seal meat, which her dog ate gratefully before finally putting her head down onto her fur bed. Arora took out her knife and began preparing food for them to share; she would then join Keroua on the furs to sleep for the night.

Arora's older sister, Halla, arranged several rocks in a circle and started a small fire; soon, the orange glow lit the entire cave, bouncing off of Keroua's black, white, and silver coat. Halla gently touched Arora's shoulder, motioning over to Keroua. Arora nodded and straightened her coat, removing her hood and letting her blonde hair fall to her sides. She remembered when Halla's dog, Haza, had shown the same decline; she knew that if Arora couldn't keep up with the group that she would be left behind. She turned her face from the firelight as tears stung her eyes, and then busied herself with rationing out food portions. Her other family members did the same, and then they sat around the fire, breaking off pieces of meat, intermittently feeding themselves and their dogs.

Afterward, Arora joined Keroua on the furs, pulling them all around. They would take turns stoking the fire, but most of their heat came from co-sleeping with their canines. Arora rested her head against her large dog's side, her head rising

DOI: 10.1079/9781800624566.0005

Fig. 5.1. Keroua and Arora. Sketch courtesy of Arianne Taylor, artist.

and falling as Keroua's sides did so. Soon they were all asleep; Keroua's breathing became labored, but Arora did not stir. As morning came, Arora gently patted Keroua, but this time, it was her dog who did not stir. She had passed only moments before, warm, and loved, and comforted by the presence of her family. Arora was relieved that Keroua had left on her own terms; she hated leaving a dog behind. Halla helped Arora remove the furs from beneath her dog—they would need these for their continued journey—and then gently placed her dog against the cave wall.

One of Keroua's bones would be found 10,000 years later in what is now called "Lawyer's Cave." The intrepid researchers who discovered the bone specimen would marvel over the find: the discovery revealed that those humans who ventured from Asia to the Americas were accompanied by their domesticated dogs (da Silva Coelho *et al.*, 2021). Primary researcher Charlotte Lindqvist of the University of Buffalo, New York, and her team would perform DNA analysis on the bone, discovering that Keroua's ancestors split off from a lineage of East Siberian dogs approximately 17,000 years ago.

"A graduate student and I found the bone specimen," Dr Lindqvist recalled. "This bone wasn't like the big megafauna, such as the mammoths; it was so small that we couldn't say much about it at first." The specimen was initially classified as an "unidentified mammal."

"We didn't find complete skeletons like Dr Perri found in Illinois, at the Stilwell and Koster II sites,"

Lindqvist said, "but this bone is of a similar age, and it was the oldest bone found in Lawyer's Cave. It provides evidence for the coastal route people took as they first migrated to the Americas. A domesticated dog also tells us something about humans, and the peopling of the Americas after the Ice Age. This find also tells us about human history. We don't know very much about this region—it may have been very crucial for the peopling after the Ice Age. Our goal is to try to reconstruct the mammalian history of the area. We want to know: were there any ice-free areas in this region during the last Ice Age?"

These finds—including the bones of bears, seals, foxes, and even humans—can help reconstruct the history of the vertebrates' diversity through time and how they have adapted to climatic changes. After human bones were discovered, the cave was shut down to repatriate the remains.

"This dog is 10,000 years old, and while the human bones and artifacts are much younger, they have used this cave for at least that many years," Lindqvist said. "And if there was one dog, there must have been more."

Keroua, and likely many, many other dogs like her, arrived in North America with their humans, sharing their food and shouldering their responsibilities. Ten thousand years ago, Keroua and her canine family would already have distinct differences from other lines of dogs—adaptations for their diet, vocation, and environment. Prior examinations of Lawyer's Cave had unearthed the bones of many other animals, including fish, birds, and mammals, as well as human artifacts such as shell beads, a bone spear point, obsidian flakes, and a bone awl.

According to Lindqvist, Lawyer's Cave has openings on each end. "It's almost like a crawl space," she said. "It's a smallish cave; it may have been a resting place for people as they traveled. There's no artwork in it—probably because it wouldn't have held up to the elements, possibly because people were just passing through. There are springs nearby, though, so it is also possible that people hung around in that area for a while, as well."

While we have outlined many of the distinct differences between dogs, for the evolutionary biologist, a dog is just a dog, is just a dog. But is it really that simple? Variation can arise within populations, but it usually takes thousands of years—or longer—for speciation to occur. Most dog breeds originated during the Victorian era, falling far short of that typical speciation timeline, though humans have accelerated the process. During speciation, a slow accumulation of mutations cause heritable changes to the phenotype. Those mutations that confer an advantage for individuals will be selected for, while those individuals who don't have advantageous mutations would have lower rates of survival and, subsequently, their lineages would have had decreased rates of reproductive success. For the domesticated dog, breeders often chose individuals with the most exaggerated traits and bred them with other dogs who were similarly unusual. When a mutation arose that owners found novel or aesthetically pleasing, they ensured the proliferation of this trait through selective breeding. Many of these traits, such as brachycephalic faces and truncated limbs, would have been deleterious for these dogs in free-living situations, but given human assistance (including Cesarean sections for many breeds), they're allowed to propagate.

We define a dog breed as: "A domestic race of dogs (selected and maintained by man) with a common gene pool and characterized appearance and function" (AKC, 2023). As a species, the range of individuals within *Canis familiaris* is vast. Dog breeds vary from one another in multiple ways, including how long they have been established. Some breeds originated as recently as this year, while others were established thousands of years ago. It makes sense, then, given the differences in the time frame when their breeds arose, how long they were geographically or artificially isolated, and how long they were selected by humans for breeding for specific tasks or appearances, that we will see a wide range of variance in this species. They vary in their physical appearance, including their skeletal and skull structure, teeth and claws, body size, coat color, length, and type, length of legs and tail, whether their ears are floppy or erect, and their face and head shape. Large dog breeds, such as the Great Dane, and small ones, such as the Chihuahua, differ in volume by as much as 40-fold—a level of diversity that applies to no other terrestrial mammal (Ostrander *et al.*, 2017). When looking at differences between body mass, the range is even larger: there is a 100-fold difference between the Chihuahua's 1 kg body mass and the Great Dane's 100 kg body mass (Sutter *et al.*, 2008). Dog breeds vary in their genetics, including their propensity to have certain diseases and physical ailments, and their level of inbreeding. Breeds vary in their senses, including their sense of smell, hearing, and vision. They vary in their temperament, including extroversion or introversion and high versus low explore drive. And they vary in their personalities, including levels of activity, adaptability, approach/withdrawal, boldness, confidence, distractibility, friendliness, independence, intensity of reaction, mood, neuroticism, persistence, security, and sensitivity. They vary in their behavior, including their social behavior, how they play (and how often), and how often they will display social dominance and aggression. They vary in how they communicate with one another and with humans. Some dogs have barks that typify their breed, and others tend to be quieter. They vary in their trainability, with some breeds learning at a faster rate than others. They vary in their vocational propensity, capability for a wide range of activities, including herding, hunting, vermin catching, sledding, and guarding. With so many physical and behavioral differences between dogs, it's unsurprising that they also differ physiologically, including differences in their digestion (Ollivier *et al.*, 2016) and reproduction, with significant interbreed differences in the speed of their sperm (Valverde *et al.*, 2019). And with dogs selected for generations to perform very specific tasks, it follows that they also differ in their neuroanatomy: those parts of the brain that correspond to the tasks that a dog was bred for, such as the olfactory area for scent hunting, are more specialized than the brains of dogs who have not been bred for this purpose. We're accustomed to a wide range of canine diversity, so perhaps it's lost its "wow" factor to many of us, but imagine having this much difference in our own species, *Homo sapiens*, or in eagles, or mosquitoes, or blue whales.

Dogs aren't the only domesticated species that exhibit significant differences between breeds. Other species, including horses and cats, have breed-specific differences. There are approximately 400 different horse breeds, which vary in multiple ways, including their physical appearances, physiology, behavior (Goodwin *et al.*, 2008), personalities (Lloyd *et al.* 2008), and even their insulin sensitivity (Bamford *et al.*, 2014). Horses, like dogs, vary greatly, from the eager to please, aptly named miniature horse to the imposingly tall, but often docile, draft breeds, and from the stocky, hardworking, robust Norwegian Fjord to the tall, slender, sensitive Turkish Akhal-Teke. In their physical mass alone, they can vary by 20-fold. The miniature horse can be as small as 34 inches at the withers (shoulder) and weigh as little as 150 pounds, while the world record holding draft horse, a Shire named Sampson, stood at 86.5 inches at the withers, and weighed approximately 3360 pounds (Whitaker and Whitelaw, 2007).

Cats also have dozens of breeds that diverge physically, behaviorally, and in personality from one another. The sleek, short-haired, medium-sized Russian Blue is intelligent, talkative, gentle, and affectionate with their people, the large, long-haired Norwegian Forest Cat is resourceful, interactive, playful, intelligent, and loving, and the short-haired, short-legged Munchkin cat is curious, affectionate, smart, and spirited (The International Cat Association, 2018). The Cat Fanciers' Association, the largest registry of cats worldwide, currently recognizes 45 pedigreed cat breeds (The Cat Fanciers Association, 2023). while The International Cat Association recognizes 73 breeds of cats for championship competition (The International Cat Association, 2018). Behavioral differences among cat breeds are an understudied topic, but one study did find evidence that behavioral differences between breeds are due to both genetic and environmental factors (Salonen *et al.*, 2019).

Breed diversity could have arisen simply due to geographic isolation and genetic mutations, but breeds have also been created by the manipulation of their breeding by humans. In this chapter, we recreate the timeline of the human–canine relationship, from estimates of the initial interactions to the creation of a number of specific breeds.

We discussed the ancestral timeline of canids in Chapter 1, so let's delve a little deeper into one of those time periods: the Quaternary period, which extends from approximately 2.5 mya to the present. We're exploring this period in particular because this is when the human–canine relationship began. This period is divided into the Pleistocene epoch, which ended approximately 11,700 years ago, and our current epoch, the Holocene, which began when the Pleistocene ended. In addition to these epochs, we also use archaeological terms to refer to our timespans. The "stone age" is divided into three periods: the Paleolithic, or old stone age, the Mesolithic, or middle stone age, and the Neolithic, or new stone age. The end of the Pleistocene corresponds to the end of the Paleolithic period. See Table 5.1.

Table 5.1. Timeline of dogs from the Upper Paleolithic to the present.

Upper Paleolithic: 50,000–15,000 BCE (52,000–17,000 years ago (ya))	
50,000 ya	The "Tirekhtyakh canid" is found in Yakutsk, Eastern Siberia; the fossil is classified as a Pleistocene wolf (Harvey, 2016).
	Denisovans (*Homo denisova*), an ancient hominid species that lived contemporaneously with *Homo sapiens* and *Homo neanderthalensis*, goes extinct. Some modern human populations (such as Melanesians, Southeast Asians, and Pacific Islanders) have up to 6% Denisovan DNA (National Library of Medicine, 2022).
47,000	Modern humans colonize Europe, Western, Southern, and Southeast Asia, Australia, and New Guinea.
40,000	Neanderthals grow closer to extinction.
	The earliest known cave paintings are created in the Cueva de El Castillo (the Cave of the Castle) in Cantabria, Spain.
	The approximate time that dogs and wolves last shared a common ancestor (Lallensack, 2017).
36,500	An ancient canid dubbed the "Goyet dog" dies in the Goyet Caves in Mozet, Belgium (Germonpré *et al.*, 2009).
35,000	A Paleolithic dog dies in Hohle Fels, Germany (Camarós *et al.*, 2016).
33,500	A Paleolithic canid dubbed the "Altai dog" dies in Razboinichya Cave in Central Asia's Altai Mountains (Ovodov *et al.*, 2011).
	The approximate age of a Paleolithic dog mandible found in the Kostyonki-8 archaeological complex in Voronezh, Russia (Germonpré *et al.*, 2015).
31,000	The approximate age of three Paleolithic dog skulls found in Predmosti, Moravia, in the Czech Republic (Mendel's homeland). One of the dogs was buried with a bone placed in their mouth (Germonpré *et al.*, 2015).
30,800	The approximate age of a fossil dubbed the "Badyarikha canid," which was found in Yakutsk, Eastern Siberia (Harvey, 2016).
28,000	The last documented evidence of Neanderthals living in the Iberian Peninsula and Gorham's Cave in Gibraltar (Muñiz *et al.*, 2019).
Mesolithic: 25,000–10,000 BCE (27,000–12,000 ya)	
26,000	The approximate age of pawprints found in the Chauvet Cave of France's Vallon-Point-d'Arc; they're morphologically similar to modern domesticated dogs, but the track line was argued to be similar to a wolf's (Ledoux and Boudadi-Maligne, 2015).
17,200	A canid skull found in Ulakhan Sular, northern Yakutia, Siberia, was determined to be that of a 17,200-year-old Paleolithic dog (Germonpré *et al.*, 2017).
17,000	The age of cave paintings of wolves found in Font-de-Gaume, France.
	Humans cross from Asia to the Americas … accompanied by their dogs (*Nature*, 2021).
14,000	A Paleolithic puppy (dubbed "Mina" in Chapter 1) dies after a prolonged bout of canine distemper in what is now modern-day Oberkassel, Bonn, Germany. Mina's was the oldest known grave containing both canine and human remains (Janssens *et al.*, 2018).
11,000	The last of the native North American horses die out (Carey, 2006).
10,000	The dire wolf, saber-toothed cat, and three-quarters of earth's species die out during the Quaternary Extinction Event (Leonard *et al.*, 2007). The dire wolf becomes an evolutionary dead end.
	The age of the remains of a domesticated dog found in Alaska. DNA tests showed that it originated from a line of dogs from Eastern Siberia that crossed the North American Arctic (da Silva Coelho *et al.*, 2021).
	The approximate age of the Stilwell II and Koster dog fossils, which were found in their own graves in what is now present-day Illinois (Bower, 2018).
Neolithic: 10,000–2000 BCE (12,000–4,000 ya)	
9500	The earliest known evidence of humans using dogs to pull sleds in Siberia (Gorman, 2020).
9000	Pigs have their first domestication events in Mekong Valley, Eurasia and Anatolia, Turkey.
	The Akita Inu emerges in Japan (National Shiba Club of America, 2023).
	Dogs are being selectively bred for different roles in Zhokov Island, Siberia (Pitulko and Kasparov, 2017).
	The age of rock carvings of dogs from the Arabian Peninsula. The depicted dogs resemble the modern-day Canaan Dog; this is the first known evidence of that breed (Guagnin *et al.*, 2018).
9000–8500	Cats are first domesticated in the fertile crescent.

Continued

Table 5.1. Continued.

8300	The approximate time that New Guinea Singing Dogs and Dingoes last shared a common ancestor; the first known recognition of these dogs as separate breeds (Zhang *et al.*, 2020).
	The Basenji emerges in Central Africa's Congo Basin (Ripley, 2016).
	The Chow Chow originates as a breed in Northern China (Yang *et al.*, 2017).
	The Afghan Hound first emerges as a breed in Afghanistan.
	The first emergence of the Saluki is documented in wall carvings from the Sumerian Empire, in what is now modern-day Iraq (Dennis-Bryan *et al.*, 2013).
7000–4000	Preserved DNA from the remains of dogs from Europe and Asia reveal they are having more copies of *Amy2B*, a gene mutation for digesting starch, which is historically seen in hominid diets, but not the canid ones (Ollivier *et al.*, 2016).
6000	The age of the oldest known New Guinea Singing Dog fossil, to date (Cairns and Wilton, 2016; San Diego Wildlife Alliance, 2023).
5300	Dogs are routinely getting leashes and collars (Mark, 2018).
5000	Images of Ibizan Hound and Pharaoh Hound adorn Egyptian tombs (Mark, 2017).
	The AKC identifies this as the age of the Elkhound (AKC, 2022a); phylogenetics indicate that it's a modern/derived breed (Parker *et al.*, 2004).
4500	Construction begins on the Sphinx in Egypt.
Bronze Age: 2000 BCE (4000 ya)	
4000	The Greyhound emerges; the age of skeletal remains of a Greyhound-like dog are found in Tell Brak, in what is modern-day Syria (Clutton-Brock, 1989).
	Construction begins on the Great Pyramid of Giza.
	Dogs first enter into literature with *The Show Dog* and *Why Dog Was Subservient to Man* (Mark, 2018).
	Some countries view dogs as deities.
3000–4000	Mastiffs are kept for hunting and guarding (Clutton-Brock, 1995).
3000	The Siberian Husky emerges in Northeast Asia/Siberia from the Chukchi Tribe (Fiszdon and Czarkowska, 2008).
	The Samoyed, named after the indigenous Samoyedic peoples, likely originates in Siberia around this time (vonHoldt *et al.*, 2010; Larson, 2012; Ahzu Samoyeds Australia, 2021).
	The Alaskan Malamute originates in Alaska's Norton Sound region (Hanford, 1998; Brown *et al.*, 2015).
3450	The age of the oldest identified Dingo fossil, to date, is found in Madura Cave, Western Australia (Balme *et al.*, 2018).
3345	King Tutankhamen dies in Egypt.
2400	The Shar-Pei originates in China (QingShen *et al.*, 2010).
2000	The Mexican Hairless Dog emerges as a breed in Mexico (Leonard *et al.*, 2002).
Iron Age: 800 BCE (approximately 2800 ya)	
43 CE:	Romans and Britons use dogs in the Seven-Year War.
47 CE	The Romans found the city of London (Londinium).
455 CE	Rome is sacked by the Vandals.
476 CE	The fall of the Roman Empire.
Middle Ages: 500–1500 CE	
1300s	The American Eskimo Dog, then referred to as the "German Spitz," emerges in Germany. It was renamed due to post-World War II anti-German sentiment.
1346	The Black Plague sweeps across Afro-Eurasia; by 1353, upwards of 200 million people perished from human history's most deadly pandemic (Philipkoski, 2001).
1493	European dogs travel to the New World (Mancini, 2018) and most native dogs disappear from the gene pool (Leathlobhair *et al.*, 2018). European dogs have been geographically and genetically isolated from the dogs in the Americas, which crossed into North America via the land bridge during the last Ice Age.
Early Modern Period: 1500–1800 CE	
1665	The Great Plague of London strikes and 100,000 people (15% of London's population) die (Harvard Library, 2023).
1700	Charles II, the King of Spain and last of the Hapsburgs, dies; this leads to the War of the Spanish Succession.

Continued

Table 5.1. Continued.

1700s	The Talbot Dog goes extinct in England (Edwards, 1800).
	The Saint John's Water Dog goes extinct in Newfoundland (Gosling and Graham, 2009).
	"Modern" dog sledding is born (Roark, 2019).
	Hounds become human trackers in North America, especially for escaped slaves.
1735	Linnaeus publishes the first edition of the *Systema Naturae* in the Netherlands.
1750s	The guide dog movement is born.
1775	The American Revolutionary War begins.
1793	An outbreak of yellow fever in Philadelphia leaves more than 5000 dead (Finger, 2023).
Late Modern Period: 1800 CE	
1800s	The French Bulldog becomes a breed in France.
1830s	The Hare Indian Dog of Canada is on the decline; the original breed, used by Hare Indians for hunting, goes extinct due to the introduction of firearms. These dogs are no longer needed and intermingle with other breeds (Bennett, 1830).
1835	The Cruelty to Animals Act is passed, leading to the extinction of the Olde English Bulldog in England.
1854	Francis Linck finds a dire wolf fossil in Indiana.
1859	Charles Darwin publishes the first edition of *On the Origin of Species* in England.
1861	The American Civil War begins.
1866	The American Society for the Prevention of Cruelty to Animals (ASPCA) is formed.
1873	The Kennel Club of the UK is founded.
1876	The Falkland Islands Wolf (*Dusicyon australis*) goes extinct; Darwin once said their friendliness would be their demise (*The Conversation*, 2013).
1877	The first Westminster Kennel Club Dog Show is held.
1878	Nine original "charter breeds" are recognized by the AKC: Pointer, Chesapeake Bay Retriever, Clumber Spaniel, Cocker Spaniel, Sussex Spaniel, Irish Water Spaniel, Irish Setter, English Setter, and Gordon Setter.
1884	The AKC begins tracking breeds.
1885	The Fox Terrier is first recognized by the AKC.
1888	The Whippet is first recognized by the AKC.
	Dogs are used to hunt for Jack the Ripper in England.
1889	The Hokkaido Wolf (*Canis lupus hattai*) goes extinct in Japan (Knight, 1997).
1891	Queen Victoria begins exhibiting her 12-pound Pomeranian, Marco; the breed standard shifts rapidly from the prior average of 30–50 pounds to Marco's size.
1897	The first study of the New Guinea Singing Dog begins at Mount Scratchley, Central Province, Papua New Guinea (Ganguly, 2020).
1899	The German Shepherd Dog is recognized as a breed in Germany.
	The earliest recognized training facility for K9s originates in Ghent, Belgium.
1903	*The Call of the Wild*, written by Jack London, is published serially by *The Saturday Evening Post* and then as a single-volume book by Macmillan & Co.
1905	The Japanese Wolf (*Canis lupus hodophilax*) goes extinct in Japan; later unconfirmed sightings are reported (Knight, 1997).
1911	The Newfoundland Wolf (*Canis lupus beothucus*) goes extinct in Newfoundland (Glover, 1942).
1914	World War I begins in Sarajevo, Bosnia, when Archduke Franz Ferdinand, the heir to the Austro-Hungarian Empire, and his wife, Sophie, are fatally shot.
1917	The Labrador Retriever is first recognized by the AKC.
1918	The Spanish flu, named for its first known outbreak in Madrid, Spain, begins. It finally subsides in 1920, after 50 million lives are lost (Centers for Disease Control and Prevention, 2018).
1920s	The Sicilian Wolf (*Canis lupus cristaldii*) goes extinct in Sicily (Angelici and Rossi, 2018).
1925	The Kenai Peninsula Wolf (*Canis lupus alces*) goes extinct in Southern Alaska (Weckworth *et al.*, 2005).
	During a deadly diphtheria outbreak, Gunnar Kaasen and his sled dogs, including Togo and Balto, carry life-saving serum to Point Safely, Alaska.
1930s	Dogs become mountain rescuers.
1937	The Portuguese Water Dog breed is "regenerated" in Portugal.

Continued

Table 5.1. Continued.

1939	"Toto," a Cairn Terrier, stars in *The Wizard of Oz*.
	World War II begins when Germany invades Poland; some dog breeds, including the Leonberger in Germany and the Lundehund in Norway, experience population bottlenecks that dramatically decrease their genetic diversity.
1943	A German Shepherd named Chips is awarded the Purple Heart, Distinguished Service Cross, and Silver Star. The awards were revoked because of an Army policy against commendations for animals. The awards were later reinstated, but with the agreement that dogs could no longer receive them (Hastings, 2018).
1952	Nixon gives his "Checkers" speech about his Cocker Spaniel who was received as a political gift (PBS, 1952).
1954	The television show *Lassie* debuts; it will run for 20 years.
1955	The Vietnam War begins; it goes on for 20 years and results in more than 1.3 million deaths (Lewy, 1978).
1956	Fred Gipson pens the young adult novel, *Old Yeller*, about a devoted dog and his boy. The book won the John Newberry Medal and became a popular film by the same name the following year.
1957	Laika the dog takes a one-way space trip on *Sputnik 2* and becomes the first animal to orbit earth.
1958	Belyaev and Sorokina begin their fox study in Russia.
1959	The Belgian Malinois is officially recognized by the AKC.
1960s	The first therapy dogs emerge.
1963	*Clifford the Big Red Dog*, written by Norman Bridwell, debuts.
1973	The Brazilian Tracker becomes extinct after a disease outbreak in Brazil (CBKC, 2020).
	The Iditarod Sled Race becomes an annual tradition to honor the Diphtheria serum race of 1925.
1974	The Tahltan Bear Dog of Northern Canada is officially declared extinct (Kercsmar, 2016) after the Canadian Kennel Club rescinds recognition (no registrations were received for more than 25 years). The Tahltan was closely related to the Hare Indian Dog.
	Benji the dog makes his first appearance.
1981	Stephen King writes *Cujo*, about a 200-pound Saint Bernard with rabies.
	HIV/AIDS is first identified.
1989	The emergence of the Labradoodle, a mix created by mating a Labrador Retriever with a Poodle. The creator, Wally Cronon, later regrets this creation (Pero, 2019).
2001	The dogs of 9/11 bring disaster search dogs to the forefront.
2003	Romeo the wolf makes his first appearance with humans.
2004	The dog genome is assembled (*Science Daily*, 2004).
2009	Romeo the wolf dies at the hands of poachers.
2010s	Dogs are increasingly put to work sniffing out diseases.
2019	COVID-19, a disease caused by a novel coronavirus, makes its appearance in the Wuhan Province of China. By mid-2020, this is a full-blown pandemic.
2020	Dogs begin sniffing out COVID in sweat, saliva, and urine.

AKC: American Kennel Club.

The More Alike, the More Different

There's enough difference between dog breeds that entire books have been devoted to describing each breed that falls within one breed group (Fig. 5.2). And the diversity between, and within, these groups arose largely due to inbreeding. To create more differences, humans bred closely related individuals with one another, resulting in increasingly exaggerated traits in the resultant offspring. When closely related individuals mate with one another, it is referred to as inbreeding; breeders often prefer to use other terms, such as line-breeding, which is just another variation of inbreeding. While breeding vastly different individuals can increase the variation within a population, breeding highly similar, closely related individuals with each other does the opposite: it decreases variety while leading to the increased expression of recessive traits. This "popular sire effect" diminishes variation in the genome (Ostrander *et al.*, 2017). Breeders looking for increasingly exaggerated phenotypic traits, such as brachycephalic faces, long fur, or a tall, lithe body, are able to produce dogs with increasingly exaggerated

Fig. 5.2. Three dogs: an Eskimo Dog, a Basenji from the Niger and Nelson, a Poodle. Thomas Musgrave Joy (1812–1866), PD-Art. This is a photograph of the original artwork and in the public domain.

features. Conversely, free-living dogs retained higher genetic variation due to fewer constraints on their breeding.

Child of Nature

While genetics is a fairly familiar concept to non-scientists, there's another important classification system that describes our evolutionary histories and relationships. Phylogenetics is the study of the evolutionary history of a taxonomic group of organisms (such as a population or species) and focuses on the relationships of an organism to other organisms according to their evolutionary similarities and differences. Phylogenetic inference evaluates DNA sequences, observable heritable traits, and morphology. Phylogenetic relationships are represented with a phylogenetic tree, which diagrams the hypothesized relatedness of extinct and extant organisms. Canidae is the most phylogenetically distinct family and diverged from the other carnivores more than 50 mya. The wolf-like canids originated more than 6 mya. The molecular phylogeny of the dog family Canidae includes 35 extant members with three primary phylogenetic groups: the fox-like canids, the South American canids, and the wolf-like canids. Domesticated dogs originated from the wolf-like canids. Phylogenetic relationships between breeds can be shown via a phylogenetic tree, precisely as we do for species, and they also reflect the amount of shared DNA sequences and inherited traits.

To examine the differences between the breeds, let's begin at the (relative) beginning, with an examination of the oldest dog breeds. Terminology pertaining to the oldest breeds varies; some prefer to use the term "ancient," while others prefer the term "primitive" or "aboriginal." An ancient breed is one that has lasted from a remote period; it's not necessarily what the first dogs would have looked or acted like, but it is a very old breed. A primitive dog breed is one that pertains to the original dog;

we can look to these dogs to see how the first dogs likely looked or acted, especially at the onset of domestication. For simplicity, we will hereafter refer to the oldest breeds as "ancient" ones, in contrast to the more recent, derived breeds, and will note the date of the breed's origin, when known. Thus, the ancient dogs include those breeds that are closest to the ancestral dog. They have the most characteristics in common with the shared ancestor of the gray wolf and contemporary dogs.

In Chapter 2, we discussed a study where 13 of the oldest dog breeds diverged from the breeds that we consider to be modern. These oldest breeds include the Afghan Hound, Akita, Alaskan Malamute, American Eskimo Dog, Basenji, Canaan Dog, Chinese Shar-Pei, Chow Chow, Dingo, New Guinea Singing Dog, Saluki, Samoyed, and Siberian Husky (vonHoldt *et al.*, 2010) (Fig. 5.3). There were also three dog groups that diverged from modern groups: an Asian group, which included admixture with Chinese wolves, and included the Akita, Chow Chow, Dingo, New Guinea Singing Dog, and Shar-Pei; a northern group, which included the Alaskan Malamute and Siberian Husky; and a Middle Eastern group, which included the Afghan Hound and Saluki (vonHoldt *et al.*, 2010).

Now let's take a closer look at the oldest breeds—the ancient ones. The ancient breeds comprise the most divergent group and includes the New Guinea Singing Dogs of Papua New Guinea, the Dingoes of Australia, the Spitz-type dogs of Russia, Scandinavia, and North America, and the Laikas of Russia. While this list might look quite different from the prior list of the 13 oldest breeds, consider that six of the 13 breeds (the Akita, Alaskan Malamute, American Eskimo Dog, Chow Chow, Samoyed, and Siberian Husky) fall under the "Spitz-type dog" umbrella, while a seventh dog, the Chinese Shar-Pei, is closely related to the Chow Chow and is also most closely related to the Spitz breeds (Yang *et al.*, 2017). Similarly, the Primitive and Aboriginal Dogs Society (PADS; https://padsociety.org) lists four groups of aboriginal dogs: the Dingo/pariah, the Nordic/Spitz, the Gazehound, and the Prick-eared hound. On the surface, this might again look very different from that prior list of the 13 oldest breeds, until you consider that each of these dogs falls into one of these four categories:

- *Dingo/pariah dogs*: Dingo, New Guinea Singing Dog, Canaan Dog (also cross-listed as Prick-eared hound)
- *Nordic/Spitz-type dogs*: Akita, Alaskan Malamute, American Eskimo Dog, Chinese Shar-Pei, Chow Chow, Samoyed, Siberian Husky
- *Gazehounds*: Afghan Hound, Saluki
- *Prick-eared hounds*: Basenji, Canaan Dog (also cross-listed as Pariah)

Thus, while dogs may be categorized with different names or group designations across studies (Spitz-type breeds versus listing specific breeds), there is a general scientific consensus on which breeds are the oldest ones.

According to PADS, the ancient dogs fit one or more of the following criteria:

1. They were present in their area of origin prior to 3000 BCE.

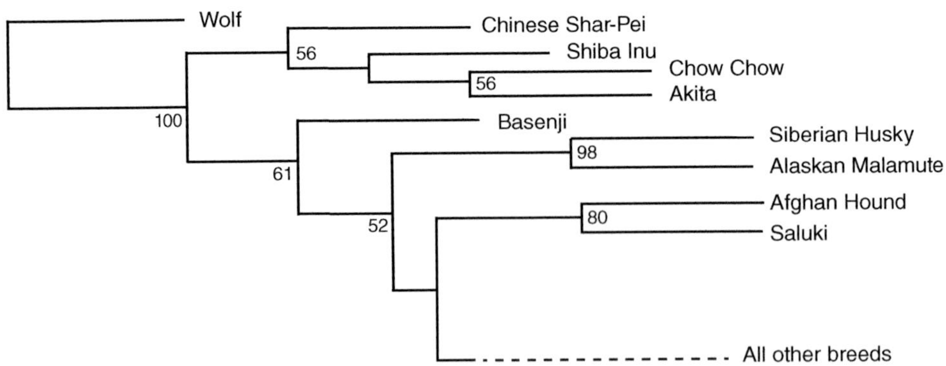

Fig. 5.3. Genetic relationships between breeds. Larger numbers on the branch reflect the proportion of analyses that support those branch relationships. From Parker *et al.* (2004). Reprinted with permission from AAAS.

2. They are documented, direct pure descendants of long-term pariahs (wild).
3. They show few, if any, derived characteristics, such as brachycephalic faces.

The ancient breeds are often used as the model for what dogs were like 10,000 15,000 or 20,000 years ago. PADS also lists "landrace" dogs—naturally occurring breed types that aren't recognized as breeds, per se, but evolved through natural selection (e.g. a single or a double coat, depending upon the climate) and are similar in appearance and often, in their behavior, as well.

Some of the oldest of the purebred dogs (bred by humans) would be those that were maintained through religious proclamations—the gazehounds of North Africa, Arabia, Middle Asia, and Europe, such as the Saluki. These dogs were considered to be the only ones that were not "unclean" by Islamic culture. Muslims do not touch "regular" dogs, but dogs such as the Saluki were historically allowed to sleep with their people in their tents and come and go as they pleased.

PADS considers the New Guinea Singing Dog and the Dingo to be the only ancient dogs that are still living in a totally wild state, and thus, likely closest in both morphology and behavior to the ancestral dog. The earliest identified Dingo fossil, to date, is 3450 years old. Dingoes are closely related to the New Guinea Singing Dog, which resides approximately 100 miles away, separated from one another by the Torres Strait. Genetic studies reveal that the two breeds last shared a common ancestor approximately 8300 years ago (Fig. 5.4). Mitochondrial genome sequences reveal that the New Guinea Singing Dog is more closely related to Dingoes that live in Southeastern Australia than those that live closer to the Malay Peninsula in the Northwest. New Guinea lies on the Australian Continental Plate. Sahul, a Pleistocene-era continent connecting New Guinea with Australia and Tasmania, was partially submerged 18,000 years ago and along with Sunda in Asia, were points of human (and other animal) migration. The Sunda shelf includes Sumatra, Borneo, Bali, Java, Melanesia, and the Malay Peninsula. The New Guinea Singing Dog is native to New Guinea and earned its moniker from its unusual "singing" vocalization.

Spitz-type dogs, also often called Northern or Nordic breeds, originated in the Arctic region or Siberia. According to the AKC, there are currently

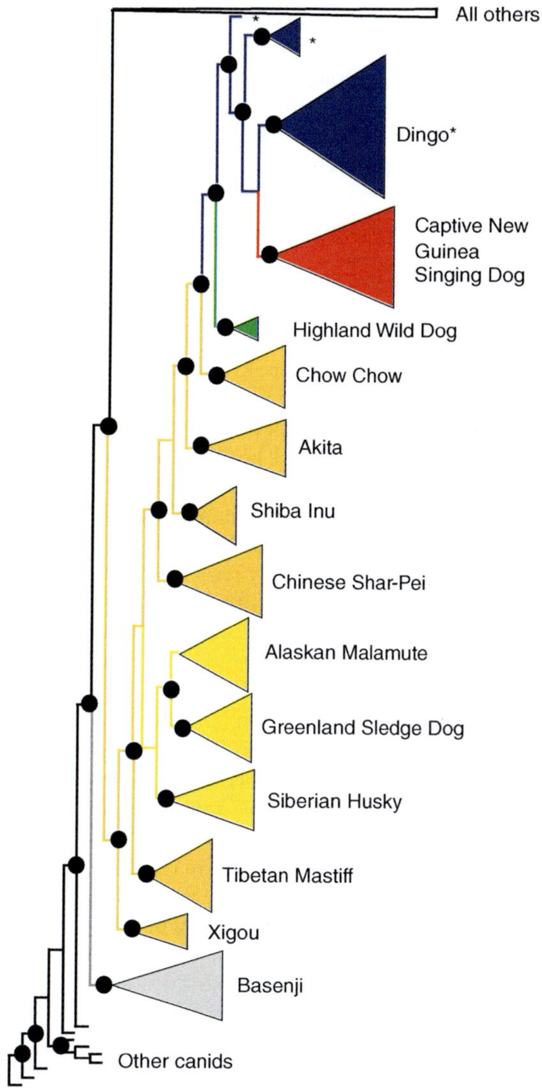

Fig. 5.4. Neighbor-joining dendrogram of dogs from 161 breeds places the Highland Water Dog (green) within the clade of East Asian (gold) and Arctic (yellow) breed dogs on a monophyletic clade with the other Oceanic dog populations (New Guinea Singing Dog in red, Dingoes in blue on three branches indicated with an asterisk). Branches with 100% bootstrap values are marked with a black dot. From Surbakti *et al.* (2020). Permission to use granted by Elaine Ostrander.

50–70 Spitz-type breeds, with the variance in numbers attributed to the difficulty in defining a Spitz (Johnstone, 2021). If the AKC struggles with this definition, it's unsurprising that dog owners do, too!

The Laika is a hunting-type dog originating from Russian Siberia and Northern Russia. According to the Federation Cynologique Internationale (FCI), the Laika currently comprises three breeds: the West Siberian Laika, the East Siberian Laika, and the Russo European Laika.

The Singing Skull

Janice Koler-Matznick smiled back serenely at the befuddled Transportation Security Administration (TSA) agent. Tall, with cropped brownish-blonde hair and an eager demeanor, she watched as the agent slowly turned the canid skull in his hands, looking from the skull back to Janice. It was 5 days after the 9/11 attacks, and Jan had been fortunate to find any flight from the US to England, where she would be participating in a conference and discussing her research on the New Guinea Singing Dog, which she affectionately refers to as "Singers."

"I had two Singer skulls with me for this conference," she recalled, "and I was the only passenger standing there, in the hollow San Francisco airport." After explaining why she was carrying two canid skulls on her European getaway, the mystified TSA agent allowed her to leave, and there she waited for 7 hours, at the beginning of a *Planes, Trains and Automobiles*-esque journey that would eventually terminate in Oxford, England, after 26 hours of travel. "There were maybe seven passengers on that entire plane," she recalled.

While the trip was taxing, it was worth it, too; by most accounts, the New Guinea Singing Dog is understudied and misunderstood, and Jan's research was the first time that many of the conference's attendees became familiar with this rare breed. According to Jan, Dingoes, which themselves are at least 12,000 years old, originated from the New Guinea Singing Dog. Jan places her confidence in this at least partially in the Singers' unique behaviors. Unlike almost all domesticated dogs, which typically have one heat, or estrus, every 6 months, Singers can have up to three estrus cycles, beginning in July. They have unique "singing vocalizations." And their cognitive tests place them squarely in between the performance of domesticated dogs and wolves.

"Dr Brian Hare did a point-gaze test with the Singers," Jan said. "He wanted to do 15 repetitions, but the Singers just wouldn't do the test this many times. They would just do five. Their test results came out exactly between domestic dogs and wolves because they just wouldn't participate. There may have been some confounding factors, as it was also breeding season, and we didn't have shade, and the males were distracted. The tests just weren't set up for this breed."

It's important to remember the cultural and social background of whomever you are testing. The classic example of culturally appropriate testing proffered to first-year psychology students is that of an IQ test that was administered to different groups of young students. The test incorporated questions with corresponding images for the answer. For example, one question asked where milk came from; the "correct" response was circling the image of the cow. When the test was administered to children in a remote Alaskan village, they all "failed" this question, insisting that milk came from a plane. They weren't wrong, though: there weren't cows in their home village, so milk came by plane.

"Despite this, Dr Hare referred to Singers as the 'Bonobos of the dog world,'" Jan said proudly. (This is no small compliment; Dr Hare has worked with bonobos extensively and considers them to be cognitively advanced.)

Jan classifies the New Guinea Singing Dog as an ancient dog, one that hasn't changed for thousands of years due to its geographic isolation, but many other breeds have also gone unchanged for millennia. "The Saluki is inbred, but they look exactly the same as they did 3500 years ago," Jan said. "The Saluki was an original landrace of the long-legged Sighthounds. They were naturally selected for their niche, so it's been harder to 'ruin' this breed. Because of the strict selection for utility at the beginning, it's harder to change them now."

Selection for utility has been a robust tradition across many cultures, whether it was for sled pulling in the Arctic or sight hunting in Africa. Perhaps it is the utilitarian value of many of these breeds that have kept them relatively "unchanged" from their founders.

From Russia, With Love

Vladimir Beregovoy surveyed his sprawling Virginia property. His three Laikas, Dinah, Pavlik, and Kiara, ran through the lushly forested spread of his 94-acre parcel, thousands of miles away from their native land. The Laika, which is the national dog breed of Russia, is considered to be an ancient breed. Vladimir has studied them since the 1960s, beginning in the taiga of Siberia.

"I'm a biologist and a zoologist, and this is the dog of my younger years in the Soviet Union," Vladimir said. "I knew them in their natural environment, in the taiga forest, in the north. I saw them in action with the native people of the north Urals."

According to Vladimir, in their natural environment, Laikas live a life full of freedom.

"They live with aboriginal people and Russian frontiersman hunters. They wander around the house, loose for most of their lives, and then the hunter takes them to hunt and gives them gifts so they aren't far away."

Vladimir, who fled from the Soviet Union to the US in 1980, brought his own Laikas a decade later. While there's a great deal of Russian pride in this breed now, that wasn't always the case.

"Strictly speaking, this isn't a 'Russian' dog," Vladimir explained. "For centuries, Russians didn't pay attention to Laikas. Only in the 19th century did they begin to care about them. These dogs were specific to the native peoples of Siberia; there were as many breeds of Laika as there were different cultural groups of people. The Russians recognized the importance of the Laika, and wanted to save this type of dog. When the Bolsheviks came to power, though, this idea was abandoned. Then, after World War II, the Russians officially established four breeds of Laika, ignoring all others. During the 1940s, the Russian government was anxious to get foreign currency, and they realized that Laikas could be a key factor in this. The other kinds of Laika that they didn't select disappeared, not because they'd been exterminated, but because of genetic mixing with the four chosen breeds."

The four chosen breeds were the West Siberian, Russo European, East Siberian, and Finnish Laikas. "The Finnish Laika is almost identical to the Finnish Spitz, so now it's recognized as the Finnish Spitz," Vladimir explained. "The West Siberian Laika was bred by hunters for other hunters; they aren't in dog shows, but their popularity is growing. The Karelian Bear Dog is similar to the Russo European Laika; without documents, you can't tell them apart. They came from the same gene pool of aboriginal dogs, which came from Finland and the most northwestern part of Russia. The East Siberian Laika is a very obedient dog—much more so than the West Siberian—and they were trained to run alongside the shore and pull their boats along. The Finnish Spitz became popular with city people, but they don't like barking, so they surgically removed their vocal cords," Vladimir said sadly.

Laikas have many behaviors that make them unique compared to other dogs. "Laikas don't need to be trained or taught. They know which animals to bark at, and which to ignore. They won't bark at the crow, magpie, or woodpecker, but they will at the turkey, which their people will eat. They are hunting dogs, but they aren't killers. They are livestock protectors. If necessary, they will kill, but that's not their first inclination. They know just how much to bark to keep prey at bay, and not scare it off. They are excellent at finding bear, and then barking to tell their people what they have found." In their native land, dogs that don't protect will get shot; only the dogs that show these behaviors—knowing when to bark, knowing how to protect livestock—live on to create more Laikas who will do this.

This focus on behavior over appearance initially confused the Europeans when they first met them. "Aboriginal dogs were referred to by Europeans as 'mongrels' because of their variation and absence of a physical standard," Vladimir explained. "They don't look 'attractive' to the European eye. They're functional, well-built, and their only differences between each other are superficial to those who live with them: a hanging or a curled tail, or upright or floppy ears. Similarly, how would we write a 'breed standard' for wolves, with their huge variation in their tails?"

Laikas live, much like they probably have for centuries, in Russia, Tajikistan, and Afghanistan, in nomadic and semi-nomadic aboriginal societies. "We need to hurry to study this breed, before it becomes extinct like the Hare Indian Dog," Vladimir warned. "The native peoples of these areas are following the same path as the natives of the US. Their ways of life are slowly changing. Native dogs are a geographic and ethnographic phenomenon—they're part of a primitive part of life. When that way of life is gone, the dog is gone, too. The way of life keeps these dogs as they were for hundreds and thousands of years."

Breed Groups

With so many different kinds of dogs, bred for different uses and appearances, kennel clubs arose to categorize them into "breed groups." Within each breed group, each breed would have breed standards for size, shape, and behavior; these standards would be the gold standard for the breed.

Unfortunately for scientists who would later study domesticated dogs, there isn't an international consensus on these groups.

The Kennel Club, the UK's official Kennel Club, is the oldest recognized kennel club; it was founded in 1873 and was granted the prefix "Royal" in 2023 to mark the 150th anniversary of its existence. The Kennel Club recognizes seven breed groups and a total of 222 breeds as of this writing. These groups include:

- The Hound Group, which comprises 39 breeds, including Beagles, Bloodhounds, Greyhounds, and Rhodesian Ridgebacks.
- The Working Group, which comprises 26 breeds, including Alaskan Malamutes, Boxers, Mastiffs, and Rottweilers.
- The Terrier Group, which comprises 27 breeds, including Airedale, Bull, Norfolk, and Staffordshire Bull Terriers.
- The Toy Group, which comprises 24 breeds, including Chihuahuas, Papillons, Pekingese, and Pugs.
- The Gundog Group, which comprises 38 breeds, including Golden Retrievers, Labrador Retrievers, Vizlas, and Irish Setters.
- The Pastoral Group, which comprises 38 breeds, including Anatolian, Australian, Belgian, and German Shepherds
- The Utility Group, which comprises 30 breeds, including Akitas, Bulldogs, and Standard and Toy Poodles.

The American Kennel Club (AKC), founded in 1884, recognizes seven breed groups and a miscellaneous class with a total of 195 breeds, as of this writing. The breed groups recognized by the AKC include:

- The Sporting Group, which comprises 31 breeds, including Golden Retrievers, Labrador Retrievers, Vizlas, and Irish Setters.
- The Hound Group, which comprises 32 breeds, including Beagles, Bloodhounds, Greyhounds, and Rhodesian Ridgebacks.
- The Working Group, which comprises 30 breeds, including Akitas, Boxers, Great Danes, and Portuguese Water Dogs.
- The Terrier Group, which comprises 31 breeds, including Cairn, Bull, Rat, and Russell.
- The Toy Group, which comprises 21 breeds, including Italian Greyhounds, Pugs, Havanese, and Shih Tzus.
- The Non-sporting Group, which comprises 20 breeds, including Bulldogs, Dalmatians, Chow Chows, and Poodles
- The Herding Group, which comprises 30 breeds, including Australian Shepherds, Bearded Collies, German Shepherds, and Old English Sheepdogs.

The AKC also has a Miscellaneous Class for those breeds that have been enrolled in the "AKC Foundation Stock Service®" (FSS), where enrollment is maintained until the AKC Board of Directors accepts the breed for regular status. This is a service that the AKC provides to allow purebred breeds to continue developing while providing them with an avenue to maintain their records. While FSS breeds are not AKC registration eligible, most are approved to compete in AKC Companion Events.

Of the groups in the two kennel clubs that have different names, the UK Kennel Club Pastoral Group and the AKC Herding Group are similar, as are the UK Kennel Club Gundog Group and the AKC Sporting Group and the AKC Nonsporting Group and the UK Kennel Club Utility Group. Between the AKC and the UK Kennel Club, there are different numbers of breeds recognized, different breed group names, and different numbers of breeds within the groups that have the same name. To further complicate this picture, the Fédération Cynologique Internationale (FCI), the world governing body of dog breeds, lists ten breed groups and 361 breeds. The FCI's breed groups include:

- Group 1, which includes the Sheepdogs and Cattledogs (excepting the Swiss Cattledogs).
- Group 2, which includes the Pinscher, Schnauzer, and Swiss Mountain Cattledogs.
- Group 3, which includes Terriers.
- Group 4, which includes Dachshunds.
- Group 5, which includes Spitz and "primitive types".
- Group 6, which includes Scenthounds and related breeds.
- Group 7, which includes Pointing Dogs.
- Group 8, which includes Retrievers, Flushing Dogs and Water Dogs.
- Group 9, which includes Companion and Toy Dogs.
- Group 10, which includes the Sighthounds.

This is highly problematic for animal behavior scientists who are trying to compare and contrast breeds.

We haven't even discussed the 100-plus other kennel clubs on every continent, except Antarctica, including the Federación Cinológica Argentina, the Australian National Kennel Council, the Canadian Kennel Club, the China Kennel Union, France's Société Centrale Canine, and the Kennel Union of Southern Africa. Some estimates of dog breeds, including those that are not officially recognized by an organization, are as high as 1000. Irrespective of whether it's 200, 400, or 1000 breeds, there's a lot of diversity within *Canis familiaris*. So how accurate are the classifications of breed group and breed, and what can they tell us about dogs? A lot, actually.

As your authors are from the US, we defer to the breed groups specified by the AKC (AKC, 2019). The AKC standards for each breed include the general appearance, size, proportion and substance, a description of the head, including the skull, muzzle, bite, ears, eyes, the neck, topline and body, including a description of the tail, the forequarters, including the shoulders, front legs, and feet, the hindquarters, the coat, the color, movement, and temperament. Dog breeds also differ in their levels of vocational propensity. While some modern breeds were bred for their physical attributes that then led to behavioral changes, other breeds were bred specifically for their behaviors (Mehrkam and Wynne, 2014). You can read more about the breed groups on the AKC's website, but briefly, their descriptions are listed in Table 5.2.

The Herding Group

Breeds in the Herding Group were originally bred to gather and herd livestock (Fig. 5.5a). The dogs in the Herding Group have the highest rates of trainability, average boldness, the lowest calmness, and low sociability (Turcsán *et al.*, 2011). They are known for their responsiveness, intelligence, high energy levels, high levels of trainability, and instinctual ability to herd other animals (AKC, 2022b). Dogs in the Herding Group were included in the Working Group until 1983.

Table 5.2. Breed groups (per the American Kennel Club) and their historical roles.

Breed groups	Historical roles
Sporting	Cooperative hunting and other field activities
Hound	Independent hunting of prey, vermin
Working	Guarding, sled-pulling, rescue/service
Terrier	Independent hunting and flushing vermin, fighting
Toy	Companion
Non-sporting	Varied
Herding	Driving livestock

Modified from Mehrkam and Wynne (2014), p. 14.

The Working Group

Dogs in the Working Group include some of the most ancient breeds and were originally bred to assist humans with tasks such as guarding homes or flocks, performing water rescues, pulling carts or sleds, or providing protection (Fig. 5.5b). They have high levels of boldness and calmness and lower levels of sociability (Turcsán *et al.*, 2011). They are known for their intelligence, quick rate of learning, alertness, strength, stature, propensity to be protective, and vigilance (AKC, 2022c).

The Sporting Group

Breeds in the Sporting Group were originally bred to provide assistance for hunters of feathered game (Fig. 5.5c). Dogs in this group are considered to be more trainable than non-sporting dogs. Sporting and working dogs are the most likely to follow an arm-pointing gesture. Hunting and herding dogs require extensive human cooperation (Coppinger and Schneider, 1995). The Sporting Group dogs have the highest rates of sociability of all of the dog groups and an average level of calmness (Turcsán *et al.*, 2011). The AKC describes the dogs in this group as likable, active, and alert, and desirous of regular exercise. The Sporting Group includes four basic types: Retrievers, Pointers, Setters, and Spaniels. The Retrievers are built for swimming and possess thick, water-repellent coats; the group's three other types were bred for finding game birds in grassland (AKC, 2022d).

The Hound Group

Breeds in the Hound Group were originally bred to chase warm-blooded prey and have a strong prey drive (Fig. 5.5d). Dogs in the Hound Group exhibit low levels of aggression toward their peers and have average levels of calmness and sociability (Turcsán *et al.*, 2011). This group is phenotypically diverse, and includes the tough Scenthounds, which have the most powerful noses in caninedom, and the long-legged, sleek

Fig. 5.5. (a) Border Collie (Herding Group). Photograph courtesy of Holly Cook Photography, LLC. (b) Saint Bernard (Working Group). Photograph courtesy of Wikipedia. This file is licensed under the Creative Commons Attribution 3.0 Unported license. Available at: https://commons.wikimedia.org/wiki/File:Saint-bernard-standing.jpg. (c). Labrador Retriever (Sporting Group). Photograph courtesy of Sokova Mary, Public domain, via Wikimedia Commons. Available at: https://commons.wikimedia.org/wiki/File:Labr.jpg. (d) Redbone Coonhound (Hound Group). Photograph courtesy of Public domain via Wikimedia Commons. This file is licensed under the Creative Commons Attribution-Share Alike 4.0 International license. Available at: https://commons.wikimedia.org/wiki/File:Redbone-coonhound-detail.jpg. (e) Jack Russell Terrier (Terrier Group). Photograph courtesy of Public domain via Wikipedia Commons. This file is licensed under the GNU Free Documentation License, Version 1.2 Available at: https://commons.wikimedia.org/wiki/File:Jack_Russell_Terrier_2.jpg. (f) French Bulldog (Non-sporting Group). Photograph courtesy of Public domain via Wikipedia Commons. This file is licensed under the Creative Commons Attribution-Share Alike 4.0 International license. Available at: https://commons.wikimedia.org/wiki/File:French_Bulldog_Male.jpg. (g) Bichon Frise (Toy Group). Photograph courtesy of Public domain via Wikipedia Commons. This file is licensed under Creative Commons Attribution-Share Alike 4.0 International license. Available at: https://commons.wikimedia.org/wiki/File:Bichon_Fris%C3%A9_0012.jpg.

Sighthounds, which possess a wide field of vision and explosive speed (AKC, 2022e).

The Terrier Group

Breeds in the Terrier Group were originally bred to pursue vermin, guard their property, or for bull baiting (Fig. 5.5e). Dogs in the Terrier Group have high levels of boldness and trainability, average calmness, and lower sociability (Turcsán *et al.*, 2011); they are known for being energetic, feisty, and stubborn. Terriers range in size from small breeds like the Norfolk Terrier to large ones like the Airedale Terrier (AKC, 2022f).

The Non-sporting Group

The breeds in the Non-sporting Group are perhaps the most diverse in terms of why they were bred and the diversity of their appearance (AKC, 2022g) (Fig. 5.5f). They tend to have lower trainability and boldness and higher calmness and sociability (Turcsán *et al.*, 2011).

The Toy Group

The breeds in the Toy Group were originally bred to be human companions (Fig. 5.5g). The dogs in the Toy Group have average boldness and trainability and lower calmness and sociability (Turcsán *et al.*, 2011). They are phenotypically diverse, coming in a wide range of colors and coat types, and are known to be sociable, affectionate, energetic, and protective (AKC, 2022h).

Breeding for Behavior—Within and Across AKC Breed Groups

Breeding for working: guarding

There is some science on the ability of various guarding breeds to safeguard against different predators in agricultural settings. One such study compared the common mixed breed "white dog" used for livestock guarding in the US to three European breeds, the Turkish Kangal, the Bulgarian Karakachan, and the Portuguese Transmontano, and found that the novel European dogs were more successful at reducing sheep predation than the "white dog" (Kinka and Young, 2019). But a study in Colorado which analyzed sheep losses over one season found no differences between Akbash, Great Pyrenees, or Komondor (Andelt, 1999). Still, genetic data suggests that there are clear breed differences between at least some of livestock guardian breeds, including three Italian, Sila's Dog, Maremma and Abruzzese Sheepdog, and Mannara's Dog, and three similar breeds, Cane Corso, Central Asian Shepherd Dog and Caucasian Shepherd Dog (Bigi *et al.*, 2018). Despite breeding for pet and companion use, another study found that the Great Pyrenees in Norway still retained natural instincts for guarding, including non-aggression towards people and livestock, and their tendency to chase bears (Hansen and Bakken, 1999).

Breeding for working: sled dogs

He pointed his muzzle into the great white beyond, snow blasting by each side of his face, his strides met by his co-lead. Their footfalls slowed as they reached the final relay point. Their musher, Gunnar Kaasen, strained to see through the snowstorm, but no one was there to take their place. The lead dogs looked around, and then back at Kaasen. The clock was ticking. The snow was falling. And Kaasen urged them onward—they would take an additional portion of this relay, traveling the final 25 miles to Nome, Alaska, rather than wait for a team that might never show. In the early morning hours of February 2, 1925, Kaasen and his lead dogs, Balto and Fox, arrived at Point Safely, Alaska. Carried in their sled was the serum to stop a deadly diphtheria epidemic. Without this medicine, the fatality rate for diphtheria approached 100%. The serum had originated in Anchorage, Alaska, and had been relayed by multiple mushers and their dogs. The inexperienced Balto ran the last two legs of the relay, totaling 55 miles, and received the lion's share of the fame for it. But it was another sled dog, Togo, who pulled his serum-transporting sled for 260 of the 674 miles of the Great Race for Mercy, including the most difficult portions. The heroic race is reenacted annually as the "Iditarod," in honor of these life-saving dog sled teams (Fig. 5.6).

Togo, Balto, and Fox were all "Alaskan sled dogs," a term that refers to a number of northern breeds that were interbred to perform a specific task: carrying people and supplies in a sled across the Arctic tundra (Huson *et al.*, 2010). While pulling archaic sleds was an ancient practice, this iteration, with more modernized sleds, was common from the late 1800s to the early 1900s, and then faded out with even more modern transportation options. There was a resurgence of interest in these dogs when sled dog racing began in 1930, and breeding selection changed from working to athletic performance (Huson *et al.*, 2010). Alaskan sled dogs are generally thought of as a unique breed, though the AKC does not recognize them as a "breed." A study found a significant genetic split between "sprint" and "distance" specialist sled dogs that were evaluated for their speed (associated with Pointer and Saluki breeds), endurance (associated with Alaskan Malamute and Siberian Husky breeds), and work ethic (associated with Anatolian Shepherds) (Huson *et al.*, 2010).

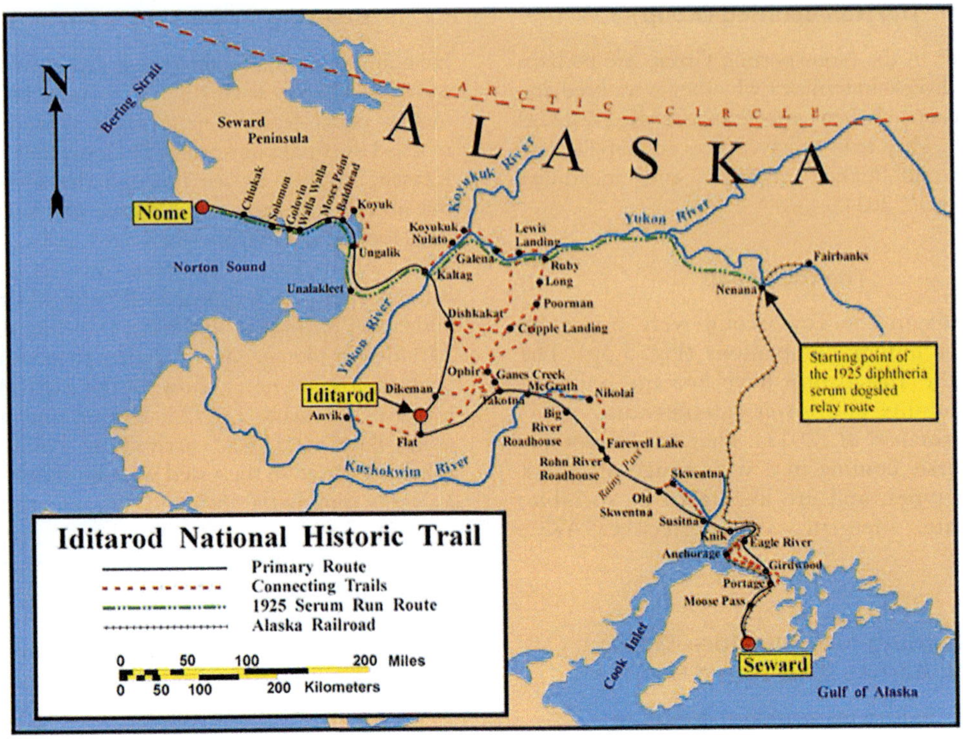

Fig. 5.6. Map of the historical Iditarod Trail and the current Iditarod National Historic Trail in Alaska. File courtesy of Public domain via Wikipedia Commons. This image is a work of a Bureau of Land Management employee, taken or made as part of that person's official duties. As a work of the US federal government, the image is in the public domain in the US. Available at: https://commons.wikimedia.org/wiki/File:Iditarod_Trail_BLM_map.jpg.

Additional evidence for an even longer history of artificial selection for sled dogs comes from a paper on the origin and evolution of a group of Greenland sled dogs. These researchers compared the DNA of ten modern Greenland sled dogs to one ancient Siberian dog (who was estimated to have lived 9500 years ago) found near sled artifacts, and an ancient Siberian wolf (estimated at 33,000 years) (Sinding *et al.*, 2020). They found significant similarity between the genetics of the modern sled dogs and the ancient Siberian dog, as well as some gene flow from ancient Siberian wolves, but not from American wolves. Together, these data suggest that humans have been using dogs to pull sleds for more than 9500 years, and that the origin of sled dogs is from Siberia (Alaska Mushing School, 2023). For centuries, sled was the primary mode of transportation for many peoples living in more remote areas, and modern traditions such as the Iditarod pay homage to this ancient way of life.

Breeding for sporting: hunting dogs

These include dogs in the Sporting, Hound, and Terrier Groups (Fig. 5.7)

One of the tasks that humans have bred dogs for is hunting with a human companion. Breeding for this task-related behavior assumes that behaviors related to hunting are genetic, and thus heritable, from the behaviors of parents. Pointing, which occurs when the dog detects a scent and assumes a stiff posture, has been found to have a high heritability in Large Munsterlanders; 40% of the variability in this behavior is associated with genetics (Schmutz and Schmutz, 1998). Large Munsterlanders are a gundog breed that isn't yet recognized by the AKC, but received recognition from the UKC in 2006. Other behaviors that have been shown to have significant heritability include: the use of the nose to track and find game, the desire to actively search for game, and a willingness to jump in and search water for game. German Shorthaired Pointers,

Fig. 5.7. English Pointer and Irish Setter. File courtesy of Public domain (PD-UK-unknown, PD-anon-70-EU) via Wikipedia Commons. This image was first published in the 1st (1876–1899), 2nd (1904–1926) or 3rd (1923–1937) edition of Nordisk familjebok.

Griffons, and Pudel-pointers were the breeds that showed 19–35% heritability in the use of the nose to track and find game (Schmutz and Schmutz, 1998). German Shorthaired Pointers and German Wirehaired Pointers showed a heritability of 48% and 31%, respectively, on searching for game. For hunting in water, German Wirehaired Pointers and Pudel-pointers showed a heritability of 32% and 31%, respectively. German Shorthaired Pointers also had a high heritability for tracking (48%) and desire to hunt (31%) (Schmutz and Schmutz, 1998). A similar study on Swedish Flatcoated Retrievers found a heritability of 23% for hunting in the water, 26% for searching, and 34% for retrieving (Lindberg *et al.*, 2004). All of these studies indicate that breeding for behaviors to help a human hunting companion are highly heritable in some breeds. Behaviors such as pointing require a human recipient to act upon this information; there would be no adaptive value for a dog to point at game if they were hunting *without* a human companion.

Breeding for vermin catching

Hounds and Terriers are well known for their vermin-catching prowess. Terriers in particular have a history of being used to catch vermin (Anonymous, 1838) and were the preferred type of dog during Victorian times in rat-filled London (Pemberton, 2014). One self-proclaimed Victorian dog breeder, a gentleman named Jack Black, applauded the rat-killing talents of his Terriers (Pemberton, 2014):

> Diminishing the numbers of rats required a new kind of dog and, as a direct consequence, terriers obtained an independent identity. There was no equal to 'that strain of black tan terriers' for the purpose of rat-killing. As a dog breeder, Black self-assuredly positioned himself as the progenitor of this breed: 'I had a little rat dog, a black tan Terrier of the name of Billy, which was the greatest stock dog in London of that day.'

There isn't much in the scientific literature about how much better Terriers are as vermin catchers than other breeds, but the idea is prevalent in historical sources, and commonly understood. Science has shown higher aggression levels in these breeds, which makes sense if they were bred to pursue vermin (Duffy *et al.*, 2008). We discuss aggression in more detail in Chapter 9.

Breeding for love

One of the relatively recent changes that humans have made in dog breeding practices has been to select them for non-working jobs, such as companionship. Studies do support that non-working breeds are lower in activity levels than their working dog peers (Asp *et al.*, 2015). This suggests that selection for companionship has changed dog behavior.

There is also some evidence that the perceived "cuteness" of a dog affects a human's perception of positive personality traits in dogs (Thorn *et al.*, 2015). Ethologist Konrad Lorenz wrote about the evolutionary basis for humans to be attracted to "cute" animals (Lorenz, 1971). A study of the morphological differences between breeds and behavioral traits in dogs suggests that shorter dogs are more likely to display negative behavioral traits such as urination and defecation when left alone, separation anxiety, dog-directed fear, non-social fear, begging for food, owner-directed aggression, urine marking, and attachment/attention seeking (McGreevy *et al.*, 2013). Dogs with lower body weights were also reported as more hyperactive and excitable. This begs the question whether in our selection of "cuteness," meaning small dogs that retain juvenile characteristics into adulthood (referred to as neoteny; see Chapter 2), we inadvertently brought along some negative traits. Alternatively, smaller Companion dogs can be more highly inbred due to selection for a subset of diminutive individuals (Irion *et al.*, 2003; Wade, 2011) and inbreeding could be the cause of these behavioral issues. At this time, we can't say what drives this pattern of smaller size and negative behaviors. Cuteness seems to hide a multitude of negative behavioral traits.

Breeding for olfaction across breed groups: which nose knows

Dogs have an amazing ability to smell. They can detect cancer, locate victims in catastrophes (Quignon *et al.*, 2012), and discriminate between identical human twins based on differences between the twins' diets (Hepper, 1988). Some breeds have been purpose-bred to use their sense of smell to work for humans, including early jobs as hunting companions to track game. Wolves and dogs would have been naturally selected to use olfaction in the wild for a variety of reasons including to hunt for food, find mates, and communicate.

Much of the science on the specifics of dog olfaction abilities is fairly recent. For example, work on the genetics of olfaction in dogs (Quignon *et al.*, 2012; Galibert *et al.*, 2016), selecting dogs for scent detection work (Beebe *et al.*, 2016), and training scent detection in dogs (Hall *et al.*, 2013) only started to be examined in 2012. Some of the delay in development is because the information on how to select dogs for detection work is not published. The information is proprietary for commercial purposes, and much of the work has been done by non-scientists. Publication is not rewarded. It is also assumed that working and sporting breeds are the best at scent detection (Beebe *et al.*, 2016). The most common breeds or breed groups used for conservation scent detection work include Retrievers, Shepherds, Collies, and Setters (Beebe *et al.*, 2016). Interestingly, Bloodhounds, with 300 million odor receptor cells, did not make the list, while Shepherds with their paltry 225 million, did (Beebe *et al.*, 2016).

The first paper to look at differences in olfaction between dog breeds and wolves was published in 2016 (Polgar *et al.*, 2016). The authors used a natural detection task that required no pre-training, and compared the olfactory ability of wolves to traditional scent dogs (Basset Hound, Beagle, German Pointer, Wirehaired Vizsla, Bracco Italiano, Grand Basset Griffon Vendéen, and Transylvanian Hound), non-scent dogs (Bichon Havanese, Chinese Crested Powder Puff, English Greyhound, Hungarian Greyhound, Whippet, Afghan Hound, Bichon Bolognese, Greyhound cross, Miniature Pinscher, and Siberian Husky), and short-nosed dogs (Cavalier King Charles Spaniel, Boston Terrier, Boxer, American Bulldog/Boxer cross, Bullmastiff, English Bulldog, and Pug) (Polgar *et al.*, 2016). The task required them to detect raw turkey in one of four ceramic pots by olfaction alone. The researchers varied the difficulty of the task by presenting pots with (i) no lid, (ii) lid with five, three, or one hole(s) in it, and (iii) a lid with no holes. The bait was only under one of the four pots, and the order of the baited container was randomized.

The researchers found that the breeds in the "scent" group were significantly better at the olfaction task than the "non-scent" and "short-nosed" groups. The short-nosed group performed the lowest across all groups. The wolves performed similarly to dogs overall, but when they were tested a second time, they improved, while the dogs did not show any improvement. The difference in improvement rates may be due to performance issues, motivation, or familiarity with humans rather than inability in the first trial.

Another study, which examined two-sample odor discrimination between five types of chemical alcohols (e.g. ethanol, propanol, etc.) among Pugs, German Shepherds, and Greyhounds, had a rather surprising result: Pugs significantly outperformed German Shepherds. And in this study, 90% of the Greyhounds failed to participate at all (Hall *et al.*, 2015). Perhaps another variable confounded the results, or perhaps the Greyhounds were unmotivated to participate. Comparing the two studies is difficult; the first includes more breeds, but not many overlap with the second study, so direct breed comparisons are not possible. The researchers also used different tasks, which required different learning times and may have introduced performance confounds. The two studies do not address the neuroanatomical differences between purpose-bred Scenthounds and other purpose-bred dogs.

Breeding for "canine cops" across breed groups

The earliest recognized training facility for K9s originated in 1899 in Ghent, Belgium, and became the model for training police dogs as far away as New York (Handy *et al.*, 1961). Much of Europe, including France, Germany, and Great Britain, started using police dogs in the early 20th century, and the US was simultaneously beginning to incorporate dogs into this specific role, as well (Pearson, 2016). Scenthounds were used in an (unsuccessful) attempt to track Jack the Ripper. Today, the breeds most often associated as police dogs are German Shepherds, or similar breeds, such as Belgian Shepherd Malinois. Another common breed used for police dog work is the Labrador Retriever. These breeds are also often used in the military (Tamimi and Wali, 2019).

Dogs trained for police work can serve quite a few different roles, including scent detection (drugs, explosives, suspect tracking, victim tracking) and attack/guard dogs. Depending on the task at hand, different traits may be favored. There is some evidence that there are breed differences between the common breeds used for these tasks. For example, a study comparing breeds on drug detection accuracy, found that German Shepherd Dogs were significantly more accurate than Labrador Retrievers, Terriers, and English Cocker Spaniels (Jezierski *et al.*, 2014). Terriers showed relatively poor performance in detection and accuracy. A study that compared German Shepherd Dogs and Labrador Retrievers on their reported performance as drug- and explosive-detection dogs found that both trainers and handlers thought Labrador Retrievers were too food motivated for both tasks, and too friendly when doing drug detection work (Adamkiewicz *et al.*, 2013). Similarly, an earlier study on working dogs with various jobs found that German Shepherd Dogs scored higher on defense drive (this is when a dog feels attacked or threatened) than Labradors (Wilsson and Sundgren, 1997). Labradors were more emotionally stable, less reactive to gunfire, and were more cooperative than German Shepherd Dogs. The researchers suggested that Labradors were more suited as guide dogs and German Shepherd Dogs as police or protection dogs. Recent work suggests that a drive or motivation to work (high arousal) and focus on the task at hand are important factors in predicting successful police dogs (Brady *et al.*, 2018). A lack of fear is also highly desired (Foyer *et al.*, 2016). However, in some dangerous tasks (e.g. explosive detection), the most important characteristic may be that of motor inhibition (impulsivity) (Tiira *et al.*, 2020).

Breeding dogs for sport fighting

When the Romans invaded Britain in 43 CE, both sides used dogs to fight in the Seven-Year War (Wikipedia, 2020). The Romans were impressed with Briton's English Mastiffs and brought some back for both war use and spectacles in the Colosseum. In the Middle Ages in Europe, Mastiff dogs were pitted against bulls or bears for spectacle and sport. This practice was banned in England in 1835, but dog–dog fighting continued because it was easier to conceal from authorities. They began to breed Terriers with Bulldogs to get the desired traits for dog–dog fighting rather than dog–bull or dog–bear matchups, and this is how they developed the Staffordshire Bull Terrier. There is evidence of dog fighting in the US from about 1750 on, but the Staffordshire Bull Terrier arrived in America in 1817. Americans bred this dog for fighting and eventually created the American Pit Bull Terrier. Dog fighting is illegal in the US, and was made a felony in 2007 (Wikipedia, 2020). Unfortunately, it still remains as an underground sport.

So, Mastiff-type breeds were originally selected for fighting, during the Roman era, but it was not until the early Victorian period when Staffordshire Bull Terriers were bred for sport fighting between dogs. The American Pit Bull Terrier is the breed most commonly used in dog–dog fighting in the US (Miller *et al.*, 2016). There is more evidence that artificial selection for fighting has changed the physical characteristics of American Pit Bull Terriers

(Kemp *et al.*, 2005) than there is for changes in behavioral traits. This is likely influenced by the fact that most breeds are no longer being bred solely for fighting (Clarke *et al.*, 2016). However, there is some data that American Pit Bull Terriers are more aggressive towards other dogs. A questionnaire study of owners found that 20% of Akitas, Jack Russell Terriers, and Pit Bull Terriers showed aggression toward unfamiliar dogs with bites or bite attempts (Duffy *et al.*, 2008). This finding may surprise some since Akitas and Jack Russell Terriers were not originally bred for dog–dog fighting. This finding underscores that while there may be an underlying predisposition for aggression apparent in American Pit Bull Terriers, the environment, individual differences, and recent breeding for pet standards are also driving factors in their characteristics.

Neuroanatomy

Because dogs have been selected for vastly different purposes, one might expect the brain of the Basset Hound, which was bred for tracking, to be different from the brain of the Pomeranian, which was bred to be a human companion. This hypothesis holds up. A 2019 study examined the neuroanatomical variation across 33 different breeds of dogs (Hecht *et al.*, 2019). Using a sample size of 62, the researchers used volumetric variation, as measured in magnetic resonance images (MRIs), to make comparisons. The researchers found significant neuroanatomical variation across breeds that correlated with the purposes that the dogs had been bred for. (While this might sound like a small data set for each breed of dog, this study provided more than enough evidence for examining differences across breeds.) Hecht *et al.* (2019) found that herding dogs showed differences in areas of the brain associated with social action and interactions, but so do many other groups bred for working with people and other animals: bird flushing and retrieving, companion, police/military/war, sport-fighting, vermin control. Areas of the brain associated with olfaction were more developed in bird flushing/retrieving, police/military/war, scent-hunting, and vermin control breed groups. Areas of the brain associated with movement, eye movement, and spatial navigation were most developed in dogs bred for bird flushing/retrieving, sight hunting, and sport fighting, but also for companion breeds. We aren't sure why that is; perhaps because our companion-bred dogs rely more heavily on social cues from their humans than dogs who have been bred to be more independent. This seems like a fruitful area for further study.

In Pure Blood: What Happens When Artificial Selection is Overdone

There are pros and cons to artificial selection, including specialized behavior (pro) and the increased prevalence of certain diseases (con). We would place appearance as neither a positive or a negative. Artificial selection can be "overdone" when inbreeding occurs, and in order to conform to official breed standards, purebred dogs are often inbred. How inbred are they?

A 2013 study found that when purebred dogs were compared to mixed breed village dogs, which exercised mate choice, the purebreds had reduced genetic diversity—particularly when they had been bred for show purposes (Pedersen *et al.*, 2013). This is likely because in trying to get an increasingly exaggerated, specific appearance in subsequent generations of dogs, breeders drew from an increasingly small potential mating pool. If Sire A and his progeny produced the desired, award-winning trait, that line would likely be used more than a genetic line that's new or that has consistently not thrown the desired trait. But in creating this artificial bottleneck, we are seeing a reduction in genetic diversity.

There's a trade-off between breeding related animals for predictable outcomes in their offspring: breeders are essentially trading uniformity, the increased prevalence of a desired trait, and an increased ability to pass on these desired traits for a slew of disadvantageous outcomes, including an increase in genetic diseases and birth defects, shorter lifespans and increased mortality rates, smaller body and litter size, and decreases in growth, vigor, and fertility (Beuchat, 2014). According to a 2018 study using data from the Swedish Kennel Club, purebred dogs lost a lot of genes between 1980 and 2012 alone (Jansson and Laikre, 2018). The study examined 12 breeds, ten of which were considered to be a concern for continued conservation. Of these ten breeds, four were "Endangered-maintained" (including the Danish-Swedish Farmdog, Drever, Hamilton Hound, and Swedish Elkhound) and six were "Critical-maintained" (including the Gotland Hound, Norbottenspitz, Schiller Hound, Smaland Hound, Swedish Lapphund, and Swedish Vallhund) according to the Food and Agriculture Organization of the United Nations (FAO) criteria for maintaining domestic animals (Scherf, 2000). The FAO's classification system includes seven categories, listed here in order of increasing genetic risk: "Unknown," "Not at risk," "Endangered-maintained," "Endangered," "Critical-maintained," "Critical," and "Extinct." For an animal to fall under the "Endangered-maintained" or "Critical-maintained" classifications, they have met the criteria for "Critical" or "Endangered," but programs are in place to conserve the breed.

According to the FAO, an individual is classified as "Critical" if: (i) the total number of breeding females is less than or equal to 100, (ii) the total number of breeding males is less than or equal to five, or (iii) if the overall population size is less than or equal to 120 and decreasing, and the percentage of females being bred to males of the same breed is below 80%.

According to the FAO, an animal will be classified as "Endangered" if they: (i) have a total number of breeding females that's greater than 100 and less than or equal to 1000, (ii) have total number of breeding males that's less than or equal to 20 and greater than five, (iii) they have an overall population size that's greater than 80 and less than 100 and increasing and the percentage of females being bred to males of the same breed is above 80%, or (iv) if the overall population size is greater than 1000 and less than or equal to 1200, and decreasing, and the percentage of females being bred to males of the same breed is below 80%.

During the 32-year timespan that this study covered, the average inbreeding coefficient (also referred to as a coefficient of inbreeding (COI)) for several breeds increased from 0.03 to 0.07 (Jansson and Laikre, 2018). Why is this problematic? According to a 2015 study by canine researcher Dr Carole Beuchat, PhD, of the Institute of Canine Biology, a COI increase of 0.04 puts these dogs firmly into genetically unhealthy territory. Beuchat contends that a COI of less than 5% (or 0.05) is ideal; above that, the breeder is running an increased risk of detrimental outcomes in subsequent litters (Beuchat, 2015). So, while a COI increase of 0.04 might seem like a small change, that increase within only 32 years, paired with its increase to above 0.07, signals big issues up ahead.

And Then There Were Eight

In the moments before the last bomb fell, all was quiet but for a hissing whoosh of air—then deafening detonation. Several buildings collapsed, and as the last beam fell, a wall of debris rose up, dusting everything in ghostly softness. Far in the distance, the whir of a plane engine faded away, and then eerie silence once again. Two large dogs clung to one another, their faces shielded from the dust and debris. For weeks afterward, they scavenged for food and fresh water. There were no people here; they had all left weeks ago. This lonely pair of dogs were the last of the Leonbergers; in the aftermath of World War II, only eight of them remained. The World Wars almost wiped out this giant, once-prolific breed: in World War I, only five Leonbergers had survived to continue the breed. As a result, every Leonberger is closely related to every other Leonberger and, because of this, the breed lacks genetic diversity (Fig. 5.8). So, what are the potential issues with all members of a closed breeding group being closely related? Well for Leonbergers, eye diseases such as entropion (inward-turned eyelid), ectropion (outward-turned eyelid), and cataracts are a common issue, as are orthopedic problems such as panosteitis, a bone disease that's characterized by painful inflammation of the shaft or outer surface on one or more of the long bones in the legs.

Leonbergers aren't the only dog that has experienced a reduced genetic load, or reduction in individual fitness due to the presence of deleterious alleles (generally recessive, but persistent, such as those that cause entropion). They aren't even the dog who has experienced the most extreme bottleneck; that dubious honor belongs to the Norwegian Lundehund, a small, Spitz-type dog. This is an old breed that got down to a population size as small as two individuals, from which all modern Lundehunds (fewer than 2000, as of this writing) descend (Stronen *et al.*, 2017). As a result, members of this breed sport a treasure trove of rather unusual physical features, including polydactyly and an unusually high level of spinal flexibility (their neck can bend so far that it almost touches their spine) (Pfahler and Distl, 2015).

From the golden age of breed creation onward, humans have created artificial bottlenecks by selecting only a few individuals to make and perpetuate a breed. While certain phenotypic traits have been strongly selected for, deleterious genes have also accumulated, resulting in higher than expected rates of anatomical abnormalities and genetic disease than if normal breeding had occurred (Ostrander *et al.*, 2017). The black

Fig. 5.8. Leonberger, male, 4 years old. File courtesy of Public domain via Wikipedia Commons. This file is licensed under the Creative Commons Attribution-Share Alike 4.0 International license. Available at: https://commons.wikimedia.org/wiki/File:Leonberger_Norway.JPG.

Standard Poodle, for example, has a higher risk of squamous cell carcinoma than white Standard Poodles, and this has been definitively linked to selecting for the darker coat color. The Shar-Pei, too, has suffered from the selection required to gain its excess of skin folds: the hyaluronan 2 (*HAS2*) gene mutation provides more skin folds, but it also leads to increased mucus production, with the combination leading to high levels of skin infections, periodic fevers and, sometimes, severe autoimmune reactions (Ostrander *et al.*, 2017).

Polydactyly, or the presence of extra digits, is a condition that occurs through a mutation on a gene, with the likelihood of this mutation occurring in future generations increasing as more related individuals with this trait or gene for this trait interbreed with one another. Lundehunds have six toes on each of their feet, instead of five, and the toes on their front feet are double- or triple-jointed. The Lundehund, which survived the smallest breeding population size one can have (one male and one female), suffers from a very low rate of heterozygosity and a high COI because every individual in this breed can be traced back to one pair of dogs (Table 5.3). While the Lundehund's inbreeding coefficient has been hovering around 0.34–0.38 since the mid-1990s, there are 33 registered individuals with a COI that is 0.5 or higher (Kettunen *et al.*, 2017).

The Lundehund is a relatively obscure breed, but more familiar breeds, such as the Doberman, also have a high COI. In 2018, the Doberman Diversity Project (www.dobermandiversityproject.org) found that, on average, Dobermans have an average COI of 0.414499 (Table 5.4), with many of the tested individuals having rates as high as 0.58. Unsurprisingly, Dobermans suffer from numerous genetic diseases;

Table 5.3. Inbreeding coefficients for various inbred relationships.

Relationship	Coefficient of inbreeding
Animal mated to its own parent (e.g. sire/daughter)	0.25
Half-sibling matings (parents have a common sire or dam)	0.125
Full-sibling matings (parents have a common sire and dam)	0.25
Animal has a single common great-grandparent	0.31

Table 5.4. Inbreeding coefficients for individuals/species/breeds.

Relationship	Coefficient of inbreeding
Cheetahs	0.99
Lundehunds	0.34–0.38
Dobermans	0.414499
Charles II	0.254
King Tut	0.25+

they are one of the breeds with the highest incidence of heart disease (familial dilated cardiomyopathy) (Meurs *et al.*, 2007) and spinal cord compression (cervical spondylomyelopathy also called Wobbler's Syndrome) (da Costa *et al.*, 2006).

The Portuguese Candidate

Bald, smiling, and wearing a smart suit, Vasco Bensaude gazed upon his dog, Leão, and then to the first litter of Leão's puppies. Over the Pacific Ocean, a tireless search for Amelia Earhart continued. The great storms of the Dust Bowl raged across the prairies of the US. The Hindenburg's explosion over New Jersey led to announcer Herbert Oglevee Morrison uttering the now famous words, "Oh, the humanity." And in Portugal, one man resurrected a dog breed that nearly went extinct. That first litter, born during turbulent 1937, would regenerate the Portuguese Water Dog. Bensaude had been introduced to Leão's breed in the early 1930s. The Portuguese Water Dog, which had been used by fishermen for centuries, had almost disappeared from the planet in large part due to technological advances, but Bensaude rejuvenated the bloodline, beginning with Leão. This dog would become the founding sire for the modern Portuguese Water Dog, and the written breed standard would be based upon Leão and his descendants. (Sadly, not much is known about the female dogs who contributed to the resurgence of this breed.)

During the 1970s, there were only 25 Portuguese Water Dogs left worldwide, but by 1982, the American Portuguese Water Dog population had grown to over 650 dogs. Because Portuguese Water Dogs, like the Leonberger, trace back to such a small founding population, they have a higher rate than would be expected of certain conditions, including cataracts, Addison's Disease (Chase *et al.*, 2006), hip dysplasia, and potentially fatal heart problems. When the Obamas received their dog, Bo, in 2009 they caused a resurgence of popularity with the breed, much like Dalmatians had a surge in popularity after Disney's *101 Dalmatians* debuted, or the surge in Northern breeds like Siberian Huskies and Alaskan Malamutes after *Game of Thrones'* dire wolves hit the small screen. But many people didn't research the behaviors and needs that tend to accompany these breeds. What inevitably ensued was the mass unloading of these breeds at area shelters several months later, much like people often unload bunnies (now rabbits) or chicks (now hens or, even worse, considering local prohibitions, roosters) after the glow of Easter has worn off.

A population genetics study reviewing 10 common breeds from the UK Kennel Club (Akita Inu, Boxer, English Bulldog, Chow Chow, Rough Collie, Golden Retriever, Greyhound, German Shepherd Dog, Labrador Retriever and English Springer Spaniel) found evidence of genetic differences between all breeds examined, and significant levels of inbreeding (Calboli *et al.*, 2008). Collies and Bulldogs had the highest levels of inbreeding, and Springer Spaniels, the lowest levels. Regardless, breeding for a breed standard has reduced the genetic variability of all of these breeds, which increases the likelihood for genetic disease.

A study on the effects of inbreeding screened 227 breeds, and found that the average inbreeding coefficient was 0.249. As noted in Table 5.3, this is similar to parent–offspring or full-sibling matings. Not surprisingly, breeds with high inbreeding coefficients had higher morbidity than breeds with low inbreeding levels (Bannasch *et al.*, 2021). The University of Cambridge Veterinary School has an online database of genetically linked dog diseases, and was created and maintained by David Sargan

(Sargan, 2004). This database tallies more than 370 diseases in specific breeds of dogs that have been partially or wholly attributed to genetics. This database is not publicly accessible, but there is a Canine Inherited Disorders Database that is accessible (Crook *et al.*, 2011).

Bottlenecks, whether they're due to environmental factors such as geographic isolation or due to artificial selection, have impacted countless species throughout our planet's evolutionary history. Around 10,000–12,000 years ago, approximately 75% of the world's large mammal species died off in the Quaternary Extinction Event. Many of the species that did survive had small populations to build upon, and some, like the cheetah, had only a few members left to attempt to rebuild their species. As a result, cheetahs (*Acinonyx jubatus*) have very low genetic diversity and suffer from low rates of reproductive success due to poor sperm quality, linked tails, and focal palatine erosion, a condition where the lower molars erode the palate until they bore a hole through the jawbone. All cheetahs are susceptible to the same diseases, as their species has no heterozygosity, or difference in their gene alleles. Cheetahs have a COI of approximately 0.99—as close to identical twins as one can get.

In the early 20th century, the Florida panther (*Puma concolor coryi*) dropped to a population size of six, including only one female. Genetic testing revealed that all members of the species can trace back to that one ancestral mother. Their modern genetic diversity is one-third of what it was during the 19th century. Another large cat, the African lion (*Panthera leo*) of the Ngorongoro crater in Tanzania, went through a population bottleneck in 1962. That year, a deluge of rain flooded the 11.8-mile radius crater. Biting flies (*Stomoxys calcitrans*) soon followed, biting the lions relentlessly. After the dual plagues of water and insect subsided, only nine females and one male lion remained. With limited gene flow into the crater, the new population of lions would be very closely related, and the males of the largely homozygous population have increased sperm abnormality rates (Packer *et al.*, 1991).

It's All Relative

Misshapen, middle-aged, and finally free from his miserable existence, the body of Charles II, King of Spain, lay in his coffin. Charles, who was referred to as "Carlos" in Spain and "El Hechizado" (the Hexed or Bewitched) behind his back, was the last of a long line of Hapsburgs. His death in 1700, just days before he turned 39, led to the War of the Spanish Succession, which lasted until 1714. Charles' family, however, had been slowly losing a genetic war for generations. While many non-human animal species have experienced population bottlenecks, most human bottlenecks have been self-imposed. For centuries, royal families throughout Europe and Northern Africa "kept it in the family" for inheritance, believing that royal blood was "pure." Over many generations, these consanguineous pairings had disastrous results. In ancient Egypt, royal brother–sister and father–daughter marriages were common. Following family tradition, King Tutankhamun (hereafter Tut) was the offspring of sibling parents. As the offspring of full siblings, his COI would have been 0.25, at a minimum, and given his family's history of within-family marriages, it was likely much higher.

Tut, too, married his own half-sister, Ankhesenamun (we'll call her Ankhe) but they had no surviving offspring. In 1922, when he discovered Tut's tomb, Howard Carter also found the mummies of two female fetuses (Carter and Mace, 1963); DNA testing revealed they were Tut and Ankhe's daughters (Hawass *et al.*, 2010). One died while Ankhe was approximately 6 months pregnant, while a second one, who was full-term, died shortly after birth (Hawass and Saleem, 2011). Had Tut and Ankhe's full-term daughter survived, she would have been plagued by spina bifida, scoliosis, and Sprengel's deformity, a rare congenital skeletal condition where one shoulder blade is higher than the other (Harrison *et al.*, 1979). It isn't clear whether these deaths were due to genetic abnormalities. The pair didn't have another chance to have a child, as Tut died in 1324 BCE. Mention of Ankhe disappears from the historical record between 1325 and 1321 BCE, signaling the likely time of her death.

As Tut's own remains would reveal, he was far from the handsome man depicted in his golden death mask. He suffered from a pronounced overbite, a club foot, and a rare, painful genetic disorder called Kohler's disease that caused him to lose bones in his foot. While many in Tut's family suffered from the effects of sanctioned incest, the Hapsburgs of Europe are the most notorious for this. The Hapsburgs, who ruled Austria for almost 650 years, even had a genetic disorder (the "Hapsburg Jaw") named after them. Charles II, who would be the last of the Hapsburgs, had a jaw that was so misshapen that he couldn't chew his food. Charles was mute until age 4 years, couldn't walk until he was 8 years old, and was barely literate; he reportedly also suffered from premature senility and was infertile. With a COI of 0.254, Charles was more inbred than the offspring of a brother–sister pairing.

Breed-specific Behavioral Disease Examples

There is evidence from a number of dog breeds that seizures with no known medical explanation (such as idiopathic epilepsy), may be genetically driven. Estimates of this suggest that it occurs in somewhere between 0.5% and 5.7% of dogs (Patterson *et al.*, 2005). Evidence for a genetic component for idiopathic epilepsy has been shown for Beagles, Belgian Tervurens, Bernese Mountain Dogs, English Springer Spaniels, German Shepherd Dogs, Golden Retrievers, Labrador Retrievers, and Vizslas. There is some evidence that Springer Spaniel Rage, a behavioral disorder that presents as unexplained episodic aggression, is also linked to seizures (Dodman *et al.*, 1992). This disorder is also found in English Cocker Spaniels (Case, 2005). We'll discuss Springer Rage further in Chapter 6.

Obsessive-compulsive disorder (OCD) is also found in dogs, and has been most commonly found in Bull Terriers and related breeds (Moon-Fanelli and Dodman, 1998). There is no evidence to suggest that there are breed differences in the occurrence of OCD in dogs, but there is evidence that the type of OCD behaviors shown varies among breeds based upon the tasks for which they were bred. For example, herding dogs were more likely to show obsessive tail chasing (Overall and Dunham, 2002), an exaggerated variation of the herding behaviors that they were bred for. There is little evidence to date of breed differences in canine cognitive dysfunction, except that smaller breeds, which live longer, seem to be more likely to show this syndrome (Kerwin *et al.*, 2017).

Conclusion

Do dog breeds differ? Yes, but it's often not in the ways we think they do. They differ in their physical appearance, temperaments, movement, motivation, predisposition for diseases, genetics, and even in their neuroanatomy. Differences among dog breeds correspond to what they have been bred for, but not all of these differences are reflected in their behavior; they also differ in their physical characteristics. The behavioral and physical diversity among dog breeds and lines of breeds is striking, but this diversity is also due to over-breeding and inbreeding. This is evidenced by rates of disease and disorders that are specific to certain breeds and lines. While we continue to breed dogs for very specific purposes and continue to see increasing differences, we must not forget the genetics behind these breeding programs.

References

Adamkiewicz, E., Jezierski, T., Walczak, M., Górecka-Bruzda, A., Sobczyńska, M. *et al.* (2013) Traits of drug and explosives detection in dogs of two breeds as evaluated by their handlers and trainers. *Animal Science Papers and Reports* 3, 205–217.

Ahzu Samoyeds Australia (2021) History of the Samoyeds. Available at: www.ahzusamoyeds.com/history-of-the-samoyed.html (accessed 2 November 2023).

AKC (2019) The 7 AKC dog breed groups explained. Available at: www.akc.org/expert-advice/lifestyle/7-akc-dog-breed-groups-explained (accessed 2 November 2023).

AKC (2022a) Norwegian Elkhound. Available at: www.akc.org/dog-breeds/norwegian-elkhound (accessed 2 November 2023).

AKC (2022b) Herding Group. Available at: www.akc.org/dog-breeds/herding/ (accessed 2 November 2023).

AKC (2022c) Working Group. Available at: www.akc.org/dog-breeds/working/ (accessed 2 November 2023).

AKC (2022d) Sporting Group. Available at: www.akc.org/dog-breeds/sporting/ (accessed 2 November 2023).

AKC (2022e) Hound Group. Available at: www.akc.org/dog-breeds/hound/ (accessed 2 November 2023).

AKC (2022f) Terrier Group. Available at: www.akc.org/dog-breeds/terrier/ (accessed 2 November 2023).

AKC (2022g) Non-sporting Group. Available at: www.akc.org/dog-breeds/non-sporting/ (accessed 9 February 2024).

AKC (2022h) Toy Group. Available at: www.akc.org/dog-breeds/toy/ (accessed 2 November 2023).

AKC (2023) Glossary. Available at: www.akc.org/about/glossary (accessed 1 November 2023).

Alaska Mushing School (2023) The history of dog sledding. Available at: www.alaskamushingschool.com/learn/history-of-dog-sledding/ (accessed 2 November 2023).

Andelt, W.F. (1999) Relative effectiveness of guarding-dog breeds to deter predation on domestic sheep in Colorado. *Wildlife Society Bulletin* 27, 706–714.

Angelici, F.M. and Rossi, L. (2018) A new subspecies of gray wolf (Carnivora, Canidae), recently extinct, from

Sicily, Italy. *Bollettino del Museo Civico di Storia Naturale di Verona* 42, 3–15. 10.1101/320655

Anonymous (1838) Rat-catchers and rat-catching. *Penny Magazine of the Society for the Diffusion of Useful Knowledge* 402(July 7), 256–259.

Asp, H.E., Fikse, W.F., Nilsson, K. and Strandberg, E. (2015) Breed differences in everyday behaviour of dogs. *Applied Animal Behaviour Science* 169, 69–77. 10.1016/j.applanim.2015.04.010

Balme, J., O'Connor, S. and Fallon, S. (2018) New dates on dingo bones from Madura cave provide oldest firm evidence for arrival of the species in Australia. *Scientific Reports* 8(1), 9933. 10.1038/s41598-018-28324-x

Bamford, N.J., Potter, S.J., Harris, P.A. and Bailey, S.R. (2014) Breed differences in insulin sensitivity and insulinemic responses to oral glucose in horses and ponies of moderate body condition score. *Domestic Animal Endocrinology* 47, 101–107. 10.1016/j.domaniend.2013.11.001

Bannasch, D., Famula, T., Donner, J., Anderson, H., Honkanen, L. *et al.* (2021) The effect of inbreeding, body size and morphology on health in dog breeds. *Canine Medicine and Genetics* 8(1), 12. 10.1186/s40575-021-00111-4

Beebe, S.C., Howell, T.J. and Bennett, P.C. (2016) Using scent detection dogs in conservation settings: a review of scientific literature regarding their selection. *Frontiers in Veterinary Science* 3, 1–96. 10.3389/fvets.2016.00096

Bennett, E.T. (1830) *The Gardens and Menagerie of the Zoological Society Delineated. Sanction of the Council, under the Superintendence of the Secretary and Vice-Secretary of the Society*. Whittingham, Chiswick, UK

Beuchat, C. (2014) The costs and benefits of inbreeding. The Institute of Canine Biology. Available at: www.instituteofcaninebiology.org/blog/the-costs-and-benefits-of-inbreeding (accessed 2 November 2023).

Beuchat, C. (2015) COI FAQS: understanding the coefficient of inbreeding. The Institute of Canine Biology. Available at: www.instituteofcaninebiology.org/blog/coi-faqs-understanding-the-coefficient-of-inbreeding (accessed 2 November 2023).

Bigi, D., Marelli, S.P., Liotta, L., Frattini, S., Talenti, A. *et al.* (2018) Investigating the population structure and genetic differentiation of livestock guard dog breeds. *Animal* 12(10), 2009–2016. 10.1017/S1751731117003573

Bower, B. (2018) Dogs lived and died with humans 10,000 years ago in the Americas. *Science News* 193(8), 22.

Brady, K., Cracknell, N., Zulch, H. and Mills, D.S. (2018) Factors associated with long-term success in working police dogs. *Applied Animal Behaviour Science* 207, 67–72. 10.1016/j.applanim.2018.07.003

Brown, S.K., Darwent, C.M., Wictum, E.J. and Sacks, B.N. (2015) Using multiple markers to elucidate the ancient, historical and modern relationships among North American Arctic dog breeds. *Heredity* 115(6), 488–495. 10.1038/hdy.2015.49

Cairns, K.M. and Wilton, A.N. (2016) New insights on the history of canids in Oceania based on mitochondrial and nuclear data. *Genetica* 144(5), 553–565. 10.1007/s10709-016-9924-z

Calboli, F.C.F., Sampson, J., Fretwell, N. and Balding, D.J. (2008) Population structure and inbreeding from pedigree analysis of purebred dogs. *Genetics* 179(1), 593–601. 10.1534/genetics.107.084954

Camarós, E., Münzel, S.C., Cueto, M., Rivals, F. and Conard, N.J. (2016) The evolution of Paleolithic hominin–carnivore interaction written in teeth: stories from the Swabian Jura (Germany). *Journal of Archaeological Science* 6, 798–809. 10.1016/j.jasrep.2015.11.010

Carey, B. (2006) Humans might have wiped out wild horses. *Live Science*. Available at: www.livescience.com/717-humans-wiped-wild-horses.html (accessed 1 November 2023).

Carter, H. and Mace, A.C. (1963) *The Tomb of Tut-Ankh-Amen*. Cooper Square Publishers, New York.

Case, L.P. (2005) *The Dog: Its Behaviour, Nutrition, and Health*, 2nd edn. Wiley-Blackwell, Ames, IA, p. 176.

CBKC (2020) Confederação Brasileira de Cinofilia. Fédération Cynologique Internationale. Available at: https://cbkc.org/application/views/docs/padroes/padrao-raca_224.pdf (accessed 2 November 2023).

Centers for Disease Control and Prevention (2018) The 1918 flu pandemic: why it matters 100 years later. Available at: https://blogs.cdc.gov/publichealthmatters/2018/05/1918-flu/ (accessed 2 November 2023).

Chase, K., Sargan, D., Miller, K., Ostrander, E.A. and Lark, K.G. (2006) Understanding the genetics of autoimmune disease: two loci that regulate late onset Addison's disease in Portuguese water dogs. *International Journal of Immunogenetics* 33(3), 179–184. 10.1111/j.1744-313X.2006.00593.x

Clarke, T., Mills, D. and Cooper, J. (2016) "Type" as central to perceptions of breed differences in behavior of domestic dog. *Society & Animals* 24(5), 467–485. 10.1163/15685306-12341422

Clutton-Brock, J. (1989) A dog and a donkey excavated at Tell Brak. *Iraq* 51, 217–224. 10.2307/4200305

Clutton-Brock, J. (1995) Origins of the dog: domestication and early history. In: Serpell, J. (ed.) *The Domestic Dog, Its Evolution, Behaviour and Interactions with People*. Cambridge University Press, Cambridge, pp. 16–20.

Coppinger, R. and Schneider, R. (1995) The evolution of working dogs. In: Serpell, J. (ed.) *The Domestic Dog, Its Evolution, Behaviour and Interactions with People*. Cambridge University Press, Cambridge, pp. 21–47.

Crook, A., Dawson, S., Côté, E., MacDonald, S. and Berry, J. (2011) Canine inherited disorders database (CIDD). Available at: https://cidd.discoveryspace.ca (accessed 2 November 2023).

da Costa, R.C., Parent, J.M., Partlow, G., Dobson, H., Holmberg, D.L. *et al.* (2006) Morphologic and morphometric magnetic resonance imaging features of Doberman Pinschers with and without clinical signs of cervical spondylomyelopathy. *American Journal of Veterinary Research* 67(9), 1601–1612. 10.2460/ajvr.67.9.1601

da Silva Coelho, F.A., Gill, S., Tomlin, C.M., Heaton, T.H. and Lindqvist, C. (2021) An early dog from Southeast Alaska supports a coastal route for the first dog migration into the Americas. *Proceedings of the Royal Society B* 288, 20203103.

Dennis-Bryan, K., Baggaley, A. and John, K. (2013) *The Dog Encyclopedia*. DK Publishing, New York, p. 13.

Dodman, N.H., Miczek, K.A., Knowles, K., Thalhammer, J.G. and Shuster, L. (1992) Phenobarbital-responsive episodic dyscontrol (rage) in dogs. *Journal of the American Veterinary Medical Association* 201(10), 1580–1583.

Duffy, D.L., Hsu, Y. and Serpell, J.A. (2008) Breed differences in canine aggression. *Applied Animal Behaviour Science* 114(3–4), 441–460. 10.1016/j.applanim.2008.04.006

Edwards, S.T. (1800) *Cynographia Britannica*. Whittingham, London.

Finger, S. (2023) Yellow fever. *The Encyclopedia of Greater Philadelphia*. Available at: https://philadelphiaencyclopedia.org/archive/yellow-fever (accessed 2 November 2023).

Fiszdon, K. and Czarkowska, K. (2008) Social behaviours in Siberian Huskies. *Annals of Warsaw University of Life Sciences – SGGW. Animal Science* 45, 19–28.

Foyer, P., Svedberg, A.-M., Nilsson, E., Wilsson, E., Faresjo, A. *et al.* (2016) Behavior and cortisol responses of dogs evaluated in a standardized temperament test for military working dogs. *Journal of Veterinary Behavior* 11, 7–12. 10.1016/j.jveb.2015.09.006

Galibert, F., Azzouzi, N., Quignon, P. and Chaudieu, G. (2016) The genetics of canine olfaction. *Journal of Veterinary Behavior* 16, 86–93. 10.1016/j.jveb.2016.06.012

Ganguly, P. (2020) Scientists use genomics to discover ancient dog species that may teach us about human vocalization. National Human Genome Research Institute. Available at: www.genome.gov/news/news-release/Scientists-use-genomics-to-discover-ancient-dog-species-that-may-teach-us-about-human-vocalization (accessed 2 November 2023).

Germonpré, M., Sablin, M.V., Stevens, R.E., Hedges, R.E.M., Hofreiter, M. *et al.* (2009) Fossil dogs and wolves from Palaeolithic sites in Belgium, the Ukraine and Russia: osteometry, ancient DNA and stable isotopes. *Journal of Archaeological Science* 36(2), 473–490. 10.1016/j.jas.2008.09.033

Germonpré, M., Lázničková-Galetová, M., Losey, R.J., Räikkönen, J. and Sablin, M.V. (2015) Large canids at the Gravettian Předmostí site, the Czech Republic: the mandible. *Quaternary International* 359–360, 261–279. 10.1016/j.quaint.2014.07.012

Germonpré, M., Fedorov, S., Danilov, P., Galeta, P., Jimenez, E.-L. *et al.* (2017) Palaeolithic and prehistoric dogs and Pleistocene wolves from Yakutia: identification of isolated skulls. *Journal of Archaeological Science* 78, 1–19. 10.1016/j.jas.2016.11.008

Glover, A. (1942) *Extinct and Vanishing Mammals of the Western Hemisphere, with the Marine Species of All the Oceans*. American Committee for International Wild Life Protection. The Intelligencer Printing Company, Lancaster, PA, pp. 205–206.

Goodwin, D., Levine, M. and McGreevy, P.D. (2008) Preliminary investigation of morphological differences between ten breeds of horses suggests selection for paedomorphosis. *Journal of Applied Animal Welfare Science* 11(3), 204–212. 10.1080/10888700802100918

Gorman, J. (2020) Dog breeding in the Neolithic age: fossils and modern DNA show the ancient roots of Arctic sled dogs. *New York Times*. Available at: www.nytimes.com/2020/06/25/science/arctic-sled-dogs-genetics.html (accessed 1 November 2023).

Gosling, S. and Graham, L. (2009) Temperament and personality in working dogs. In: Helton, W.S. (ed.) *Canine Ergonomics: The Science of Working Dogs*. CRC Press, Boca Raton, FL, pp. 63–81. 10.1201/9781420079920

Guagnin, M., Perri, A.R. and Petraglia, M.D. (2018) Pre-Neolithic evidence for dog-assisted hunting strategies in Arabia. *Journal of Anthropological Archaeology* 49, 225–236. 10.1016/j.jaa.2017.10.003

Hall, N.J., Smith, D.W. and Wynne, C.D.L. (2013) Training domestic dogs (*Canis lupus familiaris*) on a novel discrete trials odor-detection task. *Learning and Motivation* 44(4), 218–228. 10.1016/j.lmot.2013.02.004

Hall, N.J., Glenn, K., Smith, D.W. and Wynne, C.D.L. (2015) Performance of Pugs, German Shepherds, and Grey-hounds (*Canis lupus familiaris*) on an odor-discrimination task. *Journal of Comparative Psychology* 129(3), 237–246. 10.1037/a0039271

Handford, J.M. (1998) Dog sledging in the eighteenth century: North America and Siberia. *Polar Record* 34(190), 237–248. 10.1017/S0032247400025705

Handy, W.F., Harrington, M. and Pittman, D.J. (1961) The K-9 corps: the use of dogs in police work. *The Journal of Criminal Law, Criminology, and Police Science* 52(3), 328. 10.2307/1141120

Hansen, I. and Bakken, M. (1999) Livestock-guarding dogs in Norway: Part I. Interactions. *Journal of Range Management* 52(2), 6. 10.2307/4003485

Harrison, R.G., Connolly, R.C., Ahmed, S., Abdalla, A.B. and El Ghawaby, M. (1979) A mummified foetus from the tomb of Tutankhamun. *Antiquity* 53, 19–21.

Harvard Library (2023) Contagion: historical views of diseases and epidemics. Available at: https://library.harvard.edu/collections/contagion-historical-views-diseases-and-epidemics (accessed 2 November 2023).

Harvey, I. (2016) Autopsy carried out on the world's oldest dog mummified by ice. *The Vintage News*. Available at: www.thevintagenews.com/2016/08/07/sibling-tumat-dog-discovered (accessed 1 November 2023).

Hastings, D. (2018) The German shepherd-husky mix attacked a German machine gun nest in Sicily. *Inside Edition*. Available at: www.insideedition.com/chips-us-army-hero-dog-served-world-war-ii-gets-posthumous-medal-39811 (accessed 2 November 2023).

Hawass, Z. and Saleem, S.N. (2011) Mummified daughters of King Tutankhamun: archeologic and CT studies. *AJR. American Journal of Roentgenology* 197(5), W829–W836. 10.2214/AJR.11.6837

Hawass, Z., Gad, Y.Z., Ismail, Z., Khairat, R. and Fathalla, D. (2010) Ancestry and pathology in King Tutankhamun's family. *Journal of the American Medical Association* 303(7), 638–647. 10.1001/jama.2010.121

Hecht, E.E., Smaers, J.B., Dunn, W.J., Kent, M., Preuss, T.M. *et al.* (2019) Significant neuroanatomical variation among domestic dog breeds. *The Journal of Neuroscience* 39, 7748–7758. 10.1523/JNEUROSCI.0303-19.2019

Hepper, P.G. (1988) The discrimination of human odour by the dog. *Perception* 17(4), 549–554. 10.1068/p170549

Huson, H.J., Parker, H.G., Runstadler, J. and Ostrander, E.A. (2010) A genetic dissection of breed composition and performance enhancement in the Alaskan sled dog. *BMC Genetics* 11, 71. 10.1186/1471-2156-11-71

Irion, D.N., Schaffer, A.L., Famula, T.R., Eggleston, M.L. and Hughes, S.S. (2003) Analysis of genetic variation in 28 dog breed populations with 100 microsatellite markers. *Journal of Heredity* 94(1), 81–87. 10.1093/jhered/esg004

Janssens, L., Giemsch, L., Schmitz, R., Street, M., Van Dongen, S. *et al.* (2018) A new look at an old dog: Bonn-Oberkassel reconsidered. *Journal of Archaeological Science* 92, 126–138. 10.1016/j.jas.2018.01.004

Jansson, M. and Laikre, L. (2018) Pedigree data indicate rapid inbreeding and loss of genetic diversity within populations of native, traditional dog breeds of conservation concern. *PLoS One* 3(9), e0202849. 10.1371/journal.pone.0202849

Jezierski, T., Adamkiewicz, E., Walczak, M., Sobczyńska, M., Górecka-Bruzda, A. *et al.* (2014) Efficacy of drug detection by fully-trained police dogs varies by breed, training level, type of drug and search environment. *Forensic Science International* 237, 112–118. 10.1016/j.forsciint.2014.01.013

Johnstone, G. (2021) Get to know the spectacular spitz breeds. American Kennel Club. Available at: www.akc.org/expert-advice/lifestyle/spitz-dog-breeds (accessed 2 November 2023).

Kemp, T.J., Bachus, K.N., Nairn, J.A. and Carrier, D.R. (2005) Functional trade-offs in the limb bones of dogs selected for running versus fighting. *The Journal of Experimental Biology* 208, 3475–3482. 10.1242/jeb.01814

Kercsmar, J.A. (2016) Wolves at heart: how dog evolution shaped whites' perceptions of Indians in North America. *Environmental History* 21(3), 516–540. 10.1093/envhis/emw007

Kerwin, S.C., Levine, J.M., Budke, C.M., Griffin, J.F. and Boudreau, C.E. (2017) Putative cerebral microbleeds in dogs undergoing magnetic resonance imaging of the head: a retrospective study of demographics, clinical associations, and relationship to case outcome. *Journal of Veterinary Internal Medicine* 31(4), 1140–1148. 10.1111/jvim.14730

Kettunen, A., Daverdin, M., Helfjord, T. and Berg, P. (2017) Cross-breeding is inevitable to conserve the highly inbred population of puffin hunter: the Norwegian Lundehund. *PLoS One* 12(1), e0170039. 10.1371/journal.pone.0170039

Kinka, D. and Young, J.K. (2019) Evaluating domestic sheep survival with different breeds of livestock guardian dogs. *Rangeland Ecology & Management* 72(6), 923–932. 10.1016/j.rama.2019.07.002

Knight, J. (1997) On the extinction of the Japanese wolf. *Asian Folklore Studies* 56(1), 129–159. 10.2307/1178791

Lallensack, R. (2017) Ancient genomes heat up dog domestication debate. *Nature*. 10.1038/nature.2017.22320

Larson, G., Karlsson, E.K., Perri, A., Webster, M.T., Ho, S.Y.W. *et al.* (2012) Rethinking dog domestication by integrating genetics, archeology, and biogeography. *Proceedings of the National Academy of Sciences of the United States of America* 109(23), 8878–8883. 10.1073/pnas.1203005109

Leathlobhair, M.N., Perri, A.R., Irving-Pease, E.K., Witt, K.E., Linderholm, A. *et al.* (2018) The evolutionary history of dogs in the Americas. *Science* 361(6397), 81–85. 10.1126/science.aao4776

Ledoux, L. and Boudadi-Maligne, M. (2015) The contribution of geometric morphometric analysis to prehistoric ichnology: the example of large canid tracks and their implication for the debate concerning wolf domestication. *Journal of Archaeological Science* 61, 25–35. 10.1016/j.jas.2015.04.020

Leonard, J.A., Wayne, R.K., Wheeler, J., Valadez, R., Guilleen, S. *et al.* (2002) Ancient DNA evidence for old world origin of new world dogs. *Science* 298(5598), 1613–1616. 10.1126/science.1076980

Leonard, J.A., Vilà, C., Fox-Dobbs, K., Koch, P.L., Wayne, R.K. *et al.* (2007) Megafaunal extinctions and the disappearance of a specialized wolf ecomorph. *Current Biology* 17(13), 1146–1150. 10.1016/j.cub.2007.05.072

Lewy, G. (1978) *America in Vietnam*. Oxford University Press, New York, pp. 442–453.

Lindberg, S., Strandberg, E. and Swenson, L. (2004) Genetic analysis of hunting behaviour in Swedish flat-coated retrievers. *Applied Animal Behaviour Science* 88(3–4), 289–298. 10.1016/j.applanim.2004.03.007

Lloyd, A.S., Martin, J.E., Bornett-Gauci, H.L.I. and Wilkinson, R.G. (2008) Horse personality: variation between breeds. *Applied Animal Behaviour Science* 112(3–4), 369–383. 10.1016/j.applanim.2007.08.010

Lorenz, K. (1971) *Studies in Animal and Human Behavior*. Harvard University Press, Cambridge, MA.

Mancini, M. (2018) Arrival of European breeds wiped out native American dogs. *HowStuffWorks.com*. Available at: https://animals.howstuffworks.com/pets/european-breeds-wiped-out-native-american-dogs.htm (accessed 2 November 2023).

Mark, J.J. (2017) Dogs in ancient Egypt. *World History Encyclopedia*. Available at: www.ancient.eu/article/1031/dogs-in-ancient-egypt (accessed 2 November 2023).

Mark, J.J. (2018) Dogs & their collars in ancient Mesopotamia. *World History Encyclopedia*. Available at: www.ancient.eu/article/1175/dogs--their-collars-in-ancient-mesopotamia (accessed 2 November 2023).

McGreevy, P.D., Georgevsky, D., Carrasco, J., Valenzuela, M., Duffy, D.L. *et al.* (2013) Dog behavior co-varies with height, bodyweight and skull shape. *PLoS One* 8(12), e80529. 10.1371/journal.pone.0080529

Mehrkam, L.R. and Wynne, C.D.L. (2014) Behavioral differences among breeds of domestic dogs (*Canis lupus familiaris*): current status of the science. *Applied Animal Behaviour Science* 155, 12–27. 10.1016/j.applanim.2014.03.005

Meurs, K.M., Fox, P.R., Norgard, M., Spier, A.W., Lamb, A. *et al.* (2007) A prospective genetic evaluation of familial dilated cardiomyopathy in the Doberman pinscher. *Journal of Veterinary Internal Medicine* 21(5), 1016–1020. 10.1892/0891-6640(2007)21[1016:apgeof]2.0.co;2

Miller, K.A., Touroo, R., Spain, C.V., Jones, K., Reid, P. *et al.* (2016) Relationship between scarring and dog aggression in Pit Bull-type dogs involved in organized dog fighting. *Animals* 6(11), 72. 10.3390/ani6110072

Moon-Fanelli, A.A. and Dodman, N.H. (1998) Description and development of compulsive tail chasing in terriers and response to clomipramine treatment. *Journal of the American Veterinary Medical Association* 212(8), 1252–1257.

Muñiz, F., Cáceres, L.M., Rodríguez-Vidal, J., Neto de Carvalho, C., Belo, J. *et al.* (2019) Following the last Neanderthals: mammal tracks in late Pleistocene coastal dunes of Gibraltar (S Iberian Peninsula). *Quaternary Science Reviews* 217, 297–309. 10.1016/j.quascirev.2019.01.013

National Library of Medicine (2022) What does it mean to have Neanderthal or Denisovan DNA? Available at: https://medlineplus.gov/genetics/understanding/dtc-genetictesting/neanderthaldna (accessed 1 November 2023).

National Shiba Club of America (2023) Breed history. Available at: https://shibas.org/breedHistory.html (accessed 1 November 2023).

Nature (2021) An ancient Alaskan dog's DNA hints at an epic shared journey. *Nature* 591(7848), 10. 10.1038/d41586-021-00477-2

Ollivier, M., Tresset, A., Bastian, F., Lagoutte, L., Axelsson, E. *et al.* (2016) *Amy2B* copy number variation reveals starch diet adaptations in ancient European dogs. *Royal Society Open Science* 3(11), 160449. 10.1098/rsos.160449

Ostrander, E.A., Wayne, R.K., Freedman, A.H. and Davis, B.W. (2017) Demographic history, selection and functional diversity of the canine genome. *Nature Reviews Genetics* 18(12), 705–720. 10.1038/nrg.2017.67

Overall, K.L. and Dunham, A.E. (2002) Clinical features and outcome in dogs and cats with obsessive-compulsive disorder: 126 cases (1989–2000). *Journal of the American Veterinary Medical Association* 221(10), 1445–1452. 10.2460/javma.2002.221.1445

Ovodov, N.D., Crockford, S.J., Kuzmin, Y.V., Higham, T.F.G., Hodgins, G.W.L. *et al.* (2011) A 33,000-year-old incipient dog from the Altai mountains of Siberia: evidence of the earliest domestication disrupted by the last glacial maximum. *PLoS One* 6(7), e22821. 10.1371/journal.pone.0022821

Packer, C., Pusey, A.E., Rowley, H., Gilbert, D.A., Martenson, J. *et al.* (1991) Case study of a population bottleneck: lions of the Ngorongoro crater. *Conservation Biology* 5(2), 219–230. 10.1111/j.1523-1739.1991.tb00127.x

Parker, H.G., Kim, L.V., Sutter, N.B., Carlson, S., Lorentzen, T.D. *et al.* (2004) Genetic structure of the purebred domestic dog. *Science* 304(5674), 1160–1164. 10.1126/science.1097406

Patterson, E.E., Armstrong, P.J., O'Brien, D.P., Roberts, M.C., Johnson, G.S. *et al.* (2005) Clinical description and mode of inheritance of idiopathic epilepsy in English Springer Spaniels. *Journal of the American Veterinary Medical Association* 226(1), 54–58. 10.2460/javma.2005.226.54

PBS (1952) Nixon's Checkers speech. Available at: www.pbs.org/wgbh/americanexperience/features/eisenhower-checkers (accessed 2 November 2023).

Pearson, C. (2016) Between instinct and intelligence: harnessing police dog agency in early twentieth-century Paris. *Comparative Studies in Society and History* 58(2), 463–490. 10.1017/S0010417516000141

Pedersen, N., Liu, H., Theilen, G. and Sacks, B. (2013) The effects of dog breed development on genetic diversity and the relative influences of performance and conformation breeding. *Journal of Animal Breeding and Genetics* 30(3), 236–248. 10.1111/jbg.12017

Pemberton, N. (2014) The rat-catcher's prank: interspecies cunningness and scavenging in Henry Mayhew's London. *Journal of Victorian Culture* 19(4), 520–535. 10.1080/13555502.2014.967548

Pero, J. (2019) The breeder of the world's first Labradoodle says over-breeding threatens to turn into a

'Frankenstein's monster'. *MailOnline*. Available at: www.dailymail.co.uk/sciencetech/article-7500515/Original-labradoodle-breeder-says-dog-Frankenstein-monster-regrets-creating-popular-hybrid.html (accessed 2 November 2023).

Pfahler, S. and Distl, O. (2015) Effective population size, extended linkage disequilibrium and signatures of selection in the rare dog breed Lundehund. *PLoS One* 10(4), e0122680. 10.1371/journal.pone.0122680

Philipkoski, K. (2001) Black death's gene code cracked. *Wired*. Available at: www.wired.com/2001/10/black-deaths-gene-code-cracked (accessed 2 November 2023).

Pitulko, V.V. and Kasparov, A.K. (2017) Archaeological dogs from the early Holocene Zhokhov site in the Eastern Siberian Arctic. *Journal of Archaeological Science* 13, 491–515. 10.1016/j.jasrep.2017.04.003

Polgar, Z., Kinnunen, M., Ujvary, D., Miklosi, Á. and Gacsi, M. (2016) A test of canine olfactory capacity: comparing various dog breeds and wolves in a natural detection task. *PLoS One* 11(5), e0154087. 10.1371/journal.pone.0154087

QingShen, L., YanHua, S. and DingYuan, F. (2010) Analysis on mtDNA D-loop HVR-I polymorphism of the Chinese Sharpei. *Journal of South China Agricultural University* 31, 90–93.

Quignon, P., Rimbault, M., Robin, S. and Galibert, F. (2012) Genetics of canine olfaction and receptor diversity. *Mammalian Genome* 2(3), 132–143. 10.1007/s00335-011-9371-1

Ripley, K. (2016) 7 things you didn't know about the Basenji. American Kennel Club. Available at: www.akc.org/expert-advice/lifestyle/7-things-you-didnt-know-about-the-basenji (accessed 1 November 2023).

Roark, J. (2019) Dogsledding: art in chaos. *Post Register*. Available at: www.postregister.com/news/local/dog-sledding-art-in-chaos/article_82723967-b2ac-5e15-b395-905db1a1d7c3.html (accessed 2 November 2023).

Salonen, M., Vapalahti, K., Tiira, K., Mäki-Tanila, A. and Lohi, H. (2019) Breed differences of heritable behaviour traits in cats. *Scientific Reports* 9(1), 7949. 10.1038/s41598-019-44324-x

San Diego Wildlife Alliance (2023) Singing dog: *Canis hallstromi*. Available at: https://animals.sandiegozoo.org/animals/singing-dog (accessed 2 November 2023).

Sargan, D.R. (2004) IDID: Inherited Diseases in Dogs: web-based information for canine inherited disease genetics. *Mammalian Genome* 15(6), 503–506. 10.1007/s00335-004-3047-z

Scherf, B.D. (2000) *World Watch List for Domestic Animal Diversity*, 3rd edn. Food and Agriculture Organization of the United Nations, Rome.

Schmutz, S.M. and Schmutz, J.K. (1998) Heritability estimates of behaviors associated with hunting in dogs. *The Journal of Heredity* 89(3), 233–237. 10.1093/jhered/89.3.233

Science Daily (2004) Dog genome assembled: canine genome now available to research community worldwide. *Science Daily*. Available at: www.sciencedaily.com/releases/2004/07/040715080503.htm (accessed 2 November 2023).

Sinding, M.-H.S., Gopalakrishnan, S., Ramos-Madrigal, J., de Manuel, M., Pitulko, V.V. *et al.* (2020) Arctic-adapted dogs emerged at the Pleistocene–Holocene transition. *Science* 368(6498), 1495–1499. 10.1126/science.aaz8599

Stronen, A.V., Salmela, E., Baldursdóttir, B.K., Berg, P., Espelien, I.S. *et al.* (2017) Genetic rescue of an endangered domestic animal through outcrossing with closely related breeds: a case study of the Norwegian Lundehund. *PLoS One* 12(6), e0177429. 10.1371/journal.pone.0177429

Surbakti, S., Parker, H.G., McIntyre, J.K., Maury, H.K., Cairns, K.M. *et al.* (2020) New Guinea highland wild dogs are the original New Guinea singing dogs. *Proceedings of the National Academy of Sciences of the United States of America* 117(39), 24369–24376. 10.1073/pnas.2007242117

Sutter, N.B., Mosher, D.S., Gray, M.M. and Ostrander, E.A. (2008) Morphometrics within dog breeds are highly reproducible and dispute Rensch's rule. *Mammalian Genome* 19(10–12), 713–723. 10.1007/s00335-008-9153-6

Tamimi, N.S.M. and Wali, A.A. (2019) Health problems of Iraqi police dogs referred to Baghdad Veterinary Hospital during 2015-2017. *Veterinary World* 12(7), 1046–1051.

The Cat Fanciers Association (2023) CFA breeds. Available at: https://cfa.org/breeds/ (accessed 11 December 2023).

The Conversation (2013) History mystery solved: the origins of the Falkland Islands wolf. *The Conversation*. Available at: https://theconversation.com/history-mystery-solved-the-origins-of-the-falkland-islands-wolf-12637 (accessed 2 November 2023).

The International Cat Association (TICA) (2018) Browse all breeds. Available at: https://tica.org/breeds/browse-all-breeds (accessed 11 December 2023).

Thorn, P., Howell, T.J., Brown, C. and Bennett, P.C. (2015) The canine cuteness effect: owner-perceived cuteness as a predictor of human–dog relationship quality. *Anthrozoös* 28(4), 569–585. 10.1080/08927936.2015.1069992

Tiira, K., Tikkanen, A. and Vainio, O. (2020) Inhibitory control – important trait for explosive detection performance in police dogs? *Applied Animal Behaviour Science* 224, 104942.

Turcsán, B., Kubinyi, E. and Miklósi, A. (2011) Trainability and boldness traits differ between dog breed clusters based on conventional breed categories and genetic relatedness. *Applied Animal Behaviour Science* 132(1–2), 61–70. 10.1016/j.applanim.2011.03.006

Valverde, A., Arnau, S., García-Molina, A., Bompart, D., Campos, M. *et al.* (2019) Dog sperm swimming parameters analysed by computer-assisted semen analysis of motility reveal major breed differences. *Reproduction in Domestic Animals* 54(5), 795–803. 10.1111/rda.13420

vonHoldt, B.M., Pollinger, J.P., Lohmueller, K.E., Han, E., Parker, H.G. *et al.* (2010) Genome-wide SNP and haplotype analyses reveal a rich history underlying dog domestication. *Nature* 464(7290), 898–902. 10.1038/nature08837

Wade, C.M. (2011) Inbreeding and genetic diversity in dogs: results from DNA analysis. *The Veterinary Journal* 189(2), 183–188. 10.1016/j.tvjl.2011.06.017

Weckworth, B.V., Talbot, S., Sage, G.K., Person, D.K. and Cook, J. (2005) A signal for independent coastal and continental histories among North American wolves. *Molecular Ecology* 14(4), 917–931. 10.1111/j.1365-294X.2005.02461.x

Whitaker, J. and Whitelaw, I. (2007) *The Horse: A Miscellany of Equine Knowledge*. St Martin's Press, New York, p. 60.

Wikipedia (2020) Dog fighting in the United States. *Wikipedia*. Available at: https://en.wikipedia.org/wiki/Dog_fighting_in_the_United_States (accessed 5 August 2020).

Wilsson, E. and Sundgren, P.-E. (1997) The use of a behaviour test for the selection of dogs for service and breeding, I: method of testing and evaluating test results in the adult dog, demands on different kinds of service dogs, sex and breed differences. *Applied Animal Behaviour Science* 53(4), 279–295. 10.1016/S0168-1591(96)01174-4

Yang, H., Wang, G., Wang, M., Ma, Y., Yin, T. *et al.* (2017) The origin of chow chows in the light of the East Asian breeds. *BMC Genomics* 18(1), 174. 10.1186/s12864-017-3525-9

Zhang, S.J., Wang, G.D., Ma, P., Zhang, L.L., Yin, T.T. *et al.* (2020) Genomic regions under selection in the feralization of the dingoes. *Nature Communications* 11(1), 671. 10.1038/s41467-020-14515-6

6 Behavior Came Along for the Ride: Sometimes, We Breed for X, But End Up Getting Y, and Z, and …

Abstract

Chapter 6 explains how behavior was occasionally altered accidentally via artificial selection. While we might have been breeding Pugs for a phenotypic characteristic, a truncated muzzle, a behavioral characteristic, loud, laborious breathing, arose in the offspring, as well. For the breeds that have been studied, the scientific literature is examined to reveal behavioral effects, how long the breed has been isolated, when each breed originated, and how extreme the artificial selection was. Case studies demonstrate where a bad environment, such as puppy mills, abuse or abandonment, and bad genetics can lead to a predisposition for aggression.

Hands, Hagrid, and Heredity

Janna Ellison was adopted when she was only a few days old. She'd had the habit of biting her nails for as long as she could remember, but no one else in her adoptive family was a nail biter. Her adoptive mother tried to deter the behavior. While Janna would decrease the frequency of her nail biting from time to time (particularly if people were paying attention!) she never *really* stopped biting her nails. When she was stressed, or hungry, or bored, she bit her nails. It was more than a habit; it was self-soothing. It just plain "felt good" to her. But no one else in her family bit their nails when they were stressed, or hungry, or tired, or bored. Just Janna. For her entire life into middle adulthood, she only knew her adoptive parents as her family. Then in her early thirties, she finally met her birth mother. In those tense first moments, her fingers instinctively went to her mouth. As a lifelong nail biter, she habitually checked other people's hands, too. She glanced at her birth mother's hands and saw that she bit her nails, too. When she met all four of her maternal half-siblings, she noted that they also bit their nails. Five out of five close blood relatives bit their fingernails, a habit referred to by physicians and psychologists (and them alone, we'd dare say) as onychophagia. Ony … er, nail biting falls under the umbrella term of body-focused repetitive behavior (BFRB) (Teng *et al.* 2002), a group of body-grooming obsessive-compulsive disorder (OCD)-type behaviors like hair pulling (formally known as trichotillomania) and nail biting. BFRB ranges from mild to damaging (e.g. biting your nails so short that you make your fingers bleed). Nail biting has a prevalence of up to 45% for teenagers (ah, the joys of juvenescence) (Leung and Robson, 1990), but fortunately for most of us, nail biting is a fleeting trait; only 20–30% of the adult population bites their nails (Halteh *et al.*, 2017). Five (six, counting Janna), isn't a big N, but it seems far from coincidental, and the data appears to back that up. So, while Janna had been raised in a home where no one else bit their nails, all of the maternal relatives that she met did. What does this mean, exactly? Is there a "gene" for nail biting? Well, children whose parents bite their nails—even if they cease their habit before the child is born—are more likely to also be nail biters than the children of parents who have never bitten their nails. There's definitely a genetic component to this—and many other—behaviors.

While humans don't procreate to try to produce particular traits, we are often *attracted* to certain ones, like a certain height, hair or eye color, or face shape. People might say that they have a "type," but most wouldn't consider that along with these traits, other traits, including behaviors, might be associated with them. And we can see patterns of behavior among biological families, even when those family members have been separated. This

DOI: 10.1079/9781800624566.0006

happens in other species as well. And when it comes to domesticated species where we *are* actually breeding for a particular trait (e.g. for looks, for herding, or for increased milk production), we can get other traits, both morphological and behavioral, that happen to be genetically linked to the trait that was being bred for. In other words, sometimes we'll breed for trait X, but also end up with Y, and Z, and so on. So how does this all work?

To start, it's important to remember that behavior is genetic. By this, we mean that a behavior can be innate, rather than learned, to a species, breed within that species, or an individual within that species. In order to breed for a behavior, a behavior must be genetic. Examples within the animal kingdom abound. Many species, including those among the primates and felines, have an inborn trait to avoid long, skinny things that may or may not resemble snakes. This concept was captured by a viral internet sensation where people would place a pickle behind their unsuspecting cat and then alert them to the presence of it. Most of the cats would react by jumping or leaping away from the apparently benign object, and their behaviors (wide eyes, hair raised, sometimes accompanied by cries) indicated that they were very much afraid. (The authors don't condone this "experiment" as it's clearly stressful for the cats.) This is referred to in scientific circles as "snake detection theory." Among humans, there is a high visual sensitivity to snake threats, and this has been corroborated by brain scans. We have a disproportionately large portion of our brain that is used to detect a potential snake threat (Van Strien *et al.*, 2014).

Felines and primates aren't the only animals that provide clear evidence that behavior is genetic. Squirrels can open nuts without observing how to do so from another squirrel. (They can later improve their nut opening technique, however, by learning how other squirrels do this, and from practice.) Animals that were raised by another species, e.g. bottle baby orphan kittens, will still exhibit innate behaviors, such as grooming and pouncing, that they did not learn from a mother cat or from a human caregiver (although the latter would be rather amusing!)

When we say that behavior is genetic, we also mean that it can vary between individuals and certain traits can run in families, independent of learning, as exemplified by Janna Ellison and her birth family. You can intentionally breed for a behavior, or it can be linked with a physical trait that you are breeding for, or it can just happen to run in families. So, you can intentionally (this is referred to as purpose-breeding) or unintentionally breed for a behavior!

We have introduced you to the concept of epigenetics (how the environment and your behaviors can change the way that your body reads your DNA sequences) and the heritability of certain personality traits, such as neuroticism and agreeableness, but there's still a lot to discover about behavioral genetics, particularly about how genes influence behavior. One family of proteins, the SAP90/PSD95-associated protein (SAPAP), has been associated with anxiety-related behaviors. A study with laboratory mice examined individuals who had a deletion of the *SAPAP3* gene. When they did, they demonstrated behaviors associated with anxiety and excessive over-grooming. A follow-up study using 1618 human participants examined variation with the *SAPAP3* gene and its association with skin picking, nail biting, and hair pulling. Recall from Chapter 3 that a haplotype is a group of alleles on a chromosome that one inherits together from one parent. For this gene, there were three haplotypes, and all three showed a small association with grooming disorders. There was also a small association with four of the six single nucleotide polymorphisms (SNPs; or substitutions of a single nucleotide at certain positions on the genome) examined. This led the researchers to conclude that *SAPAP3* should be examined further (Bienvenu *et al.*, 2009). Follow-up work suggests that there are variants of *SAPAP3* that lead to early onset of OCDs in people (Boardman *et al.*, 2011), and there is ongoing research using mice with *SAPAP3* disorders to understand the neural mechanisms associated with the genes (Wood *et al.*, 2018).

While behavioral genetics is a fairly new field, we are seeing across multiple species that there is a genetic component of behavior. Certain behavioral traits tend to run in families. Mike Powers, a trainer at Summit Assistance Dogs in Washington State, recalled a dog named Hagrid who had a unique suite of behaviors—but these behaviors were shared by other close relatives, as well.

"He was a show dog, and he had a weird habit of 'scavenging,'" Mike said. "He would grab shoes, or anything on the floor, and move them around." Years later, Mike was working with one of Hagrid's nephews, who also displayed a propensity for ferreting things away. "Ah," he said, "you look just like your Uncle Hagrid." None of the dogs who were unrelated to Hagrid showed this behavior.

Why do some behaviors (including those that don't appear to be advantageous, like nail biting or shoe hoarding) have a genetic basis, while others don't? In 2000, behavioral geneticist Eric Turkheimer examined the relationship of genetics and behavior with humans in his paper, "The three laws of behavior genetics and what they mean." These laws were (Turkheimer, 2000):

- *First Law*: All human behavioral traits are heritable.
- *Second Law*: The effect of being raised in the same family is smaller than the effect of genes.
- *Third Law*: A substantial portion of the variation in complex human behavioral traits is not accounted for by the effects of genes or families.

In 2015, on the basis of molecular studies, researchers proposed an addition to this list (Chabris *et al.*, 2015):

- *Fourth Law*: A typical human behavioral trait is associated with very many genetic variants, each of which accounts for a very small percentage of the behavioral variability.

Behavioral geneticists were arguing that behavior did have a genetic basis, but that there were other important factors, as well, such as family (the "nurture" concept, for which Turkheimer found there was less evidence than there was for genetics) and "other" factors, such as the general environment, which have been shown to act upon one's genes, turning on or off a gene for a certain trait. Thus, the proposed paradigm for the heritability of behavior, in descending order of importance, would be: genetics, environment or "other" factors, and family. But how would this translate to studies of non-humans, particularly in the canines that we are interested in here?

Genetics can be a lot more complicated than it seems, but early researchers felt that there were clear links between behavior and genetics. In 1965, Scott and Fuller wrote, "It is obvious that breed differences in behavior are both real and important in magnitude" (Scott and Fuller, 1965). Real and important. Real, because they are observable, and testable, and quantifiable. Important, because they have far-reaching implications for all species; for dogs in particular, it will likely influence not only future breeding protocols, but policies such as breed-specific legislation, as well. After decades of artificial selection, breeds and breed groups exhibit a suite of different behaviors (Bradley, 2011; Howell and Bennett, 2011) and rates of trainability (Serpell and Hsu, 2005).

Born This Way

An animal's innate behavioral drives speak to their genetic heritage, but often, people only discuss the problems that are involved and not the positive aspects. According to a 1999 survey on reasons for relinquishing an animal to a shelter, "non-aggressive behavioral issues" (29.5% of all relinquishments) ranked second only to human housing issues (30.4%) (Scarlett *et al.*, 1999). Another study, published in 1998, found almost identical results, with "behavior" (28.8%), also ranking only below human housing issues (29.1%) (Salman *et al.*, 1998). When aggressive behavioral issues were lumped in with non-aggressive ones, the total for all behavioral issues reached 40%. Thus, four in every ten dogs are surrendered to the shelter due to a "behavioral issue." This is a relatively big animal welfare concern, considering that, by some estimates, almost 390,000 dogs are euthanized in US shelters every year (ASPCA, 2019). (We're guessing that that figure is larger, as there's neither a standardized shelter euthanasia reporting system, nor incentive to self-report the worst part of one's job.) Dogs that were relinquished due to behavioral issues are likely going to be harder to adopt out into new families than dogs who have been relinquished for non-behavioral reasons; if the original family, who (hopefully) cared about the dog, couldn't tolerate the behavioral issues, what would entice a new family to try?

So, what is it about behavior that ranks above so many other common (and often questionable) issues, including "allergies," a new baby, euthanasia requests, or specific issues with aggression? When it comes to breed types, the list of reasons for relinquishing herding dogs is often the same: they're being "mouthy," they're chasing (herding, really) other animals or people, and biting their ankles, or they're getting rough with other members of the household. But these aren't *bad* or *aggressive* dogs—they're doing precisely what they were born to do. In an apartment or house where they have inconsistent or infrequent access to a yard, they don't have an outlet to participate in their breed-specific behaviors, and they need a healthy way to do so. Across many generations, where and when it was a desirable skill, dogs have been bred for herding, with those that showed a propensity to work with large livestock (who can often be surly, or stubborn, or bold) regardless of the animal's temperament, being bred with other equally brave and perseverant individuals. Over the centuries, dogs that had a tendency to nip, herd, and run were bred with other individuals with similarly appropriate traits for herding livestock, resulting in breeds and breed groups that excel at herding, but struggle when forced into a more sedentary lifestyle. They were bred to herd, and they were born this way.

Soft Mouths and Sharp Eyes

While breed isn't the determining factor in a dog's behavior, it provides a baseline for what to expect. The McNab, while not a recognized breed, is an example of recent purpose-breeding. Much like the sled dogs of Alaska, McNabs have been bred with their jobs, rather than their good looks, in mind. Purpose-bred dogs are bred to perform a specific task, such as herding livestock, or performing life skills tasks as a service dog, or retrieving game as a hunting dog. While dogs have been purpose-bred for centuries, dog breeding is still widely considered to be "a well-established art, but a crude, unestablished science" (Rosenthal, 1991). In addition to breed and breed group, the intentional selection *within* that breed can also have an impact: for example, was the animal bred for show or were they bred for field work? Among Thoroughbred racehorses, certain genetic lines are considered to be good turf horses (running on the grass), while others are considered to be good dirt horses. Similarly, some lines are considered to be sprinters, while others are considered to be "classic distance" (a mile and a quarter) runners or "stayers"—longer-distance runners. A 2019 study on breed differences between horses found that specific personality components were more common in some breeds than in others. Saddlebreds, Tennessee Walkers, Arabians, and Thoroughbreds were considered to be the "most nervous," while Appaloosas, Paints, American Quarter Horses, and Drafts were the "least nervous." The authors concluded that genetics within the breed accounted for the largest role in these personality differences, followed by environmental factors, including social contact, housing, and interactions with humans (Sackman and Houpt, 2019). While breeds and breed groups in a number of species can provide a point of departure for behavioral genetics research, human geneticists are quick to point out that environment and family/early social experiences have important roles, as well. So how can we objectively study the heritability of behavior? In a word, cautiously. Knowing what we know—and that there's still much we don't know—about genetic blueprints and the complex interplay of genetics, the environment, and life experiences, we need to be careful about how we assess what's heritable and what's not. Is a certain behavior heritable if more than one individual in a family displays it? Maybe, maybe not.

How does that caution translate to scientific studies? Well, for one, having a large sample size, or a survey that can span multiple generations and behavioral traits, can help provide a clearer picture of what might be going on. One such study, a 2008 examination of the genetic contribution to personality, looked at 10,000 German Shepherd Dogs and Rottweilers across 16 behavioral traits. The authors found that more than 50% of the variability in these traits was explained by genetics (Saetre *et al.*, 2006). The results also pointed to the heritability of aggression and aggressive tendencies, which appeared to be heritable independently from the other characteristics. While there was a considerable heritable component to this temperament, environment and individual differences still had to be taken into account.

While scientists have honed in on studying anomalous behaviors to find genetic patterns, geneticist Dr Elaine Ostrander has found that "understanding the genetic underpinning of stereotypical breed behaviors has proven to be more difficult to study" (Ostrander *et al.*, 2017). Domestication has increased the diversity of the tasks that dogs participate in while decreasing the variation within breeds that are selected for their ability to perform these tasks. Several vocations, such as guarding, livestock supervision and herding, and hunting are cross-cultural and cross-geographic, thus, multiple breeds exhibit the traits associated with these jobs. Behaviors such as pointing are associated with a number of breeds that were selected for hunting, including the German Shorthaired Pointer, English Setter, and Brittany Spaniel. While this points to a clear genetic pattern in behaviors, some of these behaviors are also paired with other behaviors, such as swimming with the German Shorthaired Pointer, retrieving with the Brittany, and tracking in the English Setter. A genome-wide association study of 46 breeds identified a locus for canine hunting on chromosome

22 (Vaysse *et al.*, 2011), which was also associated with "boldness" on a prior study (Chase *et al.*, 2009). This makes sense, given that a fearful dog would not make a good hunting partner—if they were afraid of gunfire, or too timid to retrieve or locate a fallen animal, they wouldn't be able to get the job done.

Springer Rage

Everett and Allie Astin were playing quietly in their family room when their Springer spaniel, Bella, came tearing around the corner and raced up to them. Allie smiled happily at her dog and reached out to pet her, but Bella reciprocated with a swift, fierce bite to Allie's arm. Allie screamed as blood sprang to the surface with four large puncture wounds. Bella just as quickly raced out of the room.

"Mom!" Allie cried. "Bella ... Bella bit me!"

Allie's mom, Jennie, came racing into the room, stunned to find her young daughter holding her bloodied arm. Everett sat white-faced and slack-jawed, and then he turned to his mom. "Bella ... she just came in and ... attacked Allie," he said, choking back a sob.

This was Bella's first unprovoked bite, but it wouldn't be her last. Before this bite, Bella didn't have a known history of behavioral problems. Her family had purchased her from a breeder when she was 8 weeks old, and she had been relatively uncomplicated up until that point. But at 2.5 years old, Bella began to bite or attack her family with no provocation. The bites were increasing in frequency and severity. Jennie Astin thought that she could "handle" the issue, but she quickly found that she was out of her depth. Rather than giving up on their dog, the Astins contacted a Certified Applied Animal Behaviorist. After a 90-minute consultation, they suggested the diagnosis: Springer Rage Syndrome. The remedy? A prescription for gabapentin, an anticonvulsant medication that's typically prescribed to control seizures and to alleviate nerve pain.

Gabapentin works by influencing neurotransmitters and changing how nerves send messages in the brain. Essentially, it changes the brain's electrical activity. People who take gabapentin for pain report that the drug can work almost immediately or take 2 weeks or more for the full effects to be felt. They also report that gabapentin can make them feel calm, relaxed, and even euphoric. Because Bella's biting behavior was random, it's hard to say exactly when the drug began to take effect, but her family reported that once Bella started taking gabapentin, she never had another episode.

So what, exactly, is "Springer Rage Syndrome?" While they can be socially problematic, Springer Rage, OCD, and anxiety are all scientifically valuable, as they provide opportunities for examining the genetic basis of behavior. Rage syndrome, also referred to as sudden-onset idiopathic aggression, is a rare, severe behavioral disorder that's most commonly seen with English Springer Spaniels (Fig. 6.1) and, to a lesser extent, Cocker Spaniels, but it's not limited to these breeds. Normally, when a dog exhibits an aggressive behavior, there are indications preceding the event, but that's not the case with Rage Syndrome. Owners have reported that when their dogs have an episode, their pupils dilate and the dog becomes aggressive and attacks without any warning. Immediately preceding an episode, the dog will seem completely normal. During an episode, however, the dog will be uninhibited, disoriented, and unresponsive to efforts to cease the behavior. Rage Syndrome usually presents between the ages of 7 months to 3 years. It is genetic, rather than contagious or learned, but its rarity makes it difficult to track. Bella's breeder stated that she hadn't had any other reports of Springer Rage with the dogs that she had bred, and we still don't know what genes are involved in the expression of this disorder. Fortunately for Bella's family, a consultation and medical treatment resolved the issue completely, but this isn't always the case.

Studying behavioral diseases like Rage Syndrome and OCD has increased the awareness of behavioral genetics. With humans, OCD is a personality disorder that falls under the umbrella of anxiety disorders. It's characterized by obsessive attention to detail, repetition, excessive cleanliness, a need for control, and perfectionism. In the movie *As Good As It Gets*, Jack Nicholson portrays protagonist Melvin Udall, whose OCD behaviors include obsessive hand washing and the fear of contamination, the need for routine, and repeatedly checking door locks.

With dogs, OCD is characterized by behaviors such as circling, spinning, tail-chasing, self-injurious behaviors, pacing, vocalizing, staring, hallucinating, and pica (consuming non-nutritive substances, such as dirt or feces). By the time a professional typically see these cases, the OCD has become rather severe, resulting in injuries to the dog and damage to the home. While no OCD case is exactly

Fig. 6.1. Springer Spaniel. This file is made available by Wikipedia and is under the Creative Commons CC0 1.0 Universal Public Domain Dedication. Available at: https://commons.wikimedia.org/wiki/File:Welsh_Springer_Spaniel_1.jpg.

the same as another, the one thing that most dogs with OCD share in common is that they have been highly inbred—their coefficient of inbreeding is rather high. OCD wouldn't be a trait that was intentionally bred for, but genetics can be a lot more complicated than it seems ... remember, sometimes you breed a dog for X, but then Y comes along, too.

The Hitchhiker's Guide to the Genetiverse

In breeding dogs for a very specific look or purpose, issues such as OCD were often "hitchhikers" on that gene. This is called "genetic linkage." Genes that are close together on the same chromosome are more likely to stay linked together when the recombination of male and female DNA result in a fertilized egg. Genes that are next to one another may be coding for very different traits, such as a physical trait and a behavioral trait. This can result in some interesting inheritance patterns where particular traits tend to co-vary with one another. This linkage between behavior and morphology is exemplified in domesticated animals, where the domestication of mammals appears to produce similar differences in appearance and behavior (Trut, 1999) (Table 6.1).

One of the issues in sorting out variation in domestication is that much of this was done without careful records, and by many different breeders with different goals (Darwin, 1859). This makes sorting out the processes of domestication difficult if you have to figure out what happened retrospectively.

For canids, there is a *prospective* fox study demonstrating a linkage between morphology and behavior that pops up during the domestication process. The red fox (*Vulpes vulpes*) is the outgroup for modern canids (Trut *et al.*, 2006). Recall our discussion in Chapter 2 of Dmitri Belyaev and Nina Sorokina's classic study with a subgroup of the red fox, the silver fox (*Vulpes vulpes*). These Russian geneticists began to artificially strictly select only the tamest silver foxes to breed to one another in 1950, and this work continues on today (Wikipedia, 2020). The silver fox had never been

Table 6.1. Common traits observed in domesticated animals (Darwin, 1859; Trut, 1999; Trut *et al.*, 2009).

Trait observed in domesticated mammals	Species
Appearance of dwarf and giant varieties	All
Piebald coat color	All
Wavy or curly hair	Sheep, Poodles, donkeys, horses, pigs, goats, mice, guinea pigs
Rolled tails	Dogs, pigs
Shortened tails, fewer vertebrae	Dogs, cats, sheep
Floppy ears	Dogs, cats, pigs, horses, sheep, goats, cattle
Changes in reproductive cycle	All except sheep

Modified from Trut *et al.* (2009) and permission to use granted by Sigma Xi.

domesticated previously, so this made it a good subject for an artificial selection experiment in a species related to dogs. Keep in mind that in this breeding experiment, they were only selecting based on tameness, and not on any physical characteristics. While his work resulted in two populations of silver foxes, one that was aggressive and one that was tame (or domesticated), there were also morphological differences between the two populations (Trut *et al.*, 2006) (Table 6.2).

Table 6.2. A summary of changes seen in the domesticated silver fox.

Open eyes 1 day earlier
Respond to sound 2 days earlier
Respond to fear 3 weeks later
Piebald coat appears (after 8–10 generations)
Floppy ears appear (after 10–14 generations)
Rolled tails appear (after 10–14 generations)
Shorter tails appear (after 15–20 generations)
Shorter legs appear (after 15–20 generations)
Underbites and overbites appear (after 15–20 generations)
Reduction in production of the adrenal glands (after 12 generations, and halved again at 28–30 generations)
Higher levels of serotonin
Smaller cranium
Shorter and wider snout
Cranium sexual dimorphism decreased (males smaller like females)
Reach reproductive maturity 1 month earlier
Litter size increases by one
Mating season lengthened, and some reproduced twice per year

Note that the changes are roughly in the order that they appear during the process of domestication, and the generation they first appear in is noted when it was specified in the original publication (Trut, 1999).

Why did we get so many changes if we only selected for tameness? Some of the initial changes may have been a result of genetic linkage, but further selection may have resulted in reducing the amount of genetic variability in some traits (Trut *et al.*, 2009). Additionally, selecting for "tameness" is likely to tap into neurological and hormonal characteristics such as a reduction of fear or stress to novelty (Trut *et al.*, 2009). Changes in the brain and in hormonal systems would have a cascade of effects on other traits, both physiological and morphological. It's interesting how many of the traits that appeared unintentionally, like piebald coats, curled tails, and floppy ears, are similar to other domesticated species.

What Linkages Do We See in Domestic Dogs?

The Greyhound provides another excellent example of the link between behavior and morphology. Breeders were focused on breeding the dogs that won the most races, and this singular selection trait resulted in fast dogs, but also dogs that have a long body, limbs, head, and tail (Bradley, 2011). Another example of a link between behavior and physical appearance is that of the Korean native Jindo Dog, where dogs with a white coat are more fearful and show more submissive reactivity than dogs with a differently colored coat (Kim *et al.*, 2010). There is also evidence of a link between size and behavior. A 2016 study used trained observers to assess 67,368 dogs of 45 different breeds on a Dog Mentality Assessment, comparing that with measures of the skull length and width, bodyweight, height, and sex (Stone *et al.*, 2016). The study found that shorter dog breeds were more aggressive, while taller dogs were more affectionate, cooperative, and playful with humans. Lighter dogs were more fearful of a gunshot and a metallic noise. Besides behavior, we also see linkages between selection for physical shape (such as long bodies) and hip dysplasia (Roberts and McGreevy, 2010).

Cross-breeds: Goldendoodle and Labradoodle

This brings us to the rather interesting discussion of cross-breeds. Let's consider the popular "doodles," a designer dog mixed breed created out of the desire to have service and therapy dogs with fewer allergens, catalyzed by the erroneous belief that their offspring would be "hypoallergenic." (Spoiler alert: there's no such thing as a "hypoallergenic" dog or cat). Crossing Poodles (Miniature or Standard) with Golden Retrievers and Labrador Retrievers (and later, other breeds) to create "doodles" has become popular in the past several decades, but relatively little scientific study has been done on them. The exception to this dearth of information comes from the work of James Serpell and his colleagues, who used the Canine Behavioral Assessment & Research Questionnaire (C-BARQ) to assess owner-reported behaviors in 5141 dogs from around the world (Shouldice *et al.*, 2019). The breeds included Miniature Poodle, Standard Poodle, Labrador Retriever, "Labradoodle" (Fig. 6.2), Golden Retriever, and "Goldendoodle."

Comparing each designer to original breed parents

Serpell and his colleagues found that for most traits, the cross-breeds were generally behaviorally intermediate between the two original breeds (e.g. Goldendoodles were intermediate between Golden Retrievers and Poodle). There were a few exceptions to this tendency, though. Results for the Labradoodle indicated that it depended on whether the cross with the Labrador Retriever was with a Miniature Poodle or a Standard Poodle. There were no significant differences when the cross was with a Standard Poodle. They did find that Labradoodles showed less dog rivalry than Miniature Poodles. They defined dog rivalry as "dog shows aggressive or threatening responses to other familiar dogs in the same household" (Shouldice *et al.*, 2019).

Goldendoodles showed more dog-directed aggression than Golden Retrievers, Miniature Poodles, or Standard Poodles. The Goldendoodle also had the highest score for dog-directed fear, but this difference was only statistically significant between Standard Poodles (lower) and Goldendoodles (higher). The Goldendoodle also showed significantly higher stranger-fear than the Golden Retriever.

Across a number of concerning behaviors (touch sensitivity, non-social fear, and separation anxiety), the Miniature Poodle had higher scores. For example, the Miniature Poodle had the highest score for separation anxiety, and it was significantly higher than both Golden and Labrador Retrievers. When the "doodles" came from a cross that included Miniature Poodles, their score was higher on separation anxiety, but it was intermediate between Miniature Poodle and Golden Retriever (if Goldendoodle) and Labrador Retriever (if Labradoodle).

"Doodles" can be lovely or they can come with issues such as those described above. These issues are likely attributable to random chance based on the genetic mix, and on irresponsible breeding practices where money rather than careful selection of healthy dogs, is the motivation. More science is needed to further document and understand these differences. Other studies have found fear-related behaviors in "mixed breeds" (Bennett and Rohlf, 2007) but this isn't necessarily equivalent to a first-generation cross to create one of the "doodle" types.

But what happens when extreme inbreeding, or poorly managed breeding, and poor early environment co-occur, as in the case of puppy mill dogs? Being born and raised in a puppy mill environment combines the worst possible scenarios (inbreeding, lack of socialization, abuse, neglect, and high stress levels due to these factors). While there have been many cases where one suspects that the dog came from a puppy mill, it isn't like breeders would openly admit this. Part of this is the problem with defining a puppy mill. Although there's currently no standardized definition of a puppy mill (as of this writing, researchers are currently working on this, though), the 1984 Supreme Court case *Avenson v. Zegart* defined them as a "dog breeding operation in which the health of the dogs is disregarded in order to maintain a low overhead and maximize profits" (*Avenson v. Zegart*, 1984). This case was brought about by Lesley Zegart, who was the executive director of the Minnesota Humane Society at that time. In 1982, Zegart was investigating dog breeding enterprises in Minnesota to try to determine which ones were puppy mills. The American Society for the Prevention of Cruelty to Animals (ASPCA, 2013) concurs with this definition, defining the puppy mill as a "large-scale commercial dog breeding operation where profit is given priority over the well-being of the dogs". (Given those definitions, no breeder, and no business that sells dogs, would want to be associated with a puppy mill.) Dogs who came from puppy mills often have myriad health and behavioral problems, including anxiety disorders, aggression, and the incidence of otherwise rare and recessive traits. For the most part, puppy mill dogs are smaller breeds, such as the Bichon Frises, Corgis,

Fig. 6.2. Labradoodle. This file is provided by Wikipedia Commons and licensed under the Creative Commons Attribution-Share Alike 3.0 Unported license. Available at: https://en.m.wikipedia.org/wiki/File:Labradoodle_Brown.jpg

Papillons, Pekingese, Pomeranians, and Yorkies. You might see a few Huskies or Bernese Mountain Dogs, but typically, smaller dogs are easier for puppy millers to produce, as they can have a high selling price while not taking up an abundance of space or resources (Fig. 6.3).

Conclusion

Selection for one trait over generations can lead to an increase or exaggeration of that trait, but it can result in multiple unforeseen changes as well. Cross-breeding two different breeds does not necessarily result in an animal that has the best traits of each parental breed, but instead often results in less desirable "intermediate" levels of each trait. Additionally, given equal dominance for certain traits, it's just as likely that a cross-bred dog would have all of the worst traits of each parent. When we select for behavior, the end result can be morphological changes, and when we select for certain physical traits, the end result can be in behavior. Think of it this way: there's always going to be a stow-away or a hitchhiker, and there's no guarantee which trait is going to be at the helm of the ship when an offspring is produced. In addition to closely examining dominant and recessive traits, we also need to select

Fig. 6.3. Miniature breed dogs in a puppy mill. This work has been released into the public domain by its author, PETA.

physically and behaviorally healthy animals to ensure the best outcome. Responsible breeding doesn't happen without lots of planning and analysis; no reputable breeder should be breeding lines where hip dysplasia or cancer are common, regardless of how wonderful the animal's temperament is, just as much as no reputable breeder should breed lines where anxiety or aggression are common. And even with the best use of science, we have to remember that dog breeding is still an art, and unanticipated traits will still come along for the ride.

References

ASPCA (2013) ASPCA urges Vermont governor to sign puppy mill bill. Available at: www.aspca.org/about-us/press-releases/aspca-urges-vermont-governor-sign-puppy-mill-bill (accessed 3 November 2023).

ASPCA (2019) Pet statistics. Available at: www.aspca.org/animal-homelessness/shelter-intake-and-surrender/pet-statistics (accessed 3 November 2023).

Avenson v. Zegart (1984) 577 F. Supp. 958, 960 (United States District Court, D. Minnesota, Sixth Division January 17, 1984).

Bennett, P.C. and Rohlf, V.I. (2007) Owner-companion dog interactions: relationships between demographic variables, potentially problematic behaviours, training engagement and shared activities. *Applied Animal Behaviour Science* 102(1–2), 65–84.

Bienvenu, O.J., Wang, Y., Shugart, Y.Y., Welch, J.M., Grados, M.A. *et al.* (2009) *Sapap3* and pathological grooming in humans: results from the OCD collaborative genetics study. *American Journal of Medical Genetics, Part B* 150B(5), 710–720.

Boardman, L., van der Merwe, L., Lochner, C., Kinnear, C.J., Seedat, S. *et al.* (2011) Investigating *SAPAP3* variants in the etiology of obsessive-compulsive disorder and trichotillomania in the South African white population. *Comprehensive Psychiatry* 52(2), 181–187.

Bradley, J. (2011) *The Relevance of Breed in Selecting a Companion Dog*. National Canine Research Council, New York.

Chabris, C.F., Lee, J.J., Cesarini, D., Benjamin, D.J. and Laibson, D.I. (2015) The fourth law of behavior genetics. *Current Directions in Psychological Science* 24(4), 304–312.

Chase, K., Jones, P., Martin, A., Ostrander, E.A. and Lark, K.G. (2009) Genetic mapping of fixed phenotypes: disease frequency as a breed characteristic. *Journal of Heredity* 100(Suppl. 1), 37–41.

Darwin, C. (1859) Variation under domestication. In: *On the Origin of Species by Means of Natural Selection, or, The Preservation of Favoured Races in the Struggle for Life*. J. Murray, London.

Halteh, P., Scher, R.K. and Lipner, S.R. (2017) Onychophagia: a nail-biting conundrum for physicians. *The Journal of Dermatological Treatment* 28(2), 166–172.

Howell, T.J. and Bennett, P.C. (2011) Puppy power! Using social cognition research tasks to improve socialization practices for domestic dogs (*Canis familiaris*). *Journal of Veterinary Behavior* 6(3), 195–204.

Kim, Y.K., Lee, S.S., Oh, S.I., Kim, J.S., Suh, E.H. *et al.* (2010) Behavioural reactivity of the Korean native Jindo dog varies with coat colour. *Behavioural Processes* 84(2), 568–572.

Leung, A.K.C. and Robson, L.M. (1990) Nailbiting. *Clinical Pediatrics* 29(12), 690–692.

Ostrander, E.A., Wayne, R.K., Freedman, A.H. and Davis, B.W. (2017) Demographic history, selection and functional diversity of the canine genome. *Nature Reviews Genetics* 18(12), 705–720.

Roberts, T. and McGreevy, P.D. (2010) Selection for breed-specific long-bodied phenotypes is associated with increased expression of canine hip dysplasia. *The Veterinary Journal* 183(3), 266–272.

Rosenthal, E. (1991) Study of canine genes seeks hints on behavior. *ScienceTimes*. Available at: www.nytimes.com/1991/12/03/science/study-of-canine-genes-seeks-hints-on-behavior.html (accessed 11 December 2023).

Sackman, J.E. and Houpt, K.A. (2019) Equine personality: association with breed, use, and husbandry factors. *Journal of Equine Veterinary Science* 72, 47–55.

Saetre, P., Strandberg, E., Sundgren, P.-E., Pettersson, U., Jazin, E. *et al.* (2006) The genetic contribution to canine personality. *Genes, Brain, and Behavior* 5(3), 240–248.

Salman, M.D., New Jr, J.G., Scarlett, J.M., Kass, P.H., Ruch-Gallie, R. *et al.* (1998) Human and animal factors related to relinquishment of dogs and cats in 12 selected animal shelters in the United States. *Journal of Applied Animal Welfare Science* 1(3), 207–226.

Scarlett, J.M., Salman, M.D., New, J.G. and Kass, P.H. (1999) Reasons for relinquishment of companion animals in U.S. animal shelters: selected health and personal issues. *Journal of Applied Animal Welfare Science* 2(1), 41–57.

Scott, J.L. and Fuller, J.P. (1965) *Genetics and the Social Behavior of the Dog*. The University of Chicago Press, Chicago, IL, p. 385.

Serpell, J.A. and Hsu, Y.A. (2005) Effects of breed, sex, and neuter status on trainability in dogs. *Anthrozoös* 18(3), 196–207.

Shouldice, V.L., Edwards, A.M., Serpell, J.A., Niel, L. and Robinson, J.A.B. (2019) Expression of behavioural traits in Goldendoodles and Labradoodles. *Animals* 9(12), 1162.

Stone, H.R., McGreevy, P.D., Starling, M.J. and Forkman, B. (2016) Associations between domestic-dog morphology and behaviour scores in the dog mentality assessment. *PLoS One* 11(2), e0149403.

Teng, E.J., Woods, D.W., Twohig, M.P. and Marcks, B.A. (2002) Body-focused repetitive behavior problems: prevalence in a nonreferred population and differences in perceived somatic activity. *Behavior Modification* 26(3), 340–360.

Trut, L.N. (1999) Early canid domestication: the farm-fox experiment. *American Scientist* 87(2), 160–169.

Trut, L.N., Kharlamova, A.V., Kukekova, A.V., Acland, G.M., Carrier, D.R. *et al.* (2006) Morphology and behavior: are they coupled at the genomic level? In: Ostrander, E.A., Giger, U. and Lindblad-Toh, K. (eds) *The Dog and Its Genome*. Cold Spring Harbor Laboratory Press, Woodbury, NY, pp. 81–93.

Trut, L., Oskina, I. and Kharlamova, A. (2009) Animal evolution during domestication: the domesticated fox as a model. *BioEssays* 31(3), 349–360.

Turkheimer, E. (2000) Three laws of behavior genetics and what they mean. *Current Directions in Psychological Science* 9(5), 160–164.

Van Strien, J.W., Franken, I.H.A. and Huijding, J. (2014) Testing the snake-detection hypothesis: larger early posterior negativity in humans to pictures of snakes than to pictures of other reptiles, spiders and slugs. *Frontiers in Human Neuroscience* 8, 691.

Vaysse, A., Ratnakumar, A., Derrien, T., Axelsson, E., Rosengren, P.G. *et al.* (2011) Identification of genomic regions associated with phenotypic variation between dog breeds using selection mapping. *PLoS Genetics* 7(10), e1002316.

Wikipedia (2020) Dmitry Belyayev (zoologist). *Wikipedia*. Available at: https://en.wikipedia.org/wiki/Dmitry_Belyayev_(zoologist) (accessed 23 October 2020).

Wood, J., LaPalombara, Z. and Ahmari, S.E. (2018) Monoamine abnormalities in the *SAPAP3* knockout model of obsessive-compulsive disorder-related behaviour. *Philosophical Transactions of the Royal Society of London. Series B, Biological Sciences* 373(1742), 20170023.

7 Breed Differences in Temperament and Reactivity

Abstract

Chapter 7 reviews the scientific literature on breed differences as they pertain to temperament and reactivity, examining individual differences versus breed differences as well as variation within a breed and variation between breeds. The prevalence of mental health issues is revealed, including separation anxiety, obsessive-compulsive disorder (OCD), and depression across different breeds and breed groups, identifying which breeds have a higher occurrence of these issues, and why.

As Sick as a Dog

Gary and Gina Beaumont lived in rural Washington State with their eight dogs. While their canines represented a diverse range of breeds, including six of the AKC's seven breed groups (an Australian Cattle Dog and a Belgian Malinois, which are part of the Herding Group, a Labrador Retriever and a Pointer, which are part of the Sporting Group, a Pug, which is part of the Toy Group, a Schnauzer, which is part of the Working Group, a Rat Terrier mix, which is part of the Terrier Group, and a Beagle, which is part of the Hound Group) they lived in relative harmony with one another. That's not to say that they all had similar temperaments and rates of reactivity, though! All of the dogs had been acquired from breeders and rescues prior to 1 year of age, and had all been raised in the same home and in the same manner. But each of the eight dogs reacted differently to stimuli such as a knock on the door, a new guest in the home, or a cat traipsing through their backyard.

Who would be *most likely* to bark in response to the knock? To "disapprove" of the new guest? To chase the cat? … Well, it's in their DNA. Different dog breeds have genetic predispositions for different behaviors. The Cattle Dog would be likelier to "herd" the other dogs, the Malinois would be likelier to sound the alarm when a car came down their long driveway, the Labrador Retriever and Pointer would be likelier to chase birds off of the property, the Pug would be likelier to be content to stay on the couch, the Beagle would be likelier to bay without provocation, the Rat Terrier mix would be likelier to chase rodents, and the Schnauzer would be likelier to be happy riding "shotgun" on a trip with his family.

"Certain behaviors are 'typical' to one breed more so than they are to another. Herding dogs will herd, livestock guardian dogs will guard, and Greyhounds will run," explained animal behaviorist Wendy Dahl. "For me, it's less about looking at what the breed usually does and instead working with the communication from the dog in front of me. I like to develop a greater sense of who that dog is and understanding what it values, enjoys, and fears."

A dog's breed or breed group pertains to a host of potential differences, including temperament and reactivity. There is variation among dogs of the same breed (recall the contrasting behaviors of the Chow Chows, Annie and Charlie, and the Pomeranians, Pepper and Banjo, that we discussed in the Prologue), as well as variation between breeds and breed groups. There are also different rates in the prevalence of mental health issues, including separation anxiety, obsessive-compulsive disorder (OCD), and depression across different breeds and breed groups. It's important to identify which breeds have a higher occurrence of these issues and why.

When people are searching for a dog to join their family, "personality" is often one of the main traits driving their decision. In a UK study of more than 2000 dogs and their owners, age, size, and friendliness ranked at the top of adopters' "most important factors" lists (Siettou *et al.*, 2014). A dog's temperament has been a driving force in their evolution, as well. The ancestors of early dogs that exhibited the right combination of personality

DOI: 10.1079/9781800624566.0007

traits were given more opportunities to breed (and with similarly tempered dogs) than dogs who weren't extroverted, or friendly, or non-reactive.

Before we discuss reactivity, temperament, and personality we first need to operationally define these terms. "Reactivity" refers to a dog's sensitivity to stimuli in the environment, and may be confused with aggression. It may also be attributed to either temperament or personality. Recall that we touched upon the concept of personality when we discussed how individuals varied in Chapter 4. While "personality" and "temperament" are commonly used terms, they have slightly more specific meanings in the scientific realm. In the human psychological literature, "temperament" is the genetic component, and refers to innate characteristics such as extroversion or introversion, or high explore drive versus low explore drive, while "personality" is based upon your temperament overlaid with your life experience. Personality refers to an individual's suite of characteristics and the differences in how they behave, feel, think, and process, based upon what they have learned, been exposed to, and experienced, etc. It's the result of established cognitive capacity, including emotional reactions, which processes sensory perception as well as immediate environmental influences through learning and experience. Recall the dogs from the Prologue: Chow Chows have a genetic predisposition (temperament) to have separation anxiety; paired with the unpredictability of the negative stimulus of the stove alarm, you'll see a more reactive personality emerge in some individuals. The same terminology *can* be used across the animal kingdom, but some scientists object to using the word "personality" when it comes to non-human animals. But humans are not the only animals with personality. Despite its root in the word "person," and the typical human association with that word, this term can be used for "non-human persons," as well. Allowing non-humans to have personality acknowledges that they also have preferences and needs beyond those that serve humans. This acknowledgment is leading the way in passing landmark animal welfare legislation.

Animal personality has been a hot topic in the field of animal behavior. Here we will focus on how it is measured in dogs and provide evidence for a genetic basis to personality in dogs. Next, we will review how canine professionals view personality by breed, and how this can impact treatment in clinical cases. Finally, we'll go over what the latest scientific work on breed differences in personality suggests and reflect on the status of the field.

Measuring Personality in Dogs

The Canine Behavioral Assessment & Research Questionnaire (C-BARQ) assessment provides a standardized evaluation of canine temperament and behavior. While it has been casually referred to as a "personality assessment," the C-BARQ was originally developed by Yuying Hsu and James Serpell to specifically measure the prevalence of behavioral issues among working dogs and is now applied to pet dogs, as well. Since its development, more than 70,000 dogs representing over 300 breeds and breed mixes have been evaluated with this assessment (Serpell and The University of Pennsylvania, 2023). The C-BARQ is a survey that is now used by dog owners/guardians that contains 101 items regarding aggression, attachment, attention seeking, chasing, excitability, fear and anxiety, trainability, and separation-related behavior. The survey provides a set of numerical scores for the following 14 behavioral categories. The feline equivalent of the survey, the Fe-BARQ, used the C-BARQ as a model to provide a way to collect and measure feline behavior, as noted by cat owners/guardians. The Fe-BARQ measures 23 factors of common feline behavior, including purring, directed vocalizations, activity level, trainability, crepuscular activity, and attention seeking (Duffy *et al.*, 2017). We'll explore each of these specific areas in full, but before we delve further into the C-BARQ, let's operationally define these terms and explain why they are pertinent to measuring canine personality.

1. *Stranger-directed aggression*: Threatening or hostile responses to strangers approaching or invading the dog's or owner's personal space, territory, or home range.
2. *Owner-directed aggression*: Threatening or hostile responses to the owner or other members of the household when challenged, manhandled, stared at, stepped over, or when approached while in possession of food or objects.
3. *Dog-directed aggression*: Threatening or hostile responses when approached by unfamiliar dogs.
4. *Dog rivalry*: Threatening or hostile responses to other familiar dogs in the same household.
5. *Stranger-directed fear*: Fearful or wary responses when approached by strangers.

6. *Non-social fear*: Fearful or wary responses to sudden or loud noises, traffic, and unfamiliar objects and situations.
7. *Dog-directed fear*: Fearful or wary responses when approached by unfamiliar dogs.
8. *Separation-related behavior*: Vocalizing and/or destructiveness when separated from the owner, often accompanied or preceded by behavioral and autonomic signs of anxiety including restlessness, loss of appetite, trembling, and excessive salivation.
9. *Attachment and attention seeking*: Maintaining close proximity to the owner or other members of the household, soliciting affection or attention, and displaying agitation when the owner gives attention to third parties.
10. *Trainability*: Willingness to attend to the owner, obey simple commands, learn quickly, fetch objects, respond positively to correction, and ignore distracting stimuli.
11. *Chasing*: Chasing cats, birds, and/or other small animals, given the opportunity.
12. *Excitability*: Displaying strong reactions to potentially exciting or arousing events, such as going for walks or car trips, doorbells, arrival of visitors, and the owner arriving home; has difficulty settling down after such events.
13. *Touch sensitivity*: Fearful or wary responses to potentially painful procedures, including bathing, grooming, nail-clipping, and veterinary examinations.
14. *Energy level*: Energetic, "always on the go," and/or playful.

The C-BARQ provides information on the occurrence of an additional 22 "miscellaneous" behavior problems, from stereotypic behaviors such as tail chasing and spinning, to coprophagia (eating feces).

Using a five-point rating scale, this standardized questionnaire uses everyday scenarios to determine a dog's score for each of these areas, asking owners to indicate how their dogs responded "in the recent past" to common events or stimuli. The five-point scale is as follows: 0 for never, 1 for seldom, 2 for sometimes, 3 for usually, and 4 for always. Thus, 0 would be the complete absence of the behavior, while 4 would indicate severe levels of this behavior.

Dr Serpell discussed the C-BARQ's applicability. "This tool was originally developed to measure the prevalence of behavior problems in the working dog population, and wasn't intended to evaluate canine personality," he explained. "In an early prototype, we included positive aspects, such as sociability, but in the end, I dropped those, because they were very, very negatively correlated with problem behaviors." The C-BARQ is currently used by numerous working dog organizations, with service dogs, and for detector dogs. "Increasingly, shelters are using it to evaluate dogs who are being fostered prior to adoption. You can't really use this assessment in the shelter setting, but shelters are putting more dogs in foster homes, and it's applicable there. As a standardized assessment, the C-BARQ isn't subject to a lot of random fluctuation. It has established reliability: it's fairly reliable across individuals, has good interobserver reliability, and it has good test, re-test reliability. C-BARQ scores correlate with scores from behavioral tests. This is very important from a validation point of view because Serpell's team is demonstrating that the C-BARQ has validity when compared to direct observations of behavior. We can show that the evaluations by puppy raisers at 1 year of age are consistent. The C-BARQ has predictive and concurrent reliability, and it's consistent across geographic regions. People have compared the psychometrics across different cultures, including Japan, Taiwan, the Netherlands, Sweden, and Iran … It's very robust."

The C-BARQ is accessible to behavior professionals and laypersons alike; more than 110 peer-reviewed papers have used it. "A C-BARQ score for a particular behavior is pretty well known," Dr Serpell said. "Approximately 50,000 people have used this to assess their pet dogs, and an additional 35,000 people have assessed their working dogs." The secret of the C-BARQ's success, according to Serpell, is the specificity of its questions. "They focus on a very specific stimulus in the environment," he said, "such as how the dog acts when there's a ring at the front door. It's very specific. It asks: 'What's the dog's typical response?'" For those who are looking to use the C-BARQ to assess temperament, Dr Serpell suggests keeping the original tool intact to maintain its validity, but future researchers could add additional sections, such as friendliness or playfulness, depending upon their research interests. "It would be interesting to know how these correlate with behavioral issues," he said. "Sociability or friendliness had a very strong negative correlation with aggression—which is what you would expect."

Stuck on You

While we use terms like "attachment" quite freely, it's important to operationally define descriptors when measuring behavior. One could say a dog is "attached" to a person, but what, exactly, does this mean? In psychological terms, it refers to a close, emotional relationship between two individuals. The dog–human relationship shows a level of attachment that's similar to the human caregiver–infant relationship (Bowlby, 1958), with dogs exhibiting a "secure base effect" with their owners, much like human children do with their caregivers. A secure base effect refers to one's ability to explore their environment without stress because they feel safe that they can return to their owner, guardian, or caregiver. For both children and dogs, attachment can be healthy or unhealthy; those with healthier attachment levels tend to score higher on cognitive tests (Horn *et al.*, 2013).

So, what would a "secure base" look like with a dog? Picture a dog at the off-leash dog park who might initially hesitate to play with the other dogs. Over time, the dog would gradually venture out, increasing their distance from their owner, until they were playing with the other dogs. Periodically, they'd stop and look around to see if they could find their owner, and then continue to play. This is a behavior that's fairly easy to observe at a dog park setting. Working dogs exhibit a similar behavior, where they will "check in" with their human partner by returning to them and making eye contact and/or physical contact with them.

The DNA Behind Personality

Personality traits are heritable: a study of 123 pairs of human identical twins and 127 pairs of fraternal twins found evidence of five personality dimensions. Neuroticism had a 41% genetic effect, extraversion had 53%, openness had 61%, agreeableness had 41%, and conscientiousness had 44% (Jang *et al.*, 1996). A study that examined five dimensions of behavioral difference (Psychological interests, Psychopathology, Cognitive ability, Social attitudes, and Personality) found strong evidence that each of the domains were "moderately to substantially heritable" (Lubinski, 2000; Bouchard and McGue, 2003). A study examining global variation of the dopamine D4 receptor (DRD4) found that migratory populations of humans showed a higher proportion of long alleles for DRD4 compared to sedentary populations (Chen *et al.*, 1999). There's an undeniable correlation between personality and physiology, so if you're told that you act just like your mother or father, now you know why.

Humans aren't the only species that can inherit personality traits. Recent research has revealed that there are 131 places on a dog's DNA that may contribute to the development of 14 behavioral traits potentially influencing behavior (MacLean *et al.*, 2019). Recall the four dogs from our Prologue: The Chow Chows, Charlie and Annie, and the Pomeranians, Pepper and Banjo. Charlie and Annie were the same breed, as were Pepper and Banjo, and both pairs of dogs were raised in the same environment, but each dog in the pair had distinctly different personalities. The same could be said for human siblings in a family—even identical twins. So how does personality factor into canine research?

One of the most recent papers to use the C-BARQ examined breed *average* phenotypes instead of *individual* phenotypes (MacLean *et al.*, 2019). Researchers, including University of Arizona Tucson comparative psychologist Dr Evan MacLean, University of Washington psychologist Dr Noah Snyder-Mackler, and Dr James Serpell, used data from more than 14,000 dogs representing 101 breeds, finding a high level of among-breed heritability for 14 behavioral traits. "Breed average phenotypes were associated with certain regions of the genome," Dr Serpell explained. "There were 131 different regions of the genome that were strongly associated with the differences between the different breeds." These 131 different regions account for approximately 15% of a dog breed's personality, with the most heritable traits being trainability, aggression toward strangers, and chasing. "Most of these genome regions are known to be expressed strongly. This is exciting because it suggests that there are likely to be genes that are strongly influential, in terms of behavior differences. This is the kind of study that you can't really do in other species, because many dogs have been bred for behavior—for a particular kind of activity: barking, hunting, or fighting, all of the uses that a dog has been put to. We have this really unique animal that you can do these kinds of studies in, and reveal genetic associations that would be hard to track down [in other species]."

Dr James Serpell collaborated with colleagues specializing in genetics to try and determine the genetic drivers of breed differences in behavior (Dutrow *et al.*, 2022). To analyze this, they used over 4000 DNA samples from domestic, semi-feral, and wild canids, as well as over 46,000 C-BARQ surveys from dogs (Dutrow *et al.*, 2022). Some of the findings were that companion and toy dogs were among the breeds correlated with higher reported levels of social and non-social fear. There was also a strong correlation between herder lineages and non-social fear. These results suggest that anxiety-related behaviors in the herder lineages "may be in response to phenomena that trigger fear responses, potentially related to a hyperawareness of surroundings" (Dutrow *et al.*, 2022). The unusually large sample size and the addition of combining DNA results with behavior are major strengths of this publication. Evan MacLean, director of the Arizona Canine Cognition Center at the University of Arizona, said that the paper "... brings behavior back to the forefront when we think about where selection has acted in the evolution of dog breeds. Although it's true we have done a lot of breeding for aesthetic traits, this work makes clear that a lot of the genetic action is in pathways related to the brain (and so presumably connected to behavior and cognition)" (Callier, 2022).

Dr Monique Udell, Director of the Human–Animal Interaction Laboratory at Oregon State University, added her own thoughts to the importance of this work. "First, the authors found that genetically linked behavioral diversification was principally driven by variation that existed *prior to modern breed formation*. While this idea is consistent with predictions made by other canine biologists, such as Ray Coppinger, such demonstrations are important because they provide evidence that some behaviors may only appear to be 'breed' related because of the ongoing selection of ancestral traits within certain breeds. However, since ancestral lineages are woven through multiple breed lines and are also not present in all members of each modern breed, it is another argument against strict breed stereotypes. This complements the ideas presented in another paper published earlier this year" (Morrill *et al.*, 2022).

"The paper also makes an interesting reference to the possible origins of herding breed behavior that is different from Ray Coppinger's work, although compatible with it. Often selection for an exaugurated predatory motor pattern sequence (orient–eye–stalk–chase) is recognized as a major reason why herding breeds are so effective in their working roles. However, the authors of this study found that in addition to this predatory motor sequence, some herding breeds also have variants in a genetic region that have been associated with increased anxiety and pup-gathering behavior in mice. I found this personally fascinating, as I have not seen this connection made before. As someone with a Border Collie, it rings true that her need to have everyone in the same room is anxiety driven and not predatory (although the urge to chase moving stimuli is indeed there too). It is often recognized that herding dogs can be a bit more anxious, but the suggestion that this may have been selected for early on to aid with herding tasks (and that this genetic anxiety-driven 'herding' phenotype has been recorded and selected for scientifically in other mammals) provides a new perspective about these behaviors."

"Finally, the authors noted multiple genetic changes related to neurodevelopmental functions/developmental delays, with an overlap in genetic changes related to autism spectrum disorder (ASD) in human populations. The authors suggest that this study shows evidence of a genetic basis for neurodevelopmental diversity in domestic dogs that parallels that seen in other species, including humans."

While the C-BARQ is a powerful tool with notable strengths, it has its limitations, too. The C-BARQ, and all survey-based data collection methods, are powerful in some ways, and weak in others. They are powerful in that they allow for the relatively inexpensive collection of large amounts of data in a very efficient manner. That is unquestionable. The problem is that the data is of lower quality, unless the survey method is very well-validated. When the method is "validated," we mean that someone has made the effort of asking the survey questions of owners and then going to the homes and observing the behavior of the dogs, and perhaps the owners, themselves, and connecting the two bodies of data. This is to determine: to what degree do people tell the truth, and on which questions do they tend to be … less forthcoming? How do different people, from different backgrounds, varying levels of dog familiarity, and divergent cultures, operationally define in their minds, say, the concept of "aggressiveness?" A dog that is a 2/5 on aggression for one person might be a terrifying 4/5 for someone else. Now, C-BARQ being one of the first, and best supported, survey-type tools, has done some of this work, and is more reliable as a data collection tool than many survey systems. But *all* such survey-based tools suffer from the same benefits (data collection efficiency) and costs (data quality and interpretability). And hence, we need to be careful, and limited, in interpreting their results too broadly.

One of the greatest uses of survey-based tools is in generating new hypotheses. Ask a question, use the survey-based tools to investigate the question(s) quickly and efficiently, and if anything interesting emerges, further pursue the question with more accurate, but more costly, first-hand data collection methods. Unfortunately, many studies end at the survey-based conclusions, which, in my mind, are inherently weak without further confirmation. While there are multiple assessments of personality with non-human animals, there are also issues with agreement between measures of personality and studies of personality. Our first issues stem from our own viewpoints: we are using anthropocentric assessments of non-humans. These assessments are often questionnaires that are completed by persons who don't have a specific background in animal behavior. And then we also have limitations with breeds, with certain breeds disproportionately representing much of the data, and other breeds being omitted altogether.

In addition to the C-BARQ is the Dog Impulsivity Assessment Scale (DIAS), the Dog Personality Questionnaire (DPQ) and the Monash Canine Personality Questionnaire (MCPQ), a canine personality scale that was designed to address the limitations of these studies. But even with this goal as the primary reason to create the survey, issues still remain regarding accuracy and consistency between test-takers and results. A 2016 study examined canine personality structure using owner questionnaires, including the C-BARQ, the DIAS, and the MCPQ-Revised (Rayment *et al.*, 2016). While dog trainers and other animal professionals have long used questionnaires to assess companion animal temperament and personality, the tests themselves need to have validity, take behavioral context into account, and be used accurately, and in their entirety. This study found that three context-specific behavioral traits, neuroticism, extraversion, and behavioral regulation, had similarities with personality traits that had been identified in other species, but the questionnaire takers varied in their dog-related work experience. These differences could impact the validity of the test, and the researchers noted that we still need tools that can assess personality across a broad spectrum of skill and experience levels (Rayment *et al.*, 2016).

What all of these personality tests share in common is an issue with consistency, whether it's due to varying levels of experience with the people taking the survey, varying ages of dogs, or variance in the setting in which the survey was conducted. There is no one set "personality test" that was designed to measure canine personality and that does so successfully across breeds, cultures, and settings, but such a test does exist for cats. The feline-ality™ test, created by the American Society for the Prevention of Cruelty to Animals (ASPCA), has been widely successful in assessing the personality of cats in the shelter environment (Fig. 7.1). The test has very specific instructions for being administered and it includes 11 items, including body posture, greeting approach, cage condition, social response when the door is opened, introduction to a novel room, call and approach, open hand, stroking, play, hug, and sensitivity. Responses for each item have a numeric value. For example, for body posture, one could answer with one of three possible positions: "soft and relaxed" (add 1 point), "tense body with twitching tail" (add 1 point), and "flattened body with dilated pupils" (subtract 1 point) (ASPCA, 2013). The two former line items likely refer to the affiliative aspect and the agentic aspect of one's personality. By "agentic," we mean one's capacity, condition, or state of acting or of exerting power (Webster). Some of the items, such as body posture and greeting approach, are to be conducted simultaneously. After adding up the point value for the responses, each cat will be assigned to one of eight different "feline-alities," ranging from independent to gregarious, and measuring their valiance, adaptability, and playfulness.

The ASPCA (2013) feline-ality™ categories are as follows:

- *Private investigator*: I'm working undercover to keep an eye on you and your household. You may

Fig. 7.1. Example of cat being scored on the feline-ality™ test. Photograph courtesy of Robin Foster, PhD.

not even know you're under surveillance. I can vanish into thin air if anyone or anything interferes with my investigation. If you need a cat who knows how to stay out of trouble and will always keep your secrets, I just might take your case.

- *Secret admirer*: When it comes to relationships, I'm very level-headed. I don't leap in paws first, if you know what I mean. But give me a little time, and then I'll shower you with purrs, head-butts, and plenty of lap time. In the meantime, you may not see a lot of me but I'll be thinking a lot of you!
- *Love bug*: Do you seek affection? I do! If you also like petting, purrs, and paws kneading your lap, I think we might have a LOT in common. I'm looking for "someone who enjoys quiet times and togetherness." Could that someone be you?
- *The executive*: I have to say, I'm a busy cat. First, I've got to check out what's happening out the window. Next, I'll see if any closets or cupboards need looking into. And then there are my naps! Can't be late for those. I can fit a little socializing into my schedule. Shall we plan on breakfast and dinner? I hope you like kibbles.
- *Sidekick*: Like all sidekicks, I'm just plain good company. I like attention, and I also like my solitude. I don't go looking for trouble but I'm no scaredy-cat, either. If you are looking for a steady companion to travel with you on the road of life, look no further.
- *Personal assistant*: You're working on the computer? Let me press the keys. Reading the paper? I'll hold the pages down for you. Watching TV? I'll just plop in your lap so you can pet me. I love an orderly household, don't you? I'll help you with all your chores, and I'll help you relax when we're done. You'll wonder how you ever managed without me.
- *MVP*: I'm a savvy cat who knows the score. I'm pretty unflappable, too. I don't mind entertaining myself, but a human companion at the other end of the couch and a nice scratch behind the ears always make my day. If you're looking for a resourceful addition to your team, think about signing this Most Valuable Pussycat.
- *Party animal*: I'm a cat on a mission: PARTY! I love to play and explore and test my limits. I'd love to play with you, but I can make a toy out of anything: pencils, post-it notes, potatoes. If you're looking for some laughs and someone to liven up the party, think about inviting me.
- *Leader of the band*: I'm a cat who does everything in a big way. I not only like to be in the

middle of things—I like to lead the parade. I'm an adventurous cat, but I'll still make plenty of time to show you my affectionate side. I'm the demonstrative type, you might say. Want a cat who's brimming with confidence? That's me.

This method of personality assessment is used widely in animal shelters and has been a consistently accurate measure of feline personality. MEOW Cat Rescue, located in Kirkland, Washington, uses this assessment to describe their adoptable cats.

So why can we accurately measure the personality of cats, but struggle to do so for dogs? Domesticated dogs are a very social species, and they're also particularly accustomed to socializing with humans. While the shelter environment can be stressful for both species, it's possible to accurately ascertain feline personality in this environment. These assessments don't work nearly as well with dogs in the shelter environment because dogs have far more variability and plasticity in their behavior than cats do; dogs, far more often than cats, will adjust their behavior to the context. This should come as no surprise, given the difference in duration of domestication between dogs and cats: cats have been domesticated for 8000 years (Smith, 2017), while dogs have been domesticated for at least 15,000 years (Irving-Pease *et al.*, 2018), and for as long as 40,000 years, given genomic analyses (Botigué *et al.*, 2017) and estimates of their split from the ancestor of wolves 20,000–40,000 years ago (Thalmann and Perri, 2018). Your co-authors subscribe to the hypothesis that dogs have been domesticated for *at least* 32,000 years (Ha and Campion, 2018).

As a result, there's less of a range of behaviors and genetic variability, given that cats also only have 71 recognized breeds, compared to the several hundred to over 1000 dog breeds, depending upon the source of breed recognition. Dogs are the champions of plasticity of behavior. They will adjust their behavior to the context provided by their humans, while cats will not. Interestingly, feline personality can be measured at a fairly early age, and barring mental illness or significant trauma, will remain fairly consistent throughout a cat's lifetime. Seattle Area Feline Rescue (SAFeR), located in Seattle, Washington, uses vegetables to explain the personality of each cat. Their three vegetable categories, Radish, Squash, and Brussel sprout, are explained as follows:

- *Radish*: Cats in the Radish category are reserved and quiet. Just like a radish grows in the ground, these cats may like to shy away at first and not fully show themselves. They will appreciate a mellow environment so they have the opportunity to develop a trusting relationship with you before emerging and showing their full potential. Bonding with a shyer cat is an incredibly rewarding experience where you earn their loving affections. Radishes are so fun to watch grow!
- *Squash*: Cats in the Squash category are oh so lovable and sweet! Their affection is easily gained and they thrive pretty much anywhere. These cats are not bothered by much. With such a wonderful temperament and their desire to be loved at all times, these cats are simply squashable! Are you a first-time cat owner, or maybe a family with young children? Get yourself a Squash cat!
- *Brussel sprout*: Cats in the Brussel sprout category are a divine delicacy that aren't always for everyone. These are cats with robust spirits and hardy personalities. The Brussel sprout party is recommended for experienced cat owners who understand if you pop a hot Brussel sprout in your mouth, you get burned, but love the quirks of a strong-willed furry friend. Brussel sprouts bring zest to your life and endless entertainment!

According to SAFeR's Adoption Center Manager, Sarah Theriault, "The vegetable system is effective because it's a quick and easy way to communicate information to potential adopters. Personality categories work well for members of the public because it's a quick way to learn more about a cat when there is a selection of cats they're looking at. This helps narrow down the options of who fits what they are looking for. It's a fun way for the person to make a connection to the cat and potentially spark an interest in meeting that cat, which ultimately helps more cats find homes."

Ask the Professionals

How do professionals—behaviorists, veterinarians, and breeders—assess personality? And do they think that personality varies by breed significantly, or at least enough to make it useful to assess in behavioral treatment plans?

Behaviorist Maria Muradas has a graduate background in education psychology and an MSc in Clinical Behavior from the Royal Veterinary College, University of Edinburgh. She also holds a Certificate of Applied Animal Behavior from the

University of Washington, an Associate Nose Work Instructor Certification, and an Associated Animal Behavior Consultant Certification through the International Association of Animal Behavior Consultants: Dog Division. In these capacities, Maria has a wealth of experience with canine behavior.

During her time as a behaviorist, Maria has seen an increase in multiple unhealthy behaviors. These include hyperattachment to one person, hypervigilance and unfamiliar people aggression with guard/shepherd dogs (Saint Bernards, Great Pyrenees, German Shepherd Dogs, and Rottweilers), higher prey drive, people-directed and dog-directed aggression, excessive barking, hypervigilance, and even compulsive behaviors with hunting and herding dogs (Border Collies, Labradors, Terriers, and Pointers).

"When I refer to 'working dogs,' I'm not talking about the AKC list of working dogs; I am talking about all the breeds that were selected to fulfill a role as working companions for humans. Plenty of those breeds are just considered pets, but their genetics still have the characteristics that they were bred for. For instance, Border Collies herd and work all day long on the farm. They were selected for their stamina and ability to round up sheep. The average pet doesn't work all day long and doesn't have access to sheep to herd, so you see Border Collies chasing and barking at kids and cars, nipping heels, and herding owners at home. Even well-bred, well-socialized working dogs develop reactivity and other problem behaviors linked to the lack of opportunity to do what they were selected to do: work."

With the popularity of breeds such as the Retrievers and Border Collies, Maria has seen a corresponding increase of behavior cases with these kinds of dogs. It's unclear, however, if these increases are due to their popularity and representation in the general dog population, or if they are having an actual increase in behavioral issues.

"My experience with toy breeds is different than with working dogs, though. Even if you see the same behavioral issues, there's a big learning component: owners give less opportunities for socialization to toy puppies, they take them less to puppy class, and if they get in trouble or annoying, they just pick them up. I see a lot of cases with familiar and unfamiliar people aggression, most of the time linked to grooming, picking up, petting, moving them from spaces, etc. I don't have data, but I hypothesize that aggressive behaviors also are not culled from the breeding pool because owners don't report them to breeders and breeders don't see it as a big problem: the damage that a small dog can do is for sure less than a Labrador, for instance. Working with toy dogs always starts with working with the owners on understanding that they have a *dog*, even if she only weighs 4 pounds. From there, there's a lot of work on consent, to change the human/dog interaction."

Dog behaviorist Adam Winston (CPDT-KA, UW-AAB) has worked with hundreds of dogs in a variety of settings. "I have worked as a basic obedience dog training instructor and also served King County (Washington) as an Animal Care Technician where I handled hundreds of dogs in the shelter," he said. "No matter what, German Shepherds continue to elude me. I feel like they're fifty-fifty: half the German Shepherds that I work with are highly reactive and vocal. Conversely, the other half tend to be calm and focused on their handler. Either way, it almost always feels like it's one or the other."

It makes sense that those who work with this breed might find them to be inconsistent. Between the German Showline, American Showline, West German Working Line, and East German Working Line, as well as the fifth category comprising poorly and commercially bred, pet-purposed dogs, one would expect to find a wide range of phenotypes and behaviors. This is especially true, given the tendency to avoid mixing these lines; thus, differences will continue to become more and more distinct as these lineages continue to diverge from one another.

Thus, this fifty-fifty perspective on certain breeds, such as German Shepherd Dogs, may be an artifact of artificial selection gone awry, with breeders continuing to try to differentiate one line from another. We have found clusters of highly heritable behavioral issues with certain breeds in certain geographic regions, but not in others. For example, neuroticism, which is a highly heritable trait, might be present in more than 50% of the dogs of one breed from one area of the US, but almost non-existent in other regions. This can likely be attributed to the concentration of particular genes in those areas where the unsavory behavior or personality trait arises frequently, which occurs as a result of breeders failing to remove dogs with these psychological and behavioral issues from their breeding programs. As a result, without bringing in new genes, certain areas would have an over-representation of issues such as OCD.

Adam Winston always has to make exceptions for ancient breeds, however. "They can be difficult to train. It's also often difficult to maintain their focus on a handler," he said. "I often teach group/family classes. It can be challenging when a client brings a Chow or Shiba Inu into a group class full of Retrievers, Spaniels, and Terriers. The families with ancient breeds have difficulty incentivizing and motivating their dogs. Meanwhile, the family's classmates are swiftly moving ahead with the lesson plan by simply showing their Golden Retriever a treat or saying, 'Yes!' The ancient breeds are the only group who consistently struggle to finish my 6-week training courses."

For Adam, it's important to keep these differences in mind.

"As dogs continue to co-evolve with us in the 21st century, it's important that we recognize and consider why and how these animals were created. Sometimes, we also have to be mindful of our expectations. For example, a family might get really frustrated that their dog often ignores them when distracted. Maybe that family's dog is a Scenthound."

"I want to share with the family a little history about Bloodhounds. Bloodhounds were bred to practically ignore their handler and follow the scent of wild game. It's not a surprise that someone's youthful Bloodhound constantly wanders off while the family calls his name."

"Yes, I can help that family strengthen their Bloodhound's ability to recall. However, by understanding the dog's breed group and history I can also help the family and community by offering management and containment advice during my basic obedience training session."

Behaviorist Kevin Yeo has noticed the highest rates of reactivity in toy breeds. "These include breeds like the Maltese, Toy Poodle, and Chihuahua. Also, in Singapore, there's a growing 'poodleization' trend, where breeds are mixed with one another. Maltipoo, which is a mix of Maltese and Poodle, Cavapoo, which is a mixed of King Charles Cavalier and Poodle. You get the idea. These mixed breeds have also increasingly become part of my caseload for reactivity cases. I have no idea if unstable gene lines are a result of this and haven't read any research that suggests that this is so."

We don't just see "poodleization" in Singapore. Dog breed behavioral patterns tend to vary between different countries of origin. A 2006 study looked at the behavioral profiles of 56 purebred dogs in Japan. Researchers used a survey from prior studies that had been conducted in the UK and US and found statistically significant differences between breeds for the behavioral traits that were examined, as would be expected. But when comparing breeds from different continents (in this case, dogs from the US and dogs from the UK), there were within-breed similarities. This study found that trainability, reactivity, and aggressiveness were consistent across US and UK surveys. This suggests that the genetic basis of breed-specific temperamental traits is robust and consistent across different cultural and regional groups of owners (Takeuchi and Mori, 2006).

"As always, the second category are mongrels, or general mixed breeds. These are usually street dogs and reactivity is usually the precursor to aggression and mostly observed shortly after adoption. This usually stems from fear, which is passed down genetically, as well as a result of improper or missing socialization."

"Perhaps, the most important thing is to ensure that a particular breed group's innate instincts are satiated. Most behavior problems, such as reactivity, could stem from frustration due to an inability to perform said behaviors, or due to punishment for performing said behaviors ... I always say ... if you are trying to stop a Terrier from digging, good luck!"

For behaviorist Wendy Dahl (MA, CBATI), breed-specific education for pet owners can help them better understand their dog's behavioral nuances. "Sometimes educating the guardian about specific breed characteristics may help them understand their dog's behavior," she explained. "For instance, Great Danes can be very physical when they are playing, whether it's with other Danes or dogs from a different breed. Watching two or three Danes cavorting with each other around the yard and rearing up to box at each other can be intimidating to those not familiar with the breed."

"A Rottweiler can act like a linebacker and plow his thick body into a group of playing dogs if he thinks they are playing too rough, and they can be quite vocal about it. Dobermans are quick and can grab another dog around the neck or bite at a leg, all in the name of play."

"If a young Dane doesn't have a chance to play with other agreeable dogs or have other physical ways to channel their energy, they can become physical with the people or children in their life. They become frustrated when their guardian tries to control them in public or at home by restraining their young selves with a tight leash, asking them to sit when they are highly aroused or taking them to events where strangers want to pet them."

"There are some genetic characteristics that Dobermans display like blanket or flank sucking, and other self-soothing behaviors. A guardian should be prepared for this type of behavior, which may need some management structure when a dog's licking creates a wound. It's important that solutions are without punishment."

"These three breeds have a high desire to be near their person. They are social creatures and thrive when they are integrated into the family. Isolation can create challenging behaviors like separation anxiety, physical intimidation toward strangers and destruction from boredom."

Professional dog walker and dog trainer Lori Theis, who owns Upward Dog Walker, has been working with reactive dogs since 2011, when she purchased one of the first dog walking businesses in Seattle. "At that time, my experience with dogs was limited to being a lifelong dog owner," she said. "I'd never heard of reactivity and had never really trained a dog or found myself concerned about a dog's behavior. Within a few months of walking dogs every day, I had received a crash course on breed groups, reactivity, and the urgent need for understanding dog behavior."

"My first declaration on this subject is that *far* more dogs are anxious on some level than those that are not. Anxiety is a widespread phenomenon in the US. There seems to be no definitive research on this topic and the professional consensus pins the problem on our shelter system. That said, most dog owners don't see their dogs as anxious. It's only when a dog's behavior becomes egregious or dangerous do they seek help."

"What I have found to be completely true is that individual breeds are predisposed to perform certain behaviors. Herding breeds for instance are 'high seek' and highly reactive but very trainable. Terriers can be powerful, tenacious predators and are also highly reactive and a little more difficult to train. I have yet to meet a Boston Terrier that is not reactive and almost impossible to train. An uncommon breed, represented by five individuals over the years of my work, is the English Setter. Every one of those individuals was highly reactive to the point of their owners never taking them on walks."

"As far as training goes, breed absolutely matters, as do family dynamics. I have found Chows and Great Pyrenees to be less motivated during training than other breeds. In my experience, Chows follow their breed description as only trusting a small circle of humans. Great Pyrenees want to guard their home and territory. These are all genetic components that I help owners with when developing a behavior strategy. For instance, the owner of a Chow needs to understand that their dog is probably never going to allow every single human they meet to pet them. The owner of a Great Pyrenees cannot expect their dog to lose interest in guarding their home. These considerations affect the family dynamic which means perhaps the children in that home take a greater responsibility in making sure the Chow is not exposed to stressful situations."

"Of the hundreds of different dogs that I have walked and worked with over the last 8 years, Labs and Retrievers represent the majority of breeds, with German Shepherd-type dogs coming in second. Both of those breed clusters are supposed to be low reactivity dog breeds, but I have found that not necessarily the case, and I definitely attribute it to lack of early socialization."

Celeste Walsen (DVM) is the executive director of the Courthouse Dogs Foundation, an organization that places support dogs for youths going through trauma. While courthouse dogs are increasing in popularity, some judges have balked at having them in their courtroom, believing they would be a distraction. "I believe that reactivity is an inborn trait," Walsen explained. "Take a dog named Molly, for example. She was a three-quarters Labrador Retriever, one-quarter Golden Retriever who was from Canine Companions for Independence. She exhibited very low rates of reactivity. We were in juvenile court in Virginia with Molly and an 8-month-old puppy who was in training," Walsen recalled (Fig. 7.2). "The judge banged his gavel down several times to see if the support dogs would react, and they didn't, so he allowed them to stay." Molly and the puppy came from a long line of dogs that had low rates of reactivity. "Molly was in court once when the defendant was knocking chairs over, and she didn't even raise her head," Walsen said. So why might some breed groups, breeds, and certain lines within those breeds exhibit less reactivity than other dogs? Perhaps dogs that are intentionally bred for certain traits (like low reactivity) and exposed to a wide variety of experiences early in life, may be able to cope with stress more readily than those that are not bred for traits like lower reactivity.

Building Better Service Dogs

Marina Hall Phillips is the coordinator of the Assistance Dogs of America North American

Fig. 7.2. Molly. Photography courtesy of Ellen O'Neill-Stephens.

Breeding Cooperative (ABC). After going to school for animal science, she found her calling when she began working as a kennel tech at a guide dog school. One of her earliest interactions was teaching a blind woman how to bathe her guide dog. "Seeing that symbiosis of vulnerability, because of her disability ... I was so struck with it," she recalled.

The ABC is member-driven. Members vote and have delegates. There are currently 37 North American member schools in the cooperative, with additional partnerships with guide schools in the Netherlands and Australia.

"There are currently 38 dogs in the breeding program, plus their offspring," Marina explained.

Potential parents of service dogs are carefully chosen.

"We have a breeding evaluation work group that gathers data. When you look at it from population genetics perspective, representatives from the family line that do the job and do the job well. Indexing a lot of the dogs and using computer programs to see how related they are. We're trying to use the dogs that are more closely related with the dogs that are already doing well in the field. We keep moving toward using the population of proven dogs while doing what we can to mitigate disease and behavioral deviations."

For Marina, it helps to adopt a centralized framework to assess dogs' behavior.

"We use a system called the behavior checklist. It was created by the Guide Dogs for the Blind. They have a series of observational protocols when a puppy is young. There's more of a staged environment and they are assessed in that staged environment. There's a database that holds these results. We can run statistical analyses on and identify which dogs have greater estimated breeding values of certain behavioral traits."

"There's a difference in purpose-breeding—we don't want the most intelligent, you don't want the most heavily weighted in a trait. You want a dog who's more like a nurse or a social worker than a soldier. It's optimizing the moderate traits. If you push too far, our biggest challenge is sensitivities ... you need a dog with a certain level of social sensitivity to enjoy and willingly work with a person who's not a professional dog handler."

"You need that social and physical sensitivity for dogs to be able to do this work. If you slip too far, their sensitivity is too high, and they don't hold out under pressure, in the public, high stimulus, it's too much for them. A dog that's considered bombproof in one kind of work might make their own self-interest choices in some scenarios. A good service dog needs to have enough resilience to handle unpredictability, but enough sensitivity to work with a person with a disability."

Dogs who are being evaluated to become a part of the cooperative receive a numerical score in the behavioral system.

"When a person is assessing these scores, they decide which two would be a good pairing," she said. "Ideally, we're looking at getting more and more consistent; we're pairing similar dogs together. We also navigate the inbreeding coefficient and select the lowest to optimize the duration of the population."

Sample items on the behavioral checklist include noise sensitivity, object reactivity, underfootings (how well they can continue to stay on task), and stranger reactivity.

All of the dogs in the ABC are Labrador Retrievers, Golden Retrievers, or a mix between the two.

"We are breeding within specific, currently existing breeds of dog. While we've worked with a lot of other breeds of dogs, they just don't bring that same level of consistency to the table that the Labrador does. In the assistance dog field, we've selectively bred these dogs, and it's fairly different from your show or hunting Labrador lines, although phenotypically, they look more or less the same." But assistance dogs are bred for their personality, not to be voted "most attractive."

"In assistance dog work, no one's opposed to a dog that looks a little bit different. Sometimes, the choice is made to cross Labradors and Goldens. The Golden Retriever typically has more sensitivities than the Labrador, and that's not always a bad thing. They're typically a more emotive dog. For certain lines of work, that emotive tendency tends to draw people out. The Labrador is more stoic, less expressive. Sometimes the sensitivities are more challenging, there's a novel sound or noise. The Labrador tends to be more resilient or stoic."

Another very important behavioral trait for a successful service dog is biddability, where they have a high capacity for being led, controlled, or bid to do something. "Assistance dogs should carry a low reactivity level and a high level of biddability to work for extended periods of time with a handler who is not a professional dog handler," Marina explained. "If a medical alert dog doesn't perform well, he or she cannot simply be removed from the environment until he is up for working. We rely on selecting dogs who have a very high level of purpose-bred biddability so that we do not have to cajole, bribe, or entice them to work with their partner on a regular basis. This trait is key. You can imagine we could aim for producing a line of dogs who are very environmentally sound (think the old mastiff types, for example), but if they do not possess a desirable level of social sensitivity, then they will not be easily biddable."

Breed Differences, Temperament, and Clinical Cases

To diagnose behavior issues, we recommend taking multiple factors into account, including breed, age, social environment, and learning history (such as trauma or experience in a shelter). Of particular importance is the dog's personality—we'll use case studies to examine separation anxiety and client mismatches. For example, a Border Collie (a herding breed with a strong desire to have a job and be active) living with an owner in an apartment that limits breed-specific behavior.

In practical clinical cases, one of the most important ways to use breed differences is in the (average) differences in personality across breeds. These are, to a great degree, genetically driven, and well reflected in breed differences. A recent study using a scientifically validated online questionnaire found significant differences in personality across breeds. They note the following important point, however: "We compared the distributions of breeds whose mean scores most differed and showed that even though the distributions mostly overlapped, breeds showed differences in where the majority of individuals resided on the trait score distribution. Therefore, a dog's breed is not a predictor of its personality, but the probability of showing certain personality traits differs between breeds" (Salonen *et al.*, 2023).

We've discussed the association between Chow Chows and separation anxiety, but other patterns reveal themselves, as well. Some of these are obvious, some are less obvious, and many of them are poorly documented by science, but familiar to canine behaviorists, trainers, and owners. The high intelligence and need for both mental and physical stimulation drives many a behavioral issue in Border Collies and Australian Shepherds and their related breeds. A high demand for physical exercise is seen in many of the hunting dogs, like most of the Terriers. You can "solve" so many cases of excessive barking, inappropriate urination and defecation, and separation anxiety with prescriptions for More Exercise!, dog day care, dog walkers or joggers, or simply more walking and play time with owners. Exercise will frequently fix a problem: a tired dog is a good dog!

Breed differences don't entirely predict temperament: individual assessment of behavior and temperament is critical in these clinical cases, and the observations have to be done in their natural, comfortable habitat, which is usually the home. You can find Australian Shepherds that were *not* high energy (the genetic "show" lines are moving in that direction for many breeds, with the "working" pedigrees retaining that high energy and problem-solving/learning ability). Behavior issues can also be created by too much exercise, in the case of senior dogs who have developed arthritis or other diseases. Listening to your dog, and assessing the situation with a behaviorist, is critical in these cases.

Lifestyle mismatches have become much more common in the clinical, in-home behavior business, and are often reflected in poor choices of breeds for the owner's lifestyle. "I like the way they look" is not a good basis for choosing a lifelong companion. Behavior is far more important, and it is usually on the basis of average breed personalities that we have to make such choices. If you work long days at the office, a Border Collie, Australian Shepherd, or Cattle Dog might not be for you. Go for a nice Labrador, Sheepdog, or Toy breed: these were bred to be lower energy and require less stimulation. Labradors and many of the 'gundogs' were bred for burst activity and long periods of waiting; their stamina for the long haul is poor, and they make terrible jogging partners, as a breed. Sheepdogs and other guarding dogs were, again, bred to sit and, well, guard. They were not bred to chase, hunt afoot, or herd. Better for the small apartments and limited exercise time of many modern families. Finally, the toy breeds or lap dogs were specifically bred for royal court and very little exercise and stimulation. A better choice for Grandma than a Great Dane. No, really! We had such a case: helping Grandma, who was over 90 years old, manage her Great Dane puppy, which was given to her by her kids! This was resolved by scheduling a chat with the kids and finding a lovely new home for the Great Dane. It turned out that a rescued Pomeranian was a much better fit for Grandma.

Again, breed differences and histories explain why Labradors, Retrievers, and related breeds make great assistance or service dogs: they have been bred for hunting, and specifically a form of hunting in which they need only a burst of energy, not a constant high level of exercise. These dogs can rest and wait for long periods of time. They are very smart, easily trainable, and have been selected for very low aggression: they are supposed to retrieve the target, not attack and kill it. "Coursing" breeds, breeds that are used to chase down prey, like Deerhounds, Wolfhounds, Elkhounds, Catahoulas, and others, are bred for stamina and energy.

Assistance dogs must be low energy, bred to wait, very smart, and low aggression; pre-designed, one might say, for helping humans with their disabilities. Now, this isn't to say that individuals of other breeds don't fit the same mold: genetics is only part of the story, and even in populations of service dogs specifically bred for the work, there's a dropout rate as high as 50%. Genetics and breed characteristics only help us to narrow the field and be more efficient in our choice of a dog for the job, whether that job is hunting, helping a blind person across the street, or being a snuggly partner for an elderly owner.

Rates of Reactivity

Recognizing and measuring animals' personalities can increase the welfare of any species, in a wide range of settings, including zoos (Watters and Powell, 2012), as well as the treatment of our closest companions. Reactivity to stimuli, such as noise, is an important aspect of temperament. Not all dogs will react to the same stimuli in the same way, and noise reactivity is very common; in some surveys, up to 50% of pet dogs will have noise reactivity during some point in their lives (Storengen and Lingaas, 2015; Scheifele *et al.*, 2016). Dogs typically show the most noise reactivity with fireworks, gunfire, and thunderstorms (McCobb *et al.*, 2001), with 92.9% of dogs that have a fearful reaction to thunder also reacting to fireworks and 73.8% also exhibiting a fear of gunfire (Tiira *et al.*, 2016). They may also react with anxiety or fear to more common noises, such as vehicles, sirens, leaf blowers, vacuum cleaners (Tiira *et al.*, 2016), or stove alarms. During extreme cases, such as the case of Charlie the Chow Chow, there can be both mental and physical effects. Cases such as Charlie's are a perfect example of how noise reactivity can be comorbid with a suite of other disorders. Among humans, misophonia is a rare (less than 200,000 cases diagnosed annually) condition where someone exhibits high rates of reactivity, ranging from irritation to rage, to typically benign repetitive noises such as clicking a pen, tapping a foot, or loud chewing.

Some cases of noise reactivity may progress to noise phobia (Overall, 2013). Noise reactivity can manifest in relatively mild ways, such as pacing, panting, salivating, withdrawal, or trembling. Noise phobia, though, can result in far more severe reactions, such as Charlie's. So, what makes noise reactivity turn into a full-blown phobia? With dogs, there are certain risk factors for this progression to occur, including pain, limited vision, and hearing impairments (Overall, 2015), that might alter once-familiar, once-benign sounds. With humans, extroversion has been identified as a key predictor of noise sensitivity (Shepherd *et al.*, 2015), and with both dogs and humans, noise sensitivity, including reactivity and phobia, is often comorbid with other anxiety disorders. Charlie was already a bit of an anxious chap, and his noise reactivity, paired with other anxiety-related conditions, led to true noise phobia.

Why do such a large percentage of dogs exhibit noise reactivity during their lifetime? A big part of that is exposure. Noise reactivity with dogs can be particularly problematic over time. Each time a dog hears a loud noise, such as fireworks or thunder, their reaction often gets worse, and not better. This is because they were scared before, and now that they know what's coming, they're even more frightened. It's a form of post-traumatic stress disorder (PTSD), and certain sounds can become triggers. Then, as a dog ages and develops other health issues, such as metabolic issues like Addison's Disease, or they feel more vulnerable due to declining health, their responses to those auditory inputs become even worse. This is compounded when some dogs are scolded or punished for their fearful reactions to loud noise.

While there's a great deal of individual variation, breed differences in reactivity have also been shown in scientific studies. One study examined the presence or absence of noise reactivity with three breeds of herding dogs: Australian Shepherds, Border Collies, and German Shepherd Dogs. The researchers focused on these breeds because they are commonly used as detection, search and rescue, police/patrol, and service dogs. They asked the owners to answer a questionnaire about how their dogs reacted to storms or thunder, fireworks, gunshots, and other noises, such as a vacuum cleaner. The responses included how likely they were to react to each noise, what type of reactions they had, and how often they were exposed to each noise. If the dog showed any of the following behaviors: salivation, defecation, urination, trembling, vocalizing, destroying, pacing, escaping, freezing, panting, crouching or hiding, then they were considered "reactive." The researchers developed a weighted scale for each dog's reaction, which included the frequency and intensity of their response. They found that more Border Collies reacted to noise (68% of the dogs) than Australian Shepherds (46%) or German Shepherd Dogs (28%). Of the dogs that reacted to noises, most Australian Shepherd and Border Collie owners said they reacted regardless of the noise source, and reacted most of the time. The likelihood of reaction, however, was higher in Border Collies than Australian Shepherds. German Shepherd Dog owners reported that if they had dogs that reacted to noise, the dogs reacted about 50% of the time, regardless of the noise stimulus. While all of the dogs in the study were from the Herding Group, the Border Collies and Australian Shepherds had more noise reactivity than the German Shepherd Dogs, providing evidence that further studies should be conducted to assess the progression and comorbidity of reactivity (Overall *et al.*, 2016).

Personality Matters

Reactivity has been used as a measurement for temperament, particularly with working dog breeds. Much of the research on canine temperament and measures of research like the C-BARQ were originally created for or focused on working dogs. But with 90 million dogs in homes in the US (Megna, 2023), studying not only the temperament, but the personality of dogs (and, in particular, our pet dogs) is an increasingly important area for research.

Carly Loyer focused her PhD on personality with domesticated dogs. With a dearth of prior research and many scientists' hesitation to use the word "personality" for non-human animals, she faced a steep uphill climb. The pre-existing literature pertaining to humans and other animals provided an excellent point of departure.

"We know enough about personality in humans and other animals to say that canine personality could affect dogs' responses to various training techniques, their aptitudes, and their social behavior, including aggression," Dr Loyer explained. "We can assign mean, general, fundamental personality characteristics to breeds or breed groups. I think we can assign some general behavioral traits to dog breeds, particularly working lines or breeds that are still selected for their behavior," she explained. "The Russian silver fox experiments demonstrated that breeding for behavior can create drastically different behavioral phenotypes or temperaments."

Typically, the most rigorous personality research in animals involves direct behavioral measurements of traits in controlled settings. This kind of work is time consuming and expensive, and thus not common. Next, we review a number of scientific studies on personality in dogs, starting with Pavlov in the 1900s and a foundational study from the 1960s and then follow with more recent papers using a variety of methods and approaches.

For Whom the Bells (Don't) Toll

Even before he won a Nobel Prize, a scientist with high energy and an even more insatiable curiosity was delving into the science of animal behavior. While he's more well known for ringing bells that caused dogs to salivate, Ivan Pavlov's groundbreaking research in the early 1900s examined canine temperament, as well. In 1901 Ivan Pavlov, along with his assistant, Ivan Filippovitch Tolochinov (Ivan is a very common name in Russia, where the Ivans conducted their research, but one wonders: what did they call one another?), developed the concept of the conditional reflex, which would later be referred to as "conditioned reflex" or, as it's most commonly referred to today, "classical conditioning." Pavlov paired an auditory stimulus, such as a metronome or a buzzer (sorry, he didn't use a bell in his research, even though his name and work would become synonymous with ringing bells), with the presentation of food. After a series of consecutive sequences, the dog would salivate with the auditory stimulus alone, even when the food was absent. Pavlov characterized the dogs in his study as "Quiet," "Inhibited," "Lively," or "Excitable." Dogs in the "Quiet" category learned at a slower pace than most of their peers, but exhibited more consistency; the dogs in the "Inhibited" category also learned at a slow pace, and with less consistency; the "Lively" type learned at a rapid pace, and had the most equilibrium between excitation and inhibition; and the "Excitable" type excelled at excitatory conditioning, but performed poorly with forming inhibitory connections (Pavlov, 1906). Pavlov, who was a physiologist, became well known in psychological research for his work in classical conditioning and further research into animal temperament by people from a wide range of fields, including veterinary medicine, ethology, psychology, and zoology, provided further evidence of different personalities in animals. To date, "personality" has been attributed to more than 100 different animal species.

The Classic Study

Scott and Fuller's landmark longitudinal dog study was published in 1965 (Scott and Fuller, 1965) and used carefully controlled laboratory work with five breeds of dogs (Basenjis, Beagles, American Cocker Spaniels, Shetland Sheepdogs, and Wirehaired Fox Terriers) and explored the heritability of behavior, including reactivity. They assessed dogs at ages 17, 34, and 51 weeks in social and non-social conditions to elicit both emotional and physiological (heart rate, electromyogram from the thigh) responses to a novel situation, the approach of a stranger, a doorbell, a shock on the leg, and being addressed in a harsh voice combined with moving their muzzle forcibly side to side. The dogs had breaks, or control conditions, between the testing conditions and were measured for how long it took them to leave the testing area once released (latency). The specific behavioral responses that observers recorded included body and tail posture, tremor, investigation, attention to observer, escape behavior, lip licking, vocalization, panting, tail wagging, resistance to forced movement, biting, and elimination. The researchers rated the behaviors of each dog on a Likert scale from 1 to 5 (with 1 being a low response and 5 a higher/more energetic response).

They found that Terriers, Beagles, and Basenjis were more emotionally reactive than Shelties and Cocker Spaniels, regardless of age. There were *highly* significant differences between breeds, and only heart rate change during bell ringing and handler effect did not vary between breeds. The authors stated that "it is much more difficult to find scores which are not affected by breed differences than to find those which are affected." While these results provide strong support for breed differences in emotional reactivity, it is important to keep in mind that dogs raised in such uniform conditions as the five breeds of dogs used here, would maximize the detection of genetic effects. This is a good thing for a scientific study, so we can determine the role of genetics on behavior, but this may not translate as well to the real world where environmental differences may affect breeds and individual dogs in different ways. And, even in this highly controlled setting, they found wide differences between individuals within breeds. In fact, they suggested that there was enough variability in emotional reactivity left in Cocker Spaniels that with a few generations of selective breeding (artificial selection) they thought you could produce a litter that reacted more like the Terriers, Beagles, and Basenjis.

Two decades after Scott and Fuller published their work, researchers Hart and Hart (1985) used a dog breed database to summarize the behavioral traits of 56 breeds that had been evaluated by 48 small animal veterinarians and 48 national registered dog obedience judges. They compared breeds on 13 traits using a factor analysis to determine if there might be some basic underlying tendencies. Three factors explained most of the variation between breed groups: reactivity, aggressiveness, and trainability (Table 7.1). They argued, and we agree completely, that this was a first step in creating some objective measures of breed groups, and that this information should be used in combination with an evaluation of the dam and sire, as well as the environmental conditions in which individuals are raised.

Using the same survey that Hart and Hart (1985) used, a 1998 study ranked the 56 most popular breeds of dog in the US using those same 13 traits and compared them with a survey of the 49 most popular breeds in the UK (Bradshaw and Goodwin, 1998). Twenty-four of the 36 breeds in common showed similarities with specific traits, such as ease of housetraining, aggression, and reactivity, while nine of the 36 breeds (Airedale Terrier, Beagle, Boxer, Dalmatian, Irish Setter, Old English Sheepdog, Samoyed, Standard Poodle, and Welsh Corgi) showed distinct differences between the two countries. Three breeds (Chihuahua, Scottish Terrier, and Standard Dachshund) had smaller differences between the two countries (Bradshaw and Goodwin, 1998). The Airedale Terrier, Standard Poodle, and Beagle clustered as "high reactive" in the US study conducted by Hart and Hart, but clustered as "medium to low" in the UK. In contrast, the Samoyed and the Boxer clustered as "low reactive" in the US, but "medium" in the UK. There were also differences between the two countries in how breeds were characterized for aggression. Specifically, the Irish Setter was medium for aggression in the US and low in the UK, and the

Table 7.1. Behavioral profiles of dog breeds: cluster analysis (Hart and Hart, 1985).

	Clusters						
	1	2	3	4	5	6	7
Reactivity	High	Very low	Low	High	Low	Very low	High
Trainability	Low	Low	Low	Very high	High	Very high	Medium
Aggression	Medium	Very low	High	Medium	Low	Very high	Very high
Breeds	Lhasa Apso Pomeranian Maltese Cocker Spaniel Boston Terrier Pekinese Beagle Yorkshire Terrier Weimaraner Irish setter Pug	English Bulldog Bloodhound Elkhound English Sheepdog Basset Hound	Samoyed Malamute Husky Saint Bernard Afghan Boxer Dalmatian Great Dane Chow	Poodle Shih Tzu Shetland Sheepdog Springer Spaniel Welsh Corgi Bichon Frise	Newfoundland Brittany Spaniel Labrador Retriever Chesapeake Bay Retriever Shorthair Pointer Australian Shepherd Keeshond Collie Golden Retriever	German Shepherd Akita Doberman Rottweiler	Cairn Terrier Scottish Terrier Chihuahua Schnauzer Highland Terrier Airedale Terrier Fox Terrier Silky Terrier Dachshund

Provided with permission from the Sheridan Press, Inc.

Dalmatian was high in the US and medium or low in the UK. The Old English Sheepdog and Corgi were clustered as high aggression in the UK, but were medium or very low in the US.

Apples and Oranges

Reviews of the literature on dog temperament and personality find that most papers involved pure-bred dogs, and that the terminology and validity of the works varied widely, making it difficult to apply findings across studies (Jones and Gosling, 2005; Gartner, 2014). One review reported that reactivity, fearfulness, sociability, responsiveness to training, and aggression were the most commonly investigated traits (Jones and Gosling, 2005), while another reported that the most studied dog personality dimension is aggression, followed by anxiety (Gartner, 2014). These are the traits, of course, that are most salient to humans. Most studies did not report the reliability of their measurements, but Jones and Gosling (2005) did the calculations when available, and found that measures of dog temperament were fairly reliable, and on par with human personality standards. Validity measures were lowest for studies that looked at puppies, and highest with adult dogs. Again, these measures matched or exceeded human personality validity measures. Unfortunately, they reported that very few studies looked at temperament across breeds.

Perhaps because they have co-evolved with humans for so much longer than their feline peers, the lion's share of personality studies focus on dogs and not on cats, though both species are sociable and complex. The tide is slowly turning, though. While there's still a large disparity between canine and feline personality studies, the work that has been done on cats shows that felines have reliable personality factors and developmental differences (Gartner, 2014). Researchers like Udell are now studying cats, as well, and finding them to be an equally complex and fascinating species. Thus far, the research on cat personality reveals that issues still remain regarding study validity and reliability, and Gartner's work revealed the importance of standardization in terminology and methods.

Much of the scientific work we have on breed differences in temperament comes from surveys of owners or dog professionals such as veterinarians and obedience trial judges. Surveys have the benefit of allowing large samples to be collected relatively easily, but can lack validity, especially if the results aren't corroborated by assessing the breeds objectively to ascertain if the opinions of the humans match the behavior of the canines. Let's look at some surveys that address breed differences in temperament, and we'll follow with a summary of the recent direct observational research in this area.

The first study used the Monash Canine Personality Questionnaire (MCPQ) and surveyed 455 owners of 82 dog breeds in seven breed groups that were recognized by the Australian National Kennel Council (ANKC) (Ley *et al.*, 2009). These breed groups included Gundogs, Hounds, Non-sporting, Terriers, Toys, Working Dogs, and Utility Dogs. Utility Dogs recognized by the ANKC included Akita, Alaskan Malamute, Boxer, Doberman, Rottweiler, Samoyed, Schnauzer, and Siberian Husky Approximately one-third of the dogs represented in the questionnaire were mixed breeds or unrecognized by the ANKC. They suggested that there are *five* dimensions of personality in canines: Amicability, Extraversion, Neuroticism, Motivation, and Training Focus. Of the seven breed groups, Gundogs and Working Dogs scored the highest on Training Focus, while Terriers and Working Dogs scored the highest on Extraversion (Ley *et al.*, 2009).

In contrast, a 2011 study evaluated *four* behavioral traits (Boldness, Calmness, Dog Sociability, and Trainability) among dog breeds, and tried to assess whether the personality survey results could be related to specific breed groups or the genetic relationships between breeds (Turcsán *et al.*, 2011). The study surveyed 5733 German owners representing 98 different dog breeds and seven breed groups (including Herding Dogs, Hounds, Non-sporting Dogs, Sporting Dogs, Terriers, Toy Dogs, and Working Dogs). They assigned dogs to breed groups based upon the American Kennel Club's (AKC) definition, unless the breeds were not recognized, and then they assigned them to whichever AKC breed group most closely matched their classification by the Fédération Cynologique Internationale (FCI). This was done to try and make the results comparable. The genetic groups the researchers used were based upon a prior study (Parker *et al.*, 2007) which included categories such as "Ancient Breeds," "Mastiff/Terrier," "Herding/Sighthound," "Hunting," and "Mountain and Spaniels." When all 98 breeds were examined, the researchers found significant differences between breeds in Trainability, Boldness, Calmness, and Dog Sociability. The most popular breeds in their sample, with a minimum of 200 or more dogs, included the Beagle, German Shepherd, Golden Retriever, Jack Russell Terrier, and Labrador Retriever. These common breeds didn't differ from the overall total sample average on Trainability and Calmness. The Beagle and Labrador Retriever were significantly higher on Dog Sociability, while the German Shepherd and Jack Russell Terrier were significantly lower on this trait. Jack Russell Terriers and Labrador Retrievers also scored significantly higher on Boldness. For both traditional (AKC/FCI) and genetic (prior study) breed groups, they only found significant differences between groups in Trainability and Boldness, and not in Calmness and Dog Sociability. Specifically, in the traditional grouping category, Herding dogs were reported to be more trainable than Hounds, Working Dogs, Toys, and Non-sporting Dogs. Sporting Dogs were also reported to be more trainable than Non-sporting Dogs. In the genetic categories, Ancient Breeds were less trainable than Herding/Sighthounds or Hunting clusters. In the traditional grouping category for Boldness, the only significant difference reported was that Terriers were rated higher than Hounds and Herding Dogs. Similarly, in the genetic clusters, Mastiffs/Terriers were bolder than Ancient Breeds, Herding/Sighthounds, and Hunting Dogs (Turcsán *et al.*, 2011).

The third study developed their own questionnaire and surveyed Hungarian owners of 284 dogs across ten breed groups that are recognized by the FCI, with an additional 11th group of 86 "mongrels" (Mirkó *et al.*, 2012). Their questionnaire resulted in *four* personality traits for analysis including Activity, Aggressiveness, Stranger-directed Sociability, and Trainability. The researchers found some differences between breed groups on all traits, except for Aggressiveness, but some of the differences were unexpected (e.g. Retrievers, Flushing Dogs (birders), and Water Dogs were less trainable than mongrels). Note that "flushing" dogs are a type of hunting dog that are used to drive birds out of the grass and foliage for hunters. This finding can at least partially be explained, however, by the fact that 77% of the Labrador and Golden Retrievers in their sample were untrained, which would likely influence how the owners rated them on Trainability. Other unexpected results, such as Terriers being more sociable than all other breed groups, could be explained by the comparatively small sample sizes in their study. Overall, the sample sizes in this study may not be sufficient to overcome the variability in the dogs' housing (flat, garden, home, or combination) and training variability (from high to none), much less other differences between individual dogs and the inherent biases of owner reports.

This definitely is "apples and oranges" across these three survey-based studies, as they differed in both the number of personality dimensions they analyzed (five versus four and four), as well as the actual personality dimensions:

- *Study 1*: Amicability, Extraversion, Neuroticism, Motivation, and Training Focus (Ley *et al.*, 2009)
- *Study 2*: Boldness, Calmness, Dog Sociability, and Trainability (Turcsán *et al.*, 2011)
- *Study 3*: Activity, Aggressiveness, Stranger-directed Sociability, and Trainability (Mirkó *et al.*, 2012)

The studies also differed in their definitions of breed groups (ANC versus AKC/FCI and only FCI). This makes drawing parallels practically impossible! In addition, the dog and human populations vary, from Australia, Germany, and Hungary, so how can we generalize the results to the population of breeds around the world?

All of these studies were done in the 2000s, so they're relatively recent, but if you look at the literature in its historical entirety and try to make it comparable, you'll find some results that come up consistently.

When you draw across many more survey studies than we have shown here, including owner-reported and reports by dog experts, the overall results suggest that the Miniature Schnauzer, Scottish Terrier, and West Highland Terrier rank the highest in reactivity, while Hound Group breeds score high for inattention, and Herding Group breeds are higher than Toy and Non-sporting Group breeds on active play, constant motion, and anticipation (Mehrkam and Wynne, 2014). Labrador Retrievers, Cocker Spaniels, English Springer Spaniels, and Toy Poodles are reported to be less likely to develop fearful response to loud noises such as fireworks, and Labrador Retrievers are also reported to be the least fearful in startle tests conducted between 1 and 18 months of age, while German Shepherds are the most fearful (Mehrkam and Wynne, 2014).

Unlike survey results, direct observational research has the advantage of reducing owner bias, as it is conducted typically by a trained observer who isn't attached to the dog. This type of work is more time consuming, but is also subject to variability between observational methods, making direct comparisons across studies more difficult. Some of the best baseline work comes out of the Swedish Dog Training Centre, that specializes in the breeding of service dogs, and maintains longitudinal records of breeding and behavior. Wilsson and Sundgren (1997) evaluated differences between German Shepherds and Labrador Retrievers from the center's assessments. The researchers scored dogs on ten behavioral characteristics that the dogs showed in response to seven different testing situations. They found that German Shepherds scored higher in Sharpness (reacting with aggression) and Defensive Drive, while Labrador Retrievers scored higher in Courage, Nerve Stability (reacting appropriately to a situation), Hardness (lack of lasting effect of a pleasant or frightening experience), Affability, and Cooperation. Labrador Retrievers also reacted less to gunfire, which makes sense, given their long histories as hunting dogs (Wilsson and Sundgren, 1997). Early on during the development of this breed, less reactive dogs would have been bred with other equally non-reactive dogs. A hunting dog that ran away from gunfire wouldn't be able to earn their keep.

A 2016 study examined the behavioral differences between Golden and Labrador Retrievers that were bred as a "common type" for pet/conformation versus a "field type" for hunting (Sundman *et al.*, 2016). The study looked at 902 Goldens (698 of which were common type and 204 of which were field type) and 1672 Labradors (1023 of which were common type and 649 of which were field type). The researchers examined six behavioral characteristics: Play Interest, Curiosity, Social Curiosity, Social Greeting, Chase Proneness, and Threat Display. Type-level differences were found across both breeds, with common-type Labradors exhibiting more curiosity than their field-type counterparts. Golden Retrievers exhibited the opposite pattern, with field-type Goldens having higher rates of Play Interest, Chase Proneness, and Social Greeting than their common-type peers, but no statistically significant difference in their rates of Social Curiosity. Field-bred Labradors had higher rates of Play Interest than common Labradors, who had higher rates of Curiosity, Social Curiosity, and Social Greeting. There was no statistically significant difference between their levels of Chase Proneness. The study also found differences across categories with age and gender, with curiosity increasing with age and the other categories decreasing with age. The male dogs in the study also had higher rates of Curiosity, Play Interest, Social Greeting, and Social Curiosity than their female peers, who had higher rates of Threat Display.

Alaskan sled dogs, which have been purpose-bred, and not appearance-bred, for centuries, exemplify the difference between different genetic lines: they don't have to follow official breed standards pertaining to appearance or size. As performance-bred dogs, they're known for their endurance, pulling strength, gait, and speed; they are mixed-breed Northern dogs with ancestry coming from several breeds, including Pointers, Siberian Huskies, Alaskan Malamutes, Saluki, and Anatolian Shepherds. A 2010 study examined the genetic composition of different lines of Alaskan sled dogs (Huson *et al.*, 2010). The researchers parsed out which lines had been bred for distance and endurance versus which had been bred for sprinting, with the Siberian Husky and Alaskan Malamute showing the highest levels of endurance, Saluki and Pointer having traits for enhanced speed (sprinting), and the Anatolian Shepherd exhibiting the strongest work ethic.

A broader study done on eight of the ten FCI-based breed groups that had a sufficient sample size assessed 15,329 dogs from 164 breeds (Svartberg and Forkman, 2002). They used the Dog Mentality Assessment (DMA), which is used by the Swedish Working Dog Association (SWDA), to measure the intensity of dogs' reactions to different stimuli, including novel or potentially scary things, playing with humans and a toy, and a passive or baseline setting. Their analyses suggested the existence of *five* narrow traits: Playfulness, Curiosity/Fearlessness, Chase-proneness, Sociability and Aggressiveness. The researchers found slight, but statistically significant differences among the breed groups, particularly in the traits of Curiosity/Fearlessness, Sociability, and Aggressiveness. It's challenging to determine whether their relatively small between-breed differences suggest an absence of breed differences or reflect variability in their samples due to 201 observers conducting these tests at 235 facilities. Despite the use of a standardized assessment with trained observers, this does add some random variability into the raw data that may mask breed differences. Indeed, how could it not? That's so many different observers! High rates of interobserver reliability are hard to attain.

Another study using the DMA assessed dogs from 31 different breeds. Each dog received a score from 1 to 5 based upon the intensity of their reaction, with 1 being the lowest level of reactivity and 5 being the highest. The researchers calculated each dog's results for Aggressiveness, Curiosity/Fearlessness, Playfulness, and Sociability (we're not sure why they omitted Chase-proneness), finding statistically significant differences across all four dimensions of personality for all of the dog breeds. Of the differences, the most significant was between Labrador Retrievers and Collies. The researchers found that Labrador Retrievers exhibited the most Curiosity and least Fearfulness, while Collies exhibited the least Curiosity and most Fearfulness. Labrador Retrievers were bred to work more closely directly with humans, including humans with guns, while Collies were bred to protect livestock. The researchers also found that there were significant within-breed variations, indicating the importance of environment, including early life experiences and epigenetics, within-breed genetic variation, and interactions. The findings also provided evidence for the difficulty of this type of work, and highlight concerns such as the geographic isolation of these breeds, as they were all from Europe. Recall the differences with the same dog breeds, but from opposite sides of the pond. In the US, the Airedale Terrier, Standard Poodle, and Beagle were all rated "high reactive," but in the UK, they were rated "medium-low".

There was no relationship between traditional breed groups based on historical use and selection (e.g. Herding, Terrier, Working, and Gun) and where they fell in the *four* dimensions. Working dogs showed more Playfulness and Aggressiveness, while show dogs showed more Fearfulness and less Playfulness, Curiosity, and Aggression. Instead, the researchers suggested that recent artificial selection for other working tasks (rather than those originally bred), and show versus pet breeding has changed breed-specific behavioral characteristics (Svartberg, 2006).

A group of scientists in Finland recently performed both a standardized cognitive assessment and an observational study of 1002 dogs from 13 different breeds (Junttila *et al.*, 2022) The scientists argue that their study stands out amongst others because they were able to look at individual breeds rather than breed groups, and that lumping breeds into groups masks some of the breed-specific differences. Breeds were limited to those with a sample size of 40 or more, and sex distribution (the number of males to females) ranged between 30.6 and 69.4 for each sex in each breed, with the majority of breeds near 50–50. See Table 7.2 for the breeds they evaluated in their indoor assessments by trained observers.

Here we discuss their findings related to temperament (we'll discuss their findings on breed differences in cognition in Chapter 10). In this study, they measured how dogs responded to an unfamiliar experimenter by facing the dog and talking to them in a friendly voice. They allowed the dog to approach them, and if it was not aggressive, they attempted to pet the dog for 1–2 minutes. They rated the behavior of the dog on a Likert scale from 1 to 7, where lower scores were "fearful" and higher scores were "friendly" to "excited."

The researchers assessed activity level by attaching a FitBark, a GPS tracker that claims to provide "informed dog health monitoring" (www.fitbark.com) to the collar. Finally, they evaluated the dog's willingness to explore the unfamiliar environment with a Likert scale from 1 to 5, where 1 was "no exploration" (staying by owner's side) and 5 was "active explorative/running around." They interpreted the dog's behavior of "remaining by the owner's side" as a possible indication of fear, anxiety, or neophobia. Dogs that actively explored the novel environment were thought to be curious, bold and active.

The scientists found that the Shetland Sheepdog and the Spanish Water Dog were more fearful or aggressive when meeting a stranger in comparison to the other breeds. Five breeds (the Hovawart, Shetland Sheepdog, Finnish Lapphund, Golden

Table 7.2. Breeds and sample size in Junttila *et al.* (2022).

Breed	Sample size
Australian Kelpie	41
Australian Shepherd	49
Belgian Shepherd Malinois	49
Border Collie	106
English Cocker Spaniel	60
Finnish Lapphund	59
German Shepherd	82
Golden Retriever	74
Hovawart	50
Labrador Retriever	163
Mixed Breed	149
Shetland Sheepdog	48
Spanish Water Dog	72

Retriever, and "Mixed breed") were less active based on the FitBark scale, and the Belgian Shepherd Malinois was the most active. The Shetland Sheepdog and the Finnish Lapphund were the least exploratory (they spent more time near the owner) in this study.

We applaud the use of direct observational methods rather than owner surveys for this work. Owner surveys are inherently subjective. Despite this particular strength with this study, the temperament results are difficult to summarize in meaningful ways, but they are statistically significant and would be interesting to follow up on with more breeds.

While these observational studies showed some evidence for breed differences, a 2010 study did not find evidence of breed differences. They used yet another standardized behavioral test to examine breed differences between Belgian Shepherds (also known as the Malinois), German, and Dutch Shepherds (Sinn *et al.*, 2010). The standardized testing added validity to this study, but using very closely related breeds does not provide a representative sample of dog breeds, or dogs in general. It should come as no surprise that a nearly homogeneous sample wouldn't yield big results when it comes to breed differences.

So, what can we conclude from these studies? First, it's important to have standardized methodologies to compare breeds and behaviors. Second, when we're looking for differences in behavior between breeds, we need to be using representative, heterogeneous breeds, rather than comparing only closely related ones. Third, we need to realize that there will likely be clusters of certain behaviors or issues in particular geographical areas. If we compare dogs from one breed in one geographic area with the same breed in another, we'll have different results simply because the dogs are from different areas. With rare exceptions (such as transporting dogs from one location to another), dog breeds that are geographically distant (e.g. German Shepherds from Germany and German Shepherds from the Pacific Northwest in the US) will have a greater variance from one another than when comparing individuals all from one geographic location. Recall Darwin's mockingbirds and finches and how they speciated in the Galapagos, which is a very small area. You would find fewer differences if you compared all of the birds from one island than if you compared the birds from the two islands furthest from one another. With species or breeds that have been geographically and temporally isolated, you're likely to see more differences, but it's not a guarantee.

Try to See It My Way

So why do we continue to have such different results between studies that are trying to ask the same question? In a word: methodology. We need to consider that tests have different sample sizes and power, some studies use direct behavioral assessment, while others are owner surveys, some studies are using dogs from countries that have divergent geography and histories of dog breeding, and some studies are using different definitions for breeds. For example, some studies use *genetic* breed groups versus *traditionally recognized* breed groups (which vary by country). This all makes it very hard to compare studies. The solution to this is consistency with operational definitions, methodology, and replication, replication, replication.

In scientific studies, replication is one of the most important—but least "sexy" aspects of research. In graduate school, most students don't want to do their thesis on a research topic that has been done before; they want to do something new, novel, and exciting. But replication can be exciting, too. Replication validates and legitimizes prior findings—or calls them into question, which is also an important aspect of science! That which cannot be replicated is suspect. But replication only works when we start with consistent operational definitions and methods.

When we're creating definitions or developing measurement tools, we need to demonstrate reliability. With many studies, interobserver reliability (sometimes called "inter-rater reliability") is a key component of the methodology: will two or more persons observe the same behavior and, using the same criteria, categorize that behavior in the same way? Replication can help us solve questions about multiple species, as well. A 2010 study used domesticated dogs as a model for measuring behaviors exhibited in attention deficit hyperactive disorder (ADHD) in children, including hyperactivity, impulsivity, and variation in attention. The study slightly modified the human ADHD rating scale to determine whether this could be replicated with dogs and found reliability with inattention and with hyperactivity-impulsivity. The researchers determined that further examination of activity, attention variability, and impulsivity with domesticated dogs could help with the genetic study of attention disorders (Lit *et al.*, 2010).

It's important to measure temperament traits appropriately. Tests like the C-BARQ have high validity and reliability when they are used appropriately. The C-BARQ should only be used in its entirety, and there are issues with both validity and reliability when it's not.

While domesticated dogs are the go-to model for studying canine behavior and temperament, they aren't the only canines who have personality differences. According to Oregon State University experimental psychologist Dr Monique Udell, wolves have a wide range of temperament traits, as well.

"There are individual differences with wolves, just as there are in dogs," Dr Udell explained. "It's hard to pinpoint where those differences come from: genetics, or experiences throughout their lifetimes; just as in any species, it's likely some combination of both."

Assumptions are often made about certain subpopulations of wolves. There are isolated groups of wolves with underlying heritable factors and genetic factors, and they have certain different factors from other groups that, although they're the same species, set them apart. "Some wolves are sweet and easy going, while some have a little bit more spunk as they age," Dr Udell said.

But all of the wolves that she has worked with that are exposed to humans and hand raised at a relatively early age have some advantages when interacting with humans. They have their temperamental traits, but then they also have their early environment. "The timing of their critical window is important—they require that interaction to start with humans at 2–3 weeks at the latest," she said.

"Even if you're attempting to raise a group of wolf pups all the same, sometimes you'll do things differently or, during their early days, they'll hear a loud noise, or something will fall and startle them, so their experience will be different than the experience of other pups in their litter or other pups at the research center. Even factors such as birth order can change the interaction between wolves. Getting them early, and raising them from an early point, is important to having wolves that are truly comfortable around humans."

While the wolves have a wide range of temperaments, early life experiences are also important to the development of their personalities. "Of all of the wolves that were reared at the wolf park, none of the wolves have completely turned away humans. I've never seen one that completely wanted nothing to do with humans," Dr Udell said.

That's not to say that all of the wolves sought out human attention. Some liked to have constant companionship, while others were more relaxed about their relationship with people. We see similar trends across dogs. As a species, dogs tend to seek out human companionship, but we will still see breed group, breed, and individual differences with dogs who are more or less social or attention seeking. Some dogs are sociable with all people, while others tend to bond more closely with one or two particular people.

"Some wolves, too, have a few specific people that they are closer to, and want to be with those people, but only for a certain portion of time," Dr Udell said. "Even with dogs, their social focus, on average, is much greater toward humans than dogs. There's variability."

With dogs and wolves, the primary focus is trying to understand what aspects of their social behavior are the same and which are different.

"We're finding more and more evidence in terms of social-cognitive abilities and similar aptitude using social cues. Some research is showing that for cooperative behaviors, wolves do better than dogs do. But it's really the social focus of dogs, almost the obsessive focus, which can include people."

Part of that difference pertains to attachment. Recall that most typically socialized, healthy dogs will try to establish a secure base with their humans. Wolves, too, will have a spectrum of attachment, including insecure-avoidant, secure, and ambivalent.

"There's an overabundance of insecure-ambivalent attachment—it's the kind of dog that when they're stressed, they go to their people, but they aren't immediately calmed down by the presence of their people," Dr Udell explained. "There may have been something early in dogs' domestication history, even if it was a short-term disadvantage, where they seek that association and proximity. With dogs that are clingy, many dog owners want a dog that's loyal and that follows them around."

These "loyal" dogs tend to have a secure attachment style. They are the type of dogs that will check in with you, and then continue on for a period of time before checking back in.

"We have these traits at a high level. You want dogs that are like that, but then we also have dogs that have a high rate of separation anxiety, very loyal dogs might also have separation anxiety disorders at some point. It's very exciting that they're very socially focused, but we should also be careful to recognize that being social to a point—they also need to be able to be left alone."

For cats, it might be important to take into account their social structure versus that of the domesticated dog. Most of the species in the feline family, excepting the lion, are solitary hunters: obligate carnivores who would be very sensitive to minor changes in their environment because of their reliance on their own hunting abilities. Most of the canine species, however, are fairly social, living in family units and relying on group efforts for access to resources.

Dark Horse

With what we've learned in our crash course on genetics, it's clear that a single gene can have multiple effects, including those that effect temperament and personality. Recall that humans with red hair also have lower pain tolerance and increased heat sensitivity, and that rats, deer mice, and foxes with black coats are less active and more docile. A study with horses found that black mares are more independent than bay (brown coat and black mane and tail) ones (Jacobs *et al.*, 2016). All of these studies indicate that there's a possible link between hair pigmentation and temperament.

While we have been focusing on breed differences in personality, we would be remiss to avoid a discussion for the evidence around coat color and personality, particularly as some colors are more prominent in some breeds.

There's an age-old adage that says, "Chestnut mare, beware," warning riders to avoid red-headed horses. But are they more sensitive or reactive than horses of another color? While the scientific literature is still lacking, preliminary results reveal that there's a link between personality traits and coat color. One study found significant differences between bay and chestnut (red) horses. Chestnut horses were likelier to approach animals and objects, whether they were familiar or unfamiliar, indicating that these horses were bolder than the other horses in the study (Finn *et al.*, 2016). And anecdotal evidence about black wolves such as Romeo (Chapter 4) appears to indicate differences in health and perhaps temperament, as well.

For centuries, scientists and casual observers alike have thought that black wolves differed from gray and white colored ones not just in their coat color, but in their behavior, as well; so much so that they were once considered to be a distinct species. In his 1758 edition of the *Systema Naturae*, Linnaeus initially classified Europe's black wolves separately from lighter-coated ones, designating black-coated wolves as *Canis lycaon* (Linnaeus, 1758), a classification that was followed by French Naturalist and Zoologist Georges Cuvier (1769–1832) (Allen, 1869) and by naturalist, surgeon, Arctic explorer, and pre-Darwinian natural historian Sir John Richardson (1787–1865) in 1829 (Richardson *et al.*, 1829). American zoologist, naturalist, herpetologist, paleontologist, and physicist Richard Harlan (1796–1843) also used this binomial classification to differentiate America's black wolves, in his 1825 *Fauna Americana* (Allen, 1869). Black wolves have long had a reputation for being friendlier and more prone to interbreeding with domesticated dogs than their lighter-coated peers. In his 1839 *The Natural History of Dogs*, English artist and naturalist Lieutenant-Colonel Charles Hamilton Smith wrote that black wolves tended to be less aggressive than wolves with lighter coats (Smith and Jardine, 1839). Coat color can indicate temperament differences in a variety of species, including horses and cats (Stelow *et al.*, 2016), with research finding that darker cats are more sociable than lighter ones (Morgan, 2010). And dogs? Coat color-based differences have been found with rates of behavioral reactivity (Kim *et al.*, 2010), levels of aggression (Pérez-Guisado *et al.*, 2006), and levels of separation anxiety, agitation, excitability, fetching tendency, stereotypical behavior, and trainability (Lofgren *et al.*, 2014). Differences between rates of behavioral reactivity were found with the Korean Jindo, with fawn-colored Jindos exhibiting lower rates of submissive reactivity and fearfulness than white-colored ones (Kim *et al.*, 2010). Differences between levels of aggression were found with English Cocker Spaniels, with golden English Cocker Spaniels exhibiting more aggressive behaviors than black or multicolored Cocker Spaniels (Pérez-Guisado *et al.*, 2006). Differences between a number of traits were found among yellow, chocolate, and black Labrador Retrievers (Lofgren *et al.* 2014). This study found that yellow Labradors exhibited higher rates of separation anxiety than black Labradors did. Chocolate Labradors exhibited more agitation and excitability than black Labradors, who had a higher Fetching Tendency than chocolate Labradors.

Of the three coat colors, chocolate Labradors (Fig. 7.3) had the highest rates of Unusual Behavior (also called stereotypical behavior) and the lowest rates of trainability; they also had higher rates of excitability than black Labradors. Yellow and black Labradors also had higher noise-related fear than their chocolate Labradors. But what's behind these behavioral differences?

The brown coat color of the chocolate Labrador, and all other brown dogs, stems from a mutation in the tyrosine-related protein 1 gene (*TYRP1*) (Schmutz *et al.*, 2002). Multiple domesticated animals have this coat color mutation, including dogs, cats, horses, donkeys, cattle, goats, sheep, mice, ferrets, hamsters, and chickens (Schmutz and Berryere, 2007). While we know which gene causes the brown coat color, the linkage between *TYRP1* and possible behavioral correlations still needs further investigation.

Conclusion

Given our genetic manipulation of dogs (and cats and horses, to a lesser extent) it makes sense that differences in temperament between breeds and breed groups will exist. Dogs have been bred longer, and with stronger artificial selection, than other domesticated species, and they have been bred for a wider variety of behaviors and specific skills. While asking "what's your dog's (or cat's, or horse's) personality?" might sound like a trivial question, it actually has important implications for fit between owner and pet, between different non-human animals sharing one household, and for the

Fig. 7.3. Chocolate Labrador Retriever. This file is licensed under the Creative Commons Attribution 2.0 Generic license and provided by Wikipedia. Available at: https://commons.wikimedia.org/wiki/File:Labrador_Retriever_chocolate_Hershey_sit.jpg.

welfare of each animal. Temperament and personality can affect whether or not a dog gets surrendered; fit is important between people and their pets. An Anatolian Shepherd, with high drive, wouldn't be a good fit in an apartment building, just as much as a Chihuahua wouldn't be a good fit, temperament or size-wise, on a working cattle farm.

Measuring personality in dogs continues to be a challenge for a number of reasons, including a lack of public funding for this work, the difficulty in doing direct behavioral assessments over sufficient time and across numerous breeds, and the wide variability within breeds. Published studies have different sample sizes and power, some studies use direct behavioral assessment, while others are owner surveys, some studies are using dogs from countries that have divergent geography and histories of dog breeding, and some studies are using different definitions for breeds. Nevertheless, numerous personality studies have shown a genetic basis to these traits across a wide range of species. We've discussed a number of these papers in this chapter to demonstrate that both the variety of species where these links have been shown, as well as the more limited span of studies on dog breeds, all point to a genetic basis to dog personality.

References

Allen, J.A. (1869) *Catalogue of the Mammals of Massachusetts: With a Critical Revision of the Species*. University Press, Cambridge, MA.

ASPCA (2013) Feline-ality™ assessment. Available at: https://aspcapro.org/sites/default/files/Feline-Assessment-Forms-FINAL-05Sept13_HR.PDF (accessed 31 December 2023).

Botigué, L.R., Song, S., Scheu, A., Gopalan, S., Pendleton, A.L. *et al.* (2017) Ancient European dog genomes reveal continuity since the early Neolithic. *Nature Communications* 8(1), 16082.

Bouchard, T.J. and McGue, M. (2003) Genetic and environmental influences on human psychological differences. *Journal of Neurobiology* 54(1), 4–45.

Bowlby, J. (1958) The nature of the child's tie to his mother. *The International Journal of Psychoanalysis* 39(5), 350–373.

Bradshaw, J.W.S. and Goodwin, D. (1998) Determination of behavioural traits of purebred dogs using factor analysis and cluster analysis; a comparison of studies in the USA and UK. *Research in Veterinary Science* 66(1), 73–76.

Callier, V. (2022) Deep dive ties together dog genetics, brain physiology, and behavior to explain why collies are different from terriers. *Scientific American*. Available at: www.scientificamerican.com/article/deep-dive-ties-together-dog-genetics-brain-physiology-and-behavior-to-explain-why-collies-are-different-from-terriers (accessed 23 December 2023).

Chen, C., Burton, M., Greenberger, E. and Dmitrieva, J. (1999) Population migration and the variation of dopamine D4 receptor (DRD4) allele frequencies around the globe. *Evolution and Human Behavior* 20(5), 309–324.

Duffy, D.L., Diniz de Moura, R.T. and Serpell, J.A. (2017) Development and evaluation of the Fe-BARQ: a new survey instrument for measuring behavior in domestic cats (*Felis s. catus*). *Behavioural Processes* 141(3), 329–341.

Dutrow, E.V., Serpell, J.A. and Ostrander, E.A. (2022) Domestic dog lineages reveal genetic drivers of behavioral diversification. *Cell* 185(25), 4737–4755.

Finn, J.L., Haase, B., Willet, C.E., van Rooy, D., Chew, T. *et al.* (2016) The relationship between coat colour phenotype and equine behaviour: a pilot study. *Applied Animal Behaviour Science* 174, 66–69.

Gartner, M.C. (2014) Pet personality: a review. *Personality and Individual Differences* 75, 102–113.

Ha, J.C. and Campion, T.L. (2018) *Dog Behavior: Modern Science and Our Canine Companions*. Elsevier/Academic Press, London.

Hart, B.L. and Hart, L.A. (1985) *Canine and Feline Behavioral Therapy*. Lea & Febiger, Philadelphia, PA.

Horn, L., Huber, L. and Range, F. (2013) The importance of the secure base effect for domestic dogs – evidence from a manipulative problem-solving task. *PLoS One* 8(5), e65296.

Huson, H.J., Parker, H.G., Runstadler, J. and Ostrander, E.A. (2010) A genetic dissection of breed composition and performance enhancement in the Alaskan sled dog. *BMC Genetics* 11, 71.

Irving-Pease, E.K., Ryan, H., Jamieson, A., Dimopoulos, E.A., Larson, G. *et al.* (2018) Paleogenomics of animal domestication. In: Lindqvist, C. and Rajora, O.P. (eds) *Paleogenomics: Genome-Scale Analysis of Ancient DNA. Population Genomics*. Springer, Cham, Switzerland, pp. 225–272.

Jacobs, L.N., Staiger, E.A., Albright, J.D. and Brooks, S.A. (2016) The *MC1R* and *ASIP* coat color loci may impact behavior in the horse. *The Journal of Heredity* 107(3), 214–219.

Jang, K.L., Livesley, W.J. and Vernon, P.A. (1996) Heritability of the big five personality dimensions and their facets: a twin study. *Journal of Personality* 64(3), 577–592.

Jones, A.C. and Gosling, S.D. (2005) Temperament and personality in dogs (*Canis familiaris*): a review and evaluation of past research. *Applied Animal Behaviour Science* 95(1–2), 1–53.

Junttila, S., Valros, A., Mäki, K., Väätäjä, H., Reunanen, E. *et al.* (2022) Breed differences in social cognition, inhibitory control, and spatial problem-solving ability in the domestic dog (*Canis familiaris*). *Scientific Reports* 12(1), 22529.

Kim, Y.K., Lee, S.S., Oh, S.I., Kim, J.S., Suh, E.H. *et al.* (2010) Behavioural reactivity of the Korean native Jindo dog varies with coat colour. *Behavioural Processes* 84(2), 568–572.

Ley, J.M., Bennett, P.C. and Coleman, G.J. (2009) A refinement and validation of the Monash Canine Personality Questionnaire (MCPQ). *Applied Animal Behaviour Science* 116(2–4), 220–227.

Linnaeus, C. (1758) *Systema Naturæ per Regna Tria Naturæ, Secundum Classes, Ordines, Genera, Species, Cum Characteribus, Differentiis, Synonymis, Locis*, 10th edn, Vol. 1. Holmiæ Laurentius Salvius, Stockholm, p. 38.

Lit, L., Schweitzer, J.B., Iosif, A.-M. and Oberbauer, A.M. (2010) Owner reports of attention, activity, and impulsivity in dogs: a replication study. *Behavioral and Brain Functions* 6, 1.

Lofgren, S.E., Wiener, P., Blott, S.C., Sanchez-Molano, E., Woolliams, J.A. *et al.* (2014) Management and personality in Labrador Retriever dogs. *Applied Animal Behaviour Science* 156, 44–53.

Lubinski, D. (2000) Scientific and social significance of assessing individual differences: “sinking shafts at a few critical points.” *Annual Review of Psychology* 51, 405–444.

MacLean, E.L., Snyder-Mackler, N., vonHoldt, B.M. and Serpell, J.A. (2019) Highly heritable and functionally relevant breed differences in dog behaviour. *Proceeding of the Royal Society B: Biological Sciences* 286(1912), 20190716.

McCobb, E.C., Brown, E.A., Damiani, K. and Dodman, N.H. (2001) Thunderstorm phobia in dogs: an internet survey of 69 cases. *Journal of the American Animal Hospital Association* 37(4), 319–324.

Megna, M. (2023) Pet ownership statistics 2023. *Forbes*. Available at: www.forbes.com/advisor/pet-insurance/pet-ownership-statistics (accessed 31 October 2023).

Mehrkam, L.R. and Wynne, C.D.L. (2014) Behavioral differences among breeds of domestic dogs (*Canis lupus familiaris*): current status of the science. *Applied Animal Behaviour Science* 155, 12–27.

Mirkó, E., Kubinyi, E., Gácsi, M. and Miklosi, A. (2012) Preliminary analysis of an adjective-based dog personality questionnaire developed to measure some aspects of personality in the domestic dog (*Canis familiaris*). *Applied Animal Behaviour Science* 138(1–2), 88–98.

Morgan, R.S. (2010) Personality in domestic cats (*Felis catus*): identification and description of personality structure, predictive validity, and associations with coat color characteristics. University of California, Davis, Dissertation. ProQuest Dissertations Publishing. Available at: https://search.proquest.com/openview/c1e83fc7a4863d977d4da5cc11cd3ef7/1?pq-origsite=gscholar&cbl=18750&diss=y (accessed 1 November 2023).

Morrill, K., Hekman, J., Li, X., McClure, J., Logan, B. *et al.* (2022) Ancestry-inclusive dog genomics challenges popular breed stereotypes. *Science* 376(6592), eabk0639.

Overall, K.L. (2015) Noise, reactivity and cognition in dogs: parsing risk factors and phenotypic plasticity. In: *Tufts' Canine and Feline Breeding and Genetics Conference*. Available at: www.vin.com/apputil/content/defaultadv1.aspx?pId=12513&catId=51024&id=6976363 (accessed 1 November 2023).

Overall, K.L. (2013) *Manual of Clinical Behavioral Medicine for Dogs and Cats*. Elsevier, St Louis, MO.

Overall, K.L., Dunham, A.E. and Juarbe-Diaz, S.V. (2016) Phenotypic determination of noise reactivity in 3 breeds of working dogs: a cautionary tale of age, breed, behavioral assessment, and genetics. *Journal of Veterinary Behavior* 16, 113–125.

Parker, H.G., Kukekova, A.V., Akey, D.T., Goldstein, O., Kirkness, E.F. *et al.* (2007) Breed relationships facilitate fine-mapping studies: a 7.8-kb deletion cosegregates with Collie eye anomaly across multiple dog breeds. *Genome Research* 17(11), 1562–1571.

Pavlov, I.P. (1906) The scientific investigation of the psychical faculties or processes in the higher animals. *Science* 24(620), 613–619.

Pérez-Guisado, J., Lopez-Rodriguez, R. and Muñoz-Serrano, A. (2006) Heritability of dominant–aggressive behaviour in English Cocker Spaniels. *Applied Animal Behaviour Science* 100(3–4), 219–227.

Rayment, D.J., Peters, R.A., Marston, L.C. and De Groef, B. (2016) Investigating canine personality structure using owner questionnaires measuring pet dog behaviour and personality. *Applied Animal Behaviour Science* 180, 100–106.

Richardson, J., Swainson, W. and Kirby, W. (1829) *Fauna Boreali-americana, or, The Zoology of the Northern Parts of British America: Containing Descriptions of the Objects of Natural History Collected on the Late Northern Land Expeditions, Under Command of Captain Sir John Franklin, R.N.* J. Murray, London.

Salonen, M., Mikkola, S., Niskanen, J.E., Hakanen, E., Sulkama, S. *et al.* (2023) Breed, age, and social environment are associated with personality traits in dogs. *iScience* 26(5), 106691.

Scheifele, P.M., Sonstrom, K.E., Dunham, A.E. and Overall, K.L. (2016) Is noise reactivity reflected in auditory response variables, including those that measure cognition, in dogs? Initial findings. *Journal of Veterinary Behavior* 16, 65–75.

Schmutz, S.M. and Berryere, T.G. (2007) Genes affecting coat colour and pattern in domestic dogs: a review. *Animal Genetics* 38(6), 539–549.

Schmutz, S.M., Berryere, T.G. and Goldfinch, A.D. (2002) TYRP1 and MC1R genotypes and their effects on coat color in dogs. *Mammalian Genome* 13(7), 380–387.

Scott, J.P. and Fuller, J.L. (1965) *Genetics and the Social Behavior of the Dog*. The University of Chicago Press, Chicago, IL.

Serpell, J. and The University of Pennsylvania (2023) C-BARQ. Available at: https://vetapps.vet.upenn.edu/cbarq (accessed 31 October 2023).

Shepherd, D., Heinonen-Guzejev, M., Hautus, M.J. and Heikkilä, K. (2015) Elucidating the relationship between noise sensitivity and personality. *Noise & Health* 17(76), 165–171.

Siettou, C., Fraser, I.M. and Fraser, R.W. (2014) Investigating some of the factors that influence "consumer" choice when adopting a shelter dog in the United Kingdom. *Journal of Applied Animal Welfare Science* 17(2), 136–147.

Sinn, D.L., Gosling, S.D. and Hilliard, S. (2010) Personality and performance in military working dogs: reliability and predictive validity of behavioral tests. *Applied Animal Behaviour Science* 127(1–2), 51–65.

Smith, C. (2017) Cats domesticated themselves, ancient DNA shows. *National Geographic*. Available at: www.nationalgeographic.com/news/2017/06/domesticated-cats-dna-genetics-pets-science/#close (accessed 31 October 2023).

Smith, C.H. and Jardine, W. (1839) *The Natural History of Dogs: Canidæ or Genus Canis of Authors. Including Also the Genera Hyæna and Proteles*. W.H. Lizars, Edinburgh, UK.

Stelow, E.A., Bain, M.J. and Kass, P.H. (2016) The relationship between coat color and aggressive behaviors in the domestic cat. *Journal of Applied Animal Welfare Science* 19(1), 1–15.

Storengen, L. and Lingaas, F. (2015) Noise sensitivity in 17 dog breeds: prevalence, breed risk and correlation with fear in other situations. *Applied Animal Behaviour Science* 171, 152–160.

Sundman, A., Johnsson, M., Wright, D. and Jensen, P. (2016) Similar recent selection criteria associated with different behavioural effects in two dog breeds. *Genes, Brain, and Behavior* 15(8), 750–756.

Svartberg, K. (2006) Breed-typical behaviour in dogs – historical remnants or recent constructs? *Applied Animal Behaviour Science* 96(3–4), 293–313.

Svartberg, K. and Forkman, B. (2002) Personality traits in the domestic dog (*Canis familiaris*). *Applied Animal Behaviour Science* 79(2), 133–155.

Takeuchi, Y. and Mori, Y. (2006) A comparison of the behavioral profiles of purebred dogs in Japan to profiles of those in the United States and the United Kingdom. *The Journal of Veterinary Medical Science* 68(8), 789–796.

Thalmann, O. and Perri, A.R. (2018) Paleogenomic inferences of dog domestication. In: Lindqvist, C. and Rajora, O.P. (eds) *Paleogenomics: Genome-Scale Analysis of Ancient DNA. Population Genomics*. Springer, Cham, Switzerland, pp. 273–306.

Tiira, K., Sulkama, S. and Lohi, H. (2016) Prevalence, comorbidity, and behavioral variation in canine anxiety. *Journal of Veterinary Behavior* 16, 36–44.

Turcsán, B., Kubinyi, E. and Miklosi, Á. (2011) Trainability and boldness traits differ between dog breed clusters based on conventional breed categories and genetic relatedness. *Applied Animal Behaviour Science* 132(1–2), 61–70.

Watters, J.V. and Powell, D.M. (2012) Measuring animal personality for use in population management in zoos: suggested methods and rationale. *Zoo Biology* 31(1), 1–12.

Wilsson, E. and Sundgren, P.-E. (1997) The use of a behaviour test for the selection of dogs for service and breeding, I: method of testing and evaluating test results in the adult dog, demands on different kinds of service dogs, sex and breed differences. *Applied Animal Behaviour Science* 53(4), 279–295.

8 Social Behavior and Breed Differences

Abstract

Chapter 8 addresses the scientific evidence for breed differences in social behavior and dominance. How do multiple factors, such as genetic lineages and the environment, influence certain behaviors? There is a discussion of how "primitive" and "ancient" breeds, and breed groups, tend to show more dominant characteristics than recently created breeds with more derived characteristics. Case study examples of ancient breeds and examples of ancient and modern breed household mismatches, as well as mismatches between dogs and their humans are used to reveal the impacts of these issues on modern life.

Power Lunch

Alan Gregg gazed across the table at his dining companion, his pale eyes steady. His blonde hair was grayed and receding. His lunch guest, Clarence Cook ("C.C.") Little, 2 years his senior, was dark-eyed and distinguished, his brunette hair streaked with gray, his mustachioed mouth smiling warmly. World War II had just ended when Gregg and Little had their fateful lunch, little realizing that the future of behavioral genetics research was on the menu. Oh, to be the proverbial fly on the wall as these two elder statesmen of scientific research cracked this new field wide open! Gregg had requested the meeting, and he knew how to lay the table for a scientific feast. At only 26 years old, he had already earned his undergraduate and medical degrees from Harvard University. He then volunteered to be on the Western Front during the World War I, partnering with Great Britain's Royal Army Medical Corps and offering his newly developed medical skills to wounded soldiers. In 1919, Gregg began working for the Rockefeller Foundation, joining the International Health Division, where he was assigned to work in Brazil until 1922. Eight years later, he became the Director for the Medical Sciences at the Rockefeller Foundation. Not to be outdone, Dr C.C. Little attended Harvard, where he earned his AB, MS, and DSc in Zoology, boldly focusing on the fledgling science of genetics. Like Gregg, Little also served during World War I, where he attained the rank of Major. In 1922, Little became the president of the University of Maine. At 33 years old, he was the youngest university president in the US. Seven years later, he founded the Jackson Laboratory in Bar Harbor, Maine. Thus, this lunch was no ordinary repast. During that meal, Gregg asserted that psychiatrists and sociologists weren't focusing enough on the relationship between heredity and behavior—and he wanted to do something about it. Little concurred. With these ingredients for innovation, they cooked up what would become the Genetics and the Social Behavior of Mammals, an ambitious undertaking that aimed to use non-human animals to clarify the mysteries behind human behavior.

Genetics and the Social Behavior of Mammals was only the first course: Gregg and Little's lunch proved to be a powerful point of departure for future geneticists, including John Paul Scott and John L. Fuller. Given their sociability and storied history with humans, Scott intuited that dogs would provide a window into the behavioral problems that people faced. Scott and Fuller's seminal work, *Genetics and the Social Behavior of the Dog* (Scott and Fuller, 1965), built upon the findings of the Genetics and the Social Behavior of Mammals, examining dog breed differences in social behavior.

Dr Scott's intuition paid off: dogs were a meaningful model species for this study, both for their co-evolution with humans and for their recent artificial selection. Scott, who earned his BSc in Zoology at the University of Wyoming and his PhD

 Breed Differences in Dog Behavior: Why Tails Wag Differently. (R.L. Ha, T.L. Brad and J.C. Ha)
DOI: 10.1079/9781800624566.0008

in Psychology at the University of Chicago, worked with evolutionary scientist and geneticist Sewall Wright. He chaired the department of zoology at Wabash College from 1935 to 1945 and moved to the Jackson Laboratory in 1945. Dr Fuller, who received his PhD from the Massachusetts Institute of Technology in 1935, was an early pioneer of behavioral genetics. He was a researcher at Jackson Laboratory from 1947 to 1970. Published in 1965, Scott and Fuller's tome is no less relevant or insightful now than it was then. If anything, it has only grown in importance over the years.

These intrepid researchers delved headlong into the intersection of genetics and behavior, questioning the limited research into this topic during the 1930s and 1940s and the subsequent "scientific" atrocities at the hands of the Nazis, including the intelligence quotient (IQ) controversies that limited this work. There was a stubborn misconception that heredity had no effect on behavior—for humans or non-human animals—and Gregg and Little paved the way to dispute this. Scott and Fuller's classic is still the best study of canine breed differences in social behavior—and it's unlikely to be surpassed any time soon. So why is this work so important? Because it's experimental, rather than a survey, like most current work has been. This 13-year longitudinal study, which received major financial support from the Rockefeller Foundation, measured a wide range of potential variables, collecting data in a careful, detailed manner, by using tools such as ethograms and assessments of development, social behavior, and problem solving. They had the forethought to control extraneous environmental influences while still allowing natural social behavior to develop.

Scott and Fuller were the first to examine many areas of dog behavior and genetics, including inheritance, being born in a particular litter or to a particular dam, motor development, investigative behavior, vocalizations, agonistic behavior, and approach to human handlers (Scott and Fuller, 1965). Recall that they noted differences *between* breeds as well as those *within* breeds, which highlights the importance of individual variation. They chose 470 dogs from five breeds representing four of the six breed groups recognized at that time (Basenjis and Beagles, from the Hound Group; American Cocker Spaniels, from the Sporting Group; Shetland Sheepdogs, from the Herding Group; and Wirehaired Fox Terriers, from the Terrier Group). The only two breed groups omitted from the study were Toy and Non-sporting Group breeds. (The Working Group became officially recognized by the American Kennel Club (AKC) in 1983, bringing the total to seven breed groups.)

Scott and Fuller were trying to understand the relative influences of multiple variables, including breed, inheritance, including inheritance of behavior, from parents, being born in a particular litter, or being born to a particular dam. They analyzed behavioral differences in these purebreds and mixes between these breeds (Basenji and Cocker Spaniel) and cross-fostered puppies between litters. They also back-crossed first-generation male hybrids to their mothers and then to their sisters to study patterns of inheritance in behavior. Recall that inbreeding like this was a fairly common practice in the not-so-distant past; animal husbandry practices often included breeding closely related individuals. (And "keeping it in the family" was *en vogue* for many humans, until the past century and a half.) Across social and non-social conditions, Scott and Fuller examined responses to several stimuli, including a shock on the leg, being addressed unkindly, and novel situations and people. Across conditions, and correcting for age, Basenjis, Beagles, and Terriers were more emotionally reactive than Cocker Spaniels and Shelties. The authors found differences both between and within breeds. But they struck on something else, too: with the long co-evolutionary history of dogs and humans, they found that "anyone who wishe[d] to understand a human behavior or hereditary disease [could] usually find the corresponding condition in dogs with very little effort" (Scott and Fuller, 1965). If someone had been able to continue their work in the subsequent decades, or even in 2005 when the dog genome was complete (National Institutes of Health, 2005), we would likely today understand far more about the genetics and social behavior of dogs, but of ourselves, as well—material far exceeding the sum total of this book. It would be difficult to attain the level of funding they achieved, though: the Rockefeller Foundation's generous 13 years of support for a longitudinal study is unheard of for modern researchers. The cost of the work exceeds what the National Science Foundation can support for one effort. Due to funding limitations, most modern studies last 1 year, or perhaps up to 3 years, and have much more modest sample sizes. They also lack the input from the leaders in the field at a 3-day conference prior to the design of the work. Their findings on breed differences in behavior are summarized in our chapters dealing with social behavior and dominance, aggression, training, and reactivity.

Funding isn't the only reason why Scott and Fuller's study couldn't be replicated. It would be difficult to repeat for ethical reasons. In the decades since their study, we've become increasingly aware of animal welfare. Fueled by concerns about the treatment of laboratory animals, modern standards for animal care and welfare have changed dramatically since their book was first published in 1965, beginning with the Laboratory Animal Welfare Act (AWA) of 1966. Interestingly, the AWA doesn't cover many common laboratory animal species, including birds, mice, and rats, who are omitted from the definition of "animal" due to inspection limitations (107th Congress, H.R. 2646, 2002). Thus, the breeding and keeping of hundreds of dogs in laboratory conditions would be an ethical issue. Many studies that helped us understand learning and behavior in non-human animals couldn't ethically be replicated today, including the ape language studies of the 1960s and 1970s (Fouts, 1972; Rumbaugh *et al.*, 1973). These studies often involved taking infant chimpanzees from their mothers, as Washoe the chimpanzee was (Gardner and Gardner, 1969), raising them in laboratories or in cross-fostering experiments (Kellogg and Kellogg, 1933). Washoe learned the signs of American Sign Language, but she was denied a life with her family in West Africa. While her contributions to science were unquestionable, later ape language researchers, including those who worked with Washoe, agreed that the studies shouldn't be replicated, because chimpanzees should be with their families. These chimpanzees had to live their lives out in captivity.

Modern scientists or science students who want to conduct a study with animals have to pass the standards of the Institutional Animal Care and Use Committee (IACUC), which reviews animal experiments. The IACUC requires that animal experiments include a justification, how many animals will be used, and the species, how the animal's potential pain or discomfort will be minimized, an overview of the methods, and how the researcher is ensuring that the work is justified and doesn't simply repeat prior work.

We couldn't have come to the point of questioning whether our research projects were ethical without the foundation that was laid by so many prior researchers. While modern-day scientists may cringe at the methodology of scientists who were at the forefront of ethological research, we wouldn't have gotten here without the contributions of Scott and Fuller, the scientists at the ape language laboratory, and groundbreaking wolf researcher Erik Zimen.

Tierverhalten

Erik Zimen tilted his head back, his bearded face flanked by two wolves. Their profiles cut a soft silhouette against the skyline as Zimen let out a mournful howl. The wolves joined in, their voices commingling until they faded out into the lush Bavarian undergrowth. What might have looked, at first glance, like an odd interspecies interaction, was actually an important ethological breakthrough. And during the formative years of animal behavior research, the lion's share of these breakthroughs were being made by Germans, or occurring at German research sites, such as the forests of Bavaria. Thus, the majority of early ethological work was published in German, as well. Zimen (1971), a Swedish scientist who specialized in dog and wolf behavior, was one of these early ethologists. He conducted his research in Italy's Parco Nationale d'Abruzzo and Germany's Bavarian forests, studying wolves, Poodles, and wolf–Poodle hybrids (the original designer "doodles"). Zimen's landmark work among the wolves described their language, behavioral repertoire, and social ecology. His best-known work is his magnum opus, *Der Wolf*, published in English in 1981 (Zimen, 1981). We say "best known," but don't mistake that for "widely known." Zimen spent thousands of hours in the company of canines, but his findings still aren't widely known because he published most of it in German. Words like "Tierverhalten" halted many an intrepid researcher in their tracks. (It's German for animal behavior.) It rolls off the tongue if you say it aloud, but at first glance, it's rather unwieldy, especially if you aren't familiar with the German language. Trailblazers such as Karl von Frisch, Konrad Lorenz, and Nikolaas (Niko) Tinbergen published in German. In 1973, von Frisch, Lorenz, and Tinbergen received the Nobel Prize in Medicine "for their discoveries concerning organization and elicitation of individual and social behavior patterns" (The Nobel Prize, 1973) In his Nobel Prize speech, Professor Börje Cronholm of the Karolinska Medico-Chirurgical Institute clarified why these discoveries were so important to the field of animal behavior research. He stated, in part: "... this year's Nobel prize laureates have been pioneers. They have collected numerous data about animal behavior both in natural settings and in experimental situations. Being biological scholars, they have also studied the functions of behavior patterns, their role in the individual struggle for life and for the continuation of the species. Thus, behavior patterns have stood out as results of natural selection just as morphological characteristics and physiological functions …"

Zimen's contribution to the science of dog behavior was in his work on the parallels, or lack thereof, in the behavior of captive-reared wolves and captive-reared (as per usual) dogs. The specific insight provided to later applied animal behaviorists is his discovery of the lack of overlap, the almost complete change, we have created in how (recently derived, highly artificially selected) dogs use body language to communicate. Many dog breeds have lost their ability to effectively communicate with wolves. Of greater and more practical concern, they've lost the ability to communicate with the oldest dog breeds, including Malamutes, Huskies, Akitas, and the other ancient breeds that retain many wolf-like body language capabilities.

Zimen's work was among the first to explore these changes created by the domestication of the wolf-ancestor into the modern dog. He followed in the footsteps of the founders of ethology, like Konrad Lorenz, who discussed these ideas very early on in his seminal book, *Man Meets Dog* (Lorenz, 1949). Zimen later became a director and cinematographer, making contributions to *Beobachtun an Wolfen* (*Observations on Wolves*) (1980), *Wolf* (1978), and *Living With Wolves* (2004) which posthumously made its debut after his 2003 death.

Zimen found major differences between the species in communication, which he labeled as "expressive behavior." In particular, he noticed patterns in the use of the ears, face, and eyes, as well as differences with the leg, head, body, and tail (Fig. 8.1). There were also interesting differences in the initiation of play and in chasing play.

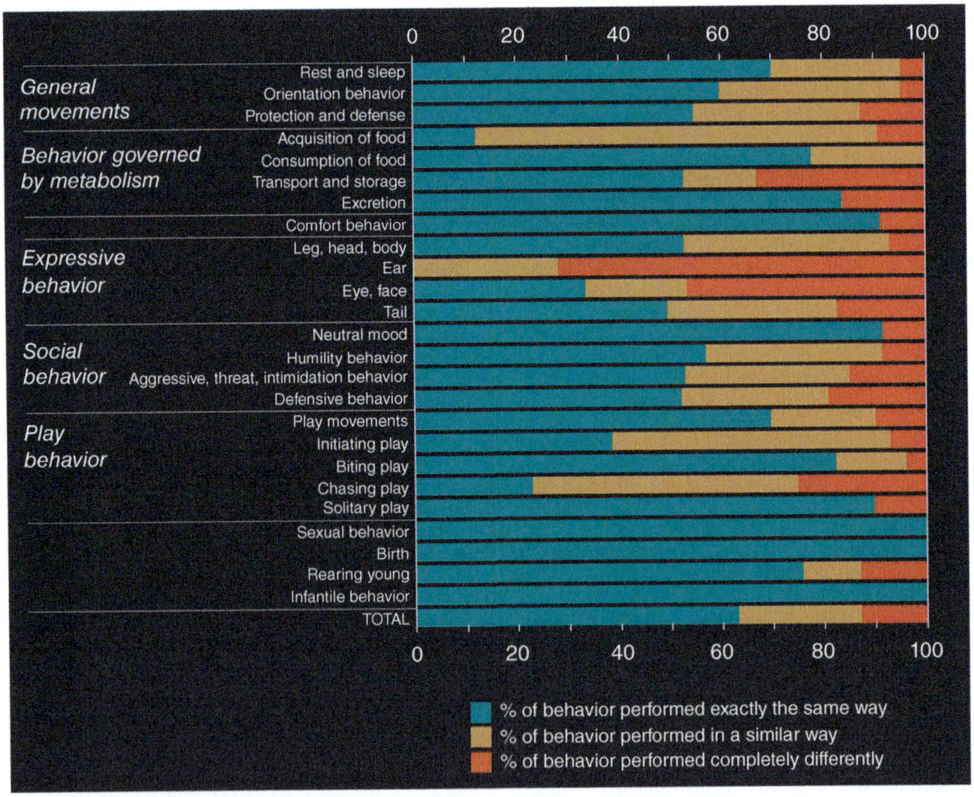

Fig. 8.1. Wolf–Poodle example. Design by Sola Dog Behavior and Karma Design, and provided courtesy to the authors. Modified from Zimen (1971).

While Zimen was studying behavioral differences between two species, they are closely enough related that some may be surprised at all of the changes that artificial selection has wrought between wild and domesticated canids. Imagine if he'd studied the same thing between a modern toy breed and an ancient breed. Would he have found similar results?

Breed versus Breed

Brenna Marose lived on a ranch in a rural town where she'd always owned Alaskan Malamutes. She was familiar with the breed's unique behaviors, provided her dogs with plenty of room to roam, and felt like she truly understood them. Her most recent Malamute, Elke, accompanied her everywhere she went. But when Brenna's father passed away, he left behind his diminutive Yorkshire Terrier, Buddy—a dog breed that Brenna was unfamiliar with—and a breed that her Malamute was uncomfortable with. Brenna had never interacted with a dog as small as Buddy before, much less cohabitated with one. And Buddy's diminutive size wasn't the only issue.

Malamutes, which got their name from the Mahlemut people of the Arctic, are an ancient dog breed (Fig. 8.2). By most estimates, the first Malamutes emerged 3000 years ago or more. These dogs were selected to hunt and to ward off polar bears, but they were particularly bred to pull heavily laden sleds across vast expanses of snow. These vocations required courage, strength, independence, and leadership. As a result, Malamutes, like most ancient dog breeds, retain many of their ancestral behavioral characteristics, including adhering to a dominance hierarchy with their family members, canine and human alike. Conversely, Yorkshire Terriers are a recent modern breed, originating in the 19th century in Yorkshire, England. They were bred for companionship and not "work," and don't retain any of the characteristics that are associated with ancient dog breeds, including adherence to a dominance hierarchy. Thus, from the onset, Buddy and

Fig. 8.2. Malamute with ancestral people. Artwork courtesy of Ari Taylor, artist.

Elke were at odds: small and large, ancient and modern, and with an owner who didn't realize the importance of these social dynamics.

From day one, this new living situation was fraught with miscommunication and behavioral issues. Elke and Buddy didn't understand each other's communicative styles. Buddy was afraid of Elke, displaying frequent bouts of anxiety aggression, and Elke reciprocated by snapping—a behavior that would appropriately keep other Malamutes in line, but which would potentially behead a tiny Toy dog. Brenna obliviously assumed that the two would "work it out." But after an escalating series of miscommunications, culminating in Elke biting Buddy, Brenna finally conceded that she was out of her depth. So, what went wrong?

What You're Doing

As we have discussed, dogs vary greatly from one another. The morphological variance of dogs is greater than that seen in any other known species. But what about their behavioral and communicative differences? Dogs communicate meaning and intention in multiple modalities, including vocalizations and with their body posture, eye contact, and tail and ear movement and placement. Brenna's near fatal (for Buddy, at least) mistake was failing to recognize the mechanistic breed differences in social communication due to morphological differences and selection differences in the use of the communication features. Let us explain. Differences in communication can occur between breeds due to a variety of differences, including morphological ones (ears, tails, coat, etc.). A dog with long, droopy ears, a shaggy coat, and short legs might have a hard time communicating with a long-legged, sleek-coated, erect-eared dog, because certain cues are obscured or presented differently. Note that these differences also have the potential to affect humans' interpretation of dog behavior as well. Humans raised with ancient breeds may not recognize the behavior signals of toy breeds and vice versa. Despite dogs being one species, their morphological variation is dramatic.

All of these variables—size, whether a dog's eyes are visible to the other dog, how a dog moves—are important cues. Were the other dog's eyes covered? In the instance of Elke and Buddy, Buddy's eyes were obscured by his hair. You probably communicate differently with someone who has sunglasses on versus someone who doesn't; eye contact is important for communication. Your parents may have told you at some point, "Get your hair out of your eyes ..." and there is something to that.

When mask mandates became the norm in 2020 because of COVID-19, many people realized just how much we rely upon facial features to convey context and meaning. Some people struggled to communicate without seeing their partner's entire face. Communication, for humans, is largely considered to be verbal, but in reality, it's our nonverbal cues—for humans and dogs alike—that play a crucial part in conveying meaning. Obscuring eyes or mouths can interfere with this, as can dramatic morphological differences.

Let's examine what this would look like in a controlled study. Researchers in Australia studied breed differences in communication in 115 dogs belonging to 30 breeds and mixed breeds (Kerswell *et al.*, 2010). The dogs were enrolled in puppy socialization classes and their free playtime interactions were videotaped and analyzed. The detailed morphological features of each dog were recorded and compared to the ancestral, wolf-like form, using categories such as tail, ears, coat, size, snout, eye cover, and lips. Table 8.1 provides a description of the allocation of scores, for each signaling structure, for domestic dogs.

Kerswell *et al.* (2010) found that dogs with shorter snouts sniffed the heads of other dogs more often, and were themselves subject to more pouncing, turning the head away, and bites to the head, from other dogs. For coat differences, they found that dogs with longer coats were more likely to turn their head away from other dogs. Dogs whose eyes were partially covered were subject to body biting from other dogs.

This brings us back to Buddy and Elke. Buddy's eyes were obscured, while Elke's were prominent. In addition to having issues between modern and ancient breeds, the dogs couldn't make eye contact. Thus, between their extreme variance in size, communication style, social behavior, social structure, and the ability to convey meaning with eye contact, this was a script for disaster. And it's a script that's played out again and again, across social contexts, when ancient and modern breeds are thrown together.

Understanding the reasons for these differences is beyond the scope of the study. The results are highly suggestive that communication is impacted by morphology, and by studying this in puppies the researchers minimize the influence of prior experience influencing their social interactions. It is interest-

Table 8.1. Allocation of scores, for each signaling structure, for domestic dogs (Kerswell *et al.*, 2010).

Signaling structure	0	1	2	3
Tail	Full length	Long docked	Short docked	Close docked
Ears	Erect	Partially erect	Floppy	Pendulous
Coat	Erect	Short	Medium	Long
Size	Giant	Medium	Short	Toy
Snout	Long	Medium	Short	Brachycephalic
Eye cover	Uncovered	Partially covered	Full cover	—
Lips	Tight	Drooping	—	—

A greater score indicates a lower ability of the structure to be used as a signaling structure and approximate a larger deviation from the lupine form. Reprinted from Kerswell *et al.* (2010) with permission from Elsevier.

ing, however, to ponder whether the fully adult forms of these breed-specific features might exaggerate the differences further. For example, puppy snouts tend to be shorter than adult snouts. Despite the challenges these morphological changes can produce on communication, we know that dissimilar dogs with long association can improve their communication through experience (learning).

Social Structure in Dogs: The Ancient Breed Effect

The most significant practical use of our knowledge of social structure in dogs comes about in cases involving the ancient breeds, which are largely indistinguishable (genetically) from their ancient wolf-like ancestors. In addition to genetic similarity to their wolf-like ancestors, these breeds look and act more "wolf-like" as well. These breeds exhibit stronger tendencies for social structure with each other, and often (disastrously) toward other breeds, and on rare occasions, toward human members of their social environment. In addition to the ancient breeds identified by Ostrander and colleagues, including the Chow Chow, Chinese Shar-Pei, Shiba Inu, Alaskan Malamute (Fig. 8.3), Greenland Sled Dog, Siberian Husky, and Tibetan Mastiff (Parker *et al.*, 2017), we have seen these sorts of issues most commonly in Boxers and most of all in Australian Cattle Dogs. Cattle Dogs appear to break a lot of the "rules"—they're outliers who don't fit neatly into many of the behavioral categories. Presumably, in the development of Cattle Dogs from ancestral breeds, the genetics for strong social structure rules, including dominance, remained strong, whereas it was diminished in more-developed, different-ancestry breeds.

The practical cases fall into a couple of types: dog–dog aggression (which is often severe) and dog–human aggression. In the case of dog–dog aggression, it's almost always the case that the trouble comes from a lack of communication: a strongly social-structured breed member is exhibiting strong social structure signals to a member of a breed that simply has had that side of its social behavior selected against over many generations. We hesitate to use that unfortunately emotionally laden term: dominance (more on dominance later in this chapter). This is primarily because this term has been misused, misappropriated, and misunderstood for so long that it often holds a negative, but inaccurate, connotation. More on this shortly. Unfortunately, when combined with common size differences (Husky, Malamute, Akita versus Toy or Teacup Poodle, Pomeranian, French Bulldog, Yorkshire Terrier, etc.), simple physical signals, like a good nip or shake, delivered to a clueless modern breed recipient can lead to serious, even fatal, damage. These signals are meant to reduce damage risks due to outright aggression, to minimize the need for fights over limited resources, but because of the size differences, and the failure of the clueless target to respond appropriately, damage can be done.

It is for this reason that we actively campaign clients to be cautious about taking small, non-ancient-breed dogs, (especially those who have not grown up with, or been very well socialized with, breeds expressing active social structure behaviors), to off-leash dog parks. All good off-leash parks have specific restricted areas for small dogs (Fig. 8.4), and this is a very wise idea! It's important to remember that, in a very practical way, social structures, hence the behaviors used to maintain social structures, if any, are simply a tool used by social species to minimize physical harm and wasted energy when it comes to dealing with limited resources. If resources are not (perceived to be) limited, whether that is food, space, or desired

Fig. 8.3. Alaskan Malamute, Zuku. Provided courtesy of Holly Cook Photography, LLC.

social contact, then social structures are not needed and those behaviors disappear.

There can be significant issues between different breeds at dog parks. A few research studies have documented this effect, but there are a few facts: we know that different breeds (or more broadly, breed groups or types) use body language, and perhaps vocalizations, differently. Think of this like regional dialects, such as the Southern US versus the Northeastern US, but with more potential consequences for misinterpretation. There's no question that ancient breeds are very wolf- or coyote-like in their use of body language, and highly derived breeds, like the Toys, Poodles, and related breeds, have had their body language use modified, in some cases dramatically, through natural selection for other features.

It's also clear that *many* significant dog park incidents (such as maulings or killings) occur between large and small breeds, and between ancient and derived breeds, to the extent that most municipal parks have exclusive small-dog-restricted areas to limit these potential interactions. The hypothesis is that some breeds still exhibit dominance and predatory behaviors, and some breeds don't act like those breeds, hence need correcting/challenging and/or being eaten. But little, or no, scientific evidence supports this concept, given few, or no, studies of this concept. There's no support, but it's not disproven either. There's just no information.

There are cases of dogs that "don't want to go to the dog park." In fact, in almost every case, when you assess the situation, these dogs are derived breeds which, for genetics or lack of early socialization, or both, are frequently faced with confused and often aggressive other dogs at the park. Our solution for these dogs is universally, "Don't take them to the parks!" In most cases, dog parks do more harm than good: if your dog is well-behaved, not showing any behavior issues, and loves the dog park … fine. Otherwise, avoid them. There are too many incompatible dogs and irresponsible owners. Dog daycares and private, stable, supervised play groups are far, far better choices for your dog to socialize.

Fig. 8.4. Dormont Park, Dormont, PA. Image shows signs for large and small dog entrances at dog park. Image licensed under the Creative Commons Attribution-Share Alike 4.0 International license and provided by Wikipedia. Available at: https://commons.wikimedia.org/wiki/File:Dormont_Park_-_dog_weight_signs.jpg.

Some breeds may benefit from high-quality doggie daycares, while others experience a smaller benefit in comparison. Almost all dogs benefit from high-quality dog daycare experiences. These opportunities, along with private "play date" groups, provide social experience at the appropriate level and with appropriately social dogs. So ancient breeds will interact with other ancient, "traditionally social" breeds and derived, modified, "non-traditional communicators" can interact with like-minded individuals. Highly intense breeds, like Terriers and Herding group breeds, can interact with the same. Laid-back breeds like Bassets and Sheepdogs interact when, and if, they feel like it. And supervision by responsible "adults" is most important.

Imagine that you are still a small child (we're using this example because dogs remain dependent upon humans, irrespective of their age) and you're taken to a playground with children who are two, ten, or even 40 times your size, and whose nonverbal behavior is foreign to you, even if you spoke the same language. Both the (relatively) very large and very small individuals are going to express some level of discomfort, at the very least. It's not a recipe for playground success. A large dog might wonder, "Is that prey?" and a small dog might ask, "Is that a predator?" We say *might* because we can only hypothesize based upon their actions.

Based on the science we have so far, and what we suspect given case studies, there are some important take home lessons about the social behavior of dogs and people. While not scientifically validated, all dog breeds *appear* to enjoy, and perhaps need, human social interaction; that is, humans have become part of the normal social environment for domestic dogs. This desire or drive varies with breed (which is less so in ancient breeds, very strongly expressed in toy or lapdog breeds, and appears to have an intermediate effect in working dogs like herders, guarders, hunters, and terriers). Of course, it also varies by individual, but less so than by breed and genetics.

It's also clear to us that a small number of breeds, usually ancient ones, with a few exceptions (such as the Cattle Dogs!), exhibit social dominance and may exhibit this social dominance towards humans. A case example is a Cattle Dog who was totally untrained, had very little structure to its life, and was severely under stimulated, both physically and mentally. He began biting his owners whenever they did something that he didn't like. This isn't an acceptable behavior for any dog. When the dog was assessed, he showed every dominance signal used by dogs towards his owner; these behaviors weren't directed toward strangers, nor to kids, only to the owners. He performed "lateral displays," which means that he was making himself look bigger.

A lateral display, also referred to as a "sideways threat" or "lateral threatening," entails presenting your side to your opponent (or in some cases, the perceived opponent) (Fig. 8.5). You may have seen this behavior between two unfamiliar cats who approached each other, their legs bouncing as though they were on springs, their heads craned to the side in an almost eerie angle. If you have grown up around animals, you may have seen this display with barn cats (where not-so-welcome Tom cats often stray) or even your own cats. We have seen this with other animals, including horses, dogs, and roosters, but it's exhibited by a wide range of species.

Besides the lateral displays, he body-blocked movement paths so he was the first to come and go, and he marked all over the house, even though he was housebroken for urination. When the owners really did something that he didn't like, he bit them: not aggressively, but making it very clear that he did not approve, and that they had ignored his previous, lower-intensity signals. This was a classic case of social dominance or control. It was explained to the owners as almost a feeling on the part of the dog that there was a vacuum of direction and stimulation, and that if the owners weren't going to exhibit *any* control over *any* resource (he was even fed ad-lib, free-feed!), that he felt that he needed to.

Most breeds have lost their social dominance drives and behaviors, and hence, the "pack leadership" concept of behavior modification taught by popular TV trainers (which doesn't use any kind of true dominance signals anyway, just punishment) fail miserably. In a few breeds and even then, only in a few individuals, social dominance is transferred to human owners, and a clear understanding of dominance and its uses and signals is required to modify the behavior.

The Unfortunately Emotionally Laden Term: Dominance

So what is dominance, as it pertains to our dogs? Dominance hierarchies are fairly common throughout the free-living animal kingdom, and are found among many species of insects, fish, birds, and mammals, including honeybees and hammerhead sharks, American crows and African lions, Nile tilapia and North American red foxes, chickadees and chimpanzees, and water monitor lizards and wolves. This type of social structure typically arises when members of a social group compete for access to limited resources such as food and mates. The dominance hierarchy is established to avoid costly aggressive encounters over these resources. Here you can think about another species such as chickens as an example. There's a "pecking order" and all individuals get along just fine so long as there

Fig. 8.5. (a) Lateral display by cat to dog. Photograph courtesy of Jenni Pfafman, UW-AAB, KPA-CTP, of her kitten, Truffle, and dog, Ky. (b) Lateral display between dogs. Photograph courtesy of Carol Harris, UW-AAB.

aren't new members introduced and the order needs to be re-assessed. Sometimes you can stir things up in the chicken coop, just by removing an individual (for health care, for example). When you try re-introducing them, there are skirmishes in the hen house.

In dogs, dominance is typically restricted to the ancient breeds, with specific breeds and breed groups tending to exhibit more dominance in social settings. Recall the Prologue and Pepper and Banjo, the Pomeranians who were displacing their owners. The dogs were trying to assert their social dominance over their people by displacing them, but Banjo took this one step further by urinating on his owner. Pomeranians themselves aren't ancient, per se, but they hail from one of the ancient breed types (the Spitzes). They became popular with royalty by the 18th century and made a special appearance in the 1788 iteration of the *Systema Naturae*: Editor Johann Friedrich Gmelin included them as *Canis pomeranus* (Linnaeus, 1792). The AKC first recognized the Pomeranian as a breed 100 years later. Pomeranians aren't the new pups on the block, though: Spitz types date back for thousands of years. Thus, Pomeranians, like Cattle Dogs, retain certain ancestral characteristics, including behaviors related to dominance and social structure.

While relatively rare, we can describe four cases of owner-dominance issues, including the peeing Pomeranian, but also a big Cattle Dog, a Corgi–Basset mix, and a Red Heeler–Cattle Dog mix. In all cases, the really big signal to the diagnostician is aggression towards the owners. It never happens otherwise, except in cases of abuse, or redirected arousal (such as reaching in to separate fighting dogs) ... otherwise, aggression toward the owners usually indicates a dog trying to take control, getting a bigger share of a resource. This usually happens in training, structure, and setting rules. Often, the owners become fearful of the dog, act "submissive" (by retreating or averting their eyes) and the dog moves in to get more food, space, time outside, whatever. Owners often respond in a challenging, violent way, and everything goes downhill fast, resulting in owner-directed aggression and bites.

We've been discussing cases of dominance with dogs and humans, but what does dog–dog dominance look like, and how common is it? Dog–dog dominance aggression cases are very rare, but when they do occur, they probably lead to a lot of the known mauling cases. For example, when you have ancient breeds versus toy breeds, such as Elke and Buddy, in areas like off-leash parks. In most breeds, though, the dogs can recognize at least some of the signals and acknowledge them appropriately enough to appease, so typically, everyone will get along.

The Big, Not So Bad Wolf

Once upon a time, dominance theory was the prevailing paradigm of science. Headed by American ethologist Lucyan David Mech (popularly known as David) and Erich Klinghammer, dominance theory was primarily based upon Mech's findings. Mech began his work with wolves in 1958 and has published hundreds of scientific pieces about them, but it was his introduction of the concept of the "alpha wolf" that earned him the most notoriety. His book, *The Wolf: Ecology and Behavior of an Endangered Species* (Mech, 1970), discussed how alphas used violence to rule their packs. This concept influenced the field of animal behavior, and in particular, canid behavior, for years to come. But the wolves in his original study didn't represent the typical wolf social structure; they instead revealed how unrelated wolves would interact when thrown into a small, shared, captive space. Almost two decades later, after mounting evidence found that the alpha was a myth, an artifact of methodology and not social mechanisms, Mech recanted this concept, even asking the publisher of his book to cease printing it, but the damage had already been done. Mech (1999) concluded that the typical wolf pack comprised parents and their dependent offspring, and his later work continued to clarify the wolf's true social structure (Mech and Boitani, 2003). The "alpha" stubbornly stayed, though; a dinosaur that never was. Even as Mech clarified that wolves form families with a mother and father wolf and their offspring, who then leave to form their own families, the myth of the alpha wolf persisted, steeping scientific and everyday literature alike. Fortunately for dogs and the humans who love them, we've had a paradigm shift in the following decades. Mech's contemporary, German-born Erich Klinghammer, perpetuated the alpha male concept (Klinghammer, 1979). Years after their initial conclusions about alphas in wolf packs, both Mech and Klinghammer retracted this stance, realizing that dominance is facultative and not obligatory. Dominance is ecosystem-based. Mech's retraction, however, was incomplete because he didn't quite understand how trainers were incorrectly using the information.

Unfortunately, as is often the case, their work was misunderstood and incorrectly extrapolated. The fundamental errors are threefold:

1. They never said that it was obligatory! (Dominance is actually facultative and depends largely on the feeding ecology of each wolf population and on the kinship structure of the pack.)
2. Further confusion arose from the fact that many extrapolated Mech's work on wolves to apply to all dogs due to the close phylogenetic relationship between wolves and dogs. But phylogeny doesn't always mean equivalence, it's just a good assumption to start with when you lack other information.
3. Additionally, many people assume that a social hierarchy is always linear. In fact, social hierarchies aren't necessarily strictly linear.

This misinformation has stuck with the public and has been hard to rewrite. Even veterinarians use this terminology. Many people still believe that you should roll a puppy over on its back and see how much it struggles to evaluate it as a good family pet. But ask yourself: what would it really measure, to dominate a young, dependent puppy in such a fashion? To make matters worse, the use of punishment techniques were used to enforce some kind of control over the supposed dominance hierarchy. This is odd because in the wild animal kingdom, it's really about controlling access to resources. If you control the resources, then you are in charge. There's no need to physically control individual animals with force.

Thanks in large part to the continued misuse and misunderstanding of the term "dominance," it has become a "bad word" in much of dog training. Cesar Millan, the self-proclaimed "Dog Whisperer" despite using some rather unquiet training techniques, was a popular proponent of using dominance theory to train dogs. He claimed that you needed to make sure your dog was subordinate to you. Unfortunately, this type of approach ignores several important factors, including the situation (is there a resource to "fight" over?) and your dog's breed and temperament. Other important factors to consider include your dog's age and thus their capacity for learning and inhibition and the length of time you have had a relationship with the dog. You don't need to dominate your dog to train it to sit or lie down; you can offer rewards for desired behaviors, such as attention, treats, or an ear scratch. Your dog's breed (think Poodle) may even prevent it from understanding the concept of a hierarchy. After all, we have artificially bred out some behaviors. In addition, you could have a naturally "shy" or "anxious" dog, and harsh corrections and efforts to dominate that individual dog could have some unfortunate side effects. For example, a condition known as "learned helplessness" occurs when an individual lacks the ability to control stressful conditions, and rather than perseverate, they "give up" or "shut down." This type of behavior has also been described as depression in animals. So, we have confusion and misuse of dominance theory that makes for some very bad outcomes. In response to this "whispering," a battle cry rose up from the positive trainers to get away from using dominance theory and punishment-based methods of dog training. This can be confusing for someone with an animal behavior background, much less, for someone who's new to cohabitating with canines.

The literature now suggests that dominance hierarchies can also form when relationships are cooperative, rather than competitive. A study focusing on packs of free-ranging dogs at three different locations in Italy found that there was a linear hierarchy based upon age (Bonanni *et al.*, 2017). The researchers in this study suggested that dominance hierarchies can still be exhibited in dogs (despite human influence on breeding and artificial selection) and that age may be a better predictor of dominance than body weight. Let's examine the concept of age as a predictor of dominance by reframing how we see "dominance" and substituting it with a word less laden with negativity, such as "leadership" or "guidance." While "dominance" has a negative connotation with many pet owners, there are evolutionary reasons for why it arises within a species. It endures in certain species and breeds or lines of species because its positive aspects outweigh its negative ones. Leadership and guidance are important in a social group.

Among many of these species, such as elephants, age is an indicator of these traits because the older female elephants—the matriarchs of the family—are the keepers of knowledge. They know where to find rare water sources during a drought. They know the best routes to seasonal grazing grounds. And they know how to relate that information to the rest of their family group—and guide them, and lead them, and keep them as safe and healthy as possible. Thus, dominance can be seen as "leadership," when we reframe it in this way.

For this particular study, the authors concluded that the hierarchy is beneficial to group members who are dependent on the experience of older members for guidance. The information on genetic relatedness was incomplete in this study, and some pack members were related. The authors argued, however, that not all of the packs were formed from relatives, and the relationship still held true in their sample. Thus, the authors concluded that dominance can still be seen in domestic dogs.

With some dog breeds retaining dominance, it's clear that there was once an evolutionary advantage for this trait. Across multiple species, dominance hierarchies can help keep a social group intact, reduce stress, and decrease rates of fighting and injury (Fournier and Festa-Bianchet, 1995; Pusey and Packer, 1997). Among horses, the dominant mare in the group is the leader and the keeper of the knowledge, much like elephants. Similarly, roosters alert the hens to a food source, such as grain or worms, and then watch as each hen had their fair share, only eating after the hens ate, and mediating any altercations. Even though roosters are "dominant," they will express "feeding tolerance" and eat last.

Feeding tolerance is exhibited differently across species. A study conducted at the Wolf Science Center in Austria compared the social tolerance of feeding in wolves versus dogs in both dyadic (pairs) and group (formed of three or more dogs) feeding scenarios (Dale *et al.*, 2017). The wolves came from North America and most of the dogs were mixed breeds from Hungarian shelters. When just two individuals were feeding, the affiliative relationship between the two predicted tolerance in both dogs and wolves. However, in the group setting, dominance was a larger factor in dogs than in wolves. Specifically, high-ranking dogs were more likely to try and monopolize the resource, while wolves were more likely to allow subordinates to feed, but they controlled access to food resources by subordinates.

It's Not All Fun and Games

In *The Descent of Man*, Darwin (1871) wrote, "Happiness is never better exhibited than by young animals, such as puppies, kittens, lambs, and company, when playing together, like our own children". Play isn't just the expression of happiness, though; it's also an important part of social and psychological development (Burghardt, 2006). For more than a century (Groos, 1898), play has been recognized as an important behavioral context for humans and non-humans alike. Studies conducted during the 1970s revealed that play supports tool use, status, and developing adaptive strategies (Vandenberg, 1978). A literature review of mammalian play conducted by Spinka *et al.* (2001) found that this behavioral context plays an important role in preparing for the unexpected, too, by teaching the players how to process stress and other strong emotions and learning how to regain one's balance after a fall. The sequences of play—including visual gestures, tactile gestures, and auditory gestures—also have an important role in understanding the attentional state of one's play partner (Campion *et al.*, 2011). Play is an important behavior for many species, from corvids to canines, but we don't all play the same way, or by the same rules. There can be subtle nuances between social or cultural groups, both within and between species. So how does this pertain to our dogs, who have such a range of physical appearances and social behaviors, but for whom play is particularly important to their development?

Mehrkam *et al.* (2017) compared Herding Group dogs, Retrievers (part of the AKC's Sporting Group), and livestock guardian dogs (part of the AKC's Working Group) in social versus solitary play using experimental treatments. Specifically, the researchers created 30 breed-matched dyads (controlling for age, sex and neuter status) and exposed them to four test conditions and two control conditions in randomized order. Four of the six conditions (owner attention, less familiar person, toy, and escape from short separation from owner/playmate) were meant to simulate either social play or solitary play in the pairs of dogs (10 pairs of dogs in each working dog group). The controls were owner control and alone control, and were designed to measure background rates of play without manipulation.

The researchers found that Retrievers played at higher levels than the other two working groups (whether social or solitary). Retrievers showed significantly more solitary play behavior than livestock guardian dogs. Retrievers also showed a trend for more solitary play than Herding Group dogs did. There was no statistically significant difference between working groups in social play, however. They argued that this effect was driven by the intact predatory motor sequence found in Retrievers, and that this pattern has also been found in domestic cats (Hall and Bradshaw, 1998).

Why is this work significant? It was the first study to show the relationship between *breed type* and *play* in domestic dogs. It did so using an experimental method, well-controlled stimuli/conditions, and with the hypothesis that there has been selection on the sequences of predatory behavior patterns that influence play behavior. These results suggest that like other behaviors, it is influenced by (breed) genetic predispositions, the environmental context, and the experience of the individual throughout its lifetime.

Speaking of experience, take a look at the play of the 1-year-old Maltese–Poodle cross and the Labrador, Frannie, shown in Fig. 8.6. The Labrador adjusted her play for the smaller dog, including adjusting her play "bites" and play bows to be smaller and gentler, and spending a lot of her time low to the ground. The smaller dog also adjusted her play style with the larger dog. This interaction between two very differently sized dogs was made possible by a Labrador that was well-socialized by an assistance dog organization and a puppy that was extremely well-socialized, including attending a high-quality doggie daycare during weekdays.

Communication Between Humans and Dogs

Breed differences in dog–human communication

One of the hottest research areas in dogs has been various studies looking at how well dogs understand us. Many of these studies were done in Hungary by Ádám Miklósi and his colleagues at the Eötvös Loránd University, Department of Ethology. The early research in this area was mostly about directional pointing and whether dogs would follow a human's pointing gestures, but has expanded to include learning, temperament and perception as well as other topics. Rather than summarize all of the literature on dog communication, we focus here on the breed differences found in the science to date.

Researchers looked at the acquisition and subsequent extinction of a dog's gaze to a human face in a scenario where food was in sight, but out of reach (Jakovcevic *et al.*, 2010). Retrievers, German Shepherds, and Poodles were compared with and without explicit training. Retrievers showed significantly more gazing towards humans than the

Fig. 8.6. Frannie and Maltipoo. The Labrador adjusts her play for the smaller dog. Photographs courtesy of Renee Robinette Ha, PhD.

other breeds. Both in their own observations, and in their review of the literature, the authors concluded that gaze toward a human is associated with sociability in these breeds and that Retrievers are significantly more social than German Shepherds or Poodles.

In a similar study (Konno *et al.*, 2016), breed group (vonHoldt *et al.*, 2010) mattered during an unsolvable task (where dogs might look to humans for "help"). This particular study used a large number of breeds to represent each genetic group, so those interested in the details can follow up based on the original publication. Ancient breeds waited longer to look to the human compared to Retrievers and Working breeds. Ancient breeds also looked at humans for shorter durations than Herding, Hound, Retriever and Working breeds. Likewise, a recent study showed that breeds that were bred to cooperate with humans on working tasks (e.g. Shepherds, Retrievers, Collies, etc.) benefited from observing a human demonstrator solving a problem, whereas breeds that were bred to do their work independently of humans (e.g. Terriers, Hounds, Huskies, etc.) did not improve after observing a human demonstration (Dobos and Pongrácz, 2023).

Udell *et al.* (2014) compared a human pointing object choice task in three breeds of dogs that vary in their predatory motor pattern sequences. For hunting dogs, they used the Airedale Terriers, which are selected for their fully intact predatory sequence, and contrasted them with Border Collies, a herding breed selected on the basis of the traditional eye–stalk–chase component of the predatory sequence, and Anatolian Shepherds, which are livestock guardian dogs, selected for the *absence* of the predatory sequence.

So what, exactly, is a "predatory sequence?" This concept was originally referred to by Canadian-born behavioral scientist (and ethological field co-founder) Wallace Craig as "appetitive behavior" (Craig, 1918). Craig referred to an appetite as a "state of agitation which continues so long as a certain stimulus, which may be called the appeted stimulus, is absent. When the appeted stimulus is at length received it stimulates a consummatory reaction, after which the appetitive behavior ceases and is succeeded by a state of relative rest." Modern definitions of the predatory sequence include searching, stalking, rushing or chasing, grabbing, killing, and then dissecting. If you've ever watched a predator stalk their prey, you might be familiar with these predatory behaviors and the unfortunate aftermath for the prey animal. And that's because the typical motive behind the predatory sequence is the final part: dissection, and then consumption. It behooves a potential prey animal, then, to take notice when a potential predator is acting predatory. While aggression serves to increase distance, predatory behavior serves to decrease distance as much as possible. Many dogs were bred to be hunters, and whether that's their current vocation or not, they'll still exhibit some parts of the predatory sequence.

There were significant differences in accuracy between all of the breeds, with Border Collies being the most accurate with using human signals, Airedale Terriers performing similarly to wolves, and Anatolian Shepherds performing at chance levels (Udell *et al.*, 2014). Udell *et al.* (2014) did find that the Anatolian Shepherds could be trained to improve in the task in relatively few trials, suggesting that this limitation was more of a predisposed tendency than a cognitive deficit. This paper is a nice contribution to the literature in that it makes a direct connection with behaviors under direct selection (the predatory motor sequence) and how that links to other behavioral tasks.

Finally, there was a study that compared different breeds of domestic dogs against New Guinea Singing Dogs for their ability to make use of various human communicative behaviors to find hidden food (Wobber *et al.*, 2009). The New Guinea Singing Dogs were raised in a human household as puppies similarly to the other dogs in the study, and demonstrated similar results to the other dogs. However, the "working" breeds (Huskies and Shepherds) were more skilled at using gestural cues than the "non-working" breeds (Basenji and Toy Poodles) were. Similarly, another study compared the likelihood of dogs in the primitive, hunting/herding and Mastiff-like breed groups to choose a smaller pile of food (over a larger one) when a human showed more interest in it (Barnard *et al.*, 2019). They found that dogs in the hunting/herding group were significantly more likely than the Mastiff-like group to choose the small pile of food over the larger one when a human reacted to it. All breed groups, however, chose the larger pile over the smaller one when they weren't directed towards the smaller one. This suggests that the ability to read and follow human gestural cues arose early in evolutionary time and has continued to be under selection, particularly in dogs that work closely on tasks with humans. These studies show that being able to observe, interpret, and react to human signals wasn't just strongly selected for across all dog breeds, though. It was particularly selected for in those dogs that worked alongside their humans and whose success depended upon this skill. Border Collies, in particular, need to be able to discern audible and visible cues to perform their traditional jobs. Historically, a Border Collie who couldn't do so could find themselves in an unsafe situation, and perhaps worse, could be deemed a non-contributor, which would result in fewer opportunities to mate or to share a home with humans. And a Husky pulling a travois or a sled would need to be able to respond to gestural cues more so than a Poodle would.

Oxytocin

Oxytocin has been implicated as a mechanism for human-directed social behavior in dogs. A 2017 review paper has summarized the findings (Kis *et al.*, 2017). Recent experimental work shows that there may be breed differences in the effects of oxytocin on dog–human social behavior and communication (Kovács *et al.*, 2016). Specifically, in a study comparing Siberian Huskies to Border Collies, the researchers examined the effect of oxytocin or placebo nasal injections on their human-directed social behavior during three tasks: impossible problem (access to food blocked by a gate when it was previously accessible), a potentially dangerous object, and tolerating prolonged eye contact. They found that in the oxytocin condition, Border Collies looked more quickly and for a longer duration to the humans for "help" with the impossible task in comparison to Siberian Huskies. Similarly, Border Collies looked more often to the owners and between the owners and the "dangerous" object. In the prolonged eye-contact task, under the influence of oxytocin, Border Collies tolerated eye contact longer than in the control condition, and Siberian Huskies decreased their tolerance under the influence of oxytocin. Taken as a whole, these studies suggest that oxytocin is under genetic control, and has been selected for in more derived breeds versus more ancient breeds. Given the importance of oxytocin on social behavior and the bond between dogs and humans, it's clear that we need more information beyond the results of what these studies have yielded so far.

Genetic mechanism and evolutionary origins

We now know that at least some breeds of dogs have genetic predispositions for social behavior and communication with humans (Persson *et al.*, 2015, 2018). Beagles were shown to have a genetic basis to solicit help from humans when faced with an impossible problem (referred to as an "unsolvable task"), and specifically that heritability estimates for related behaviors were 0.23–0.32. Let's provide some context for that. This is the proportion of the variation that seems to be correlated with genetics, and it's a moderate effect of genetics on behavior. In terms of heritability, 0.7 or 0.8 are considered to be large effects, but for a behavioral trait, 0.23–0.32 is relatively robust. However, they also found that eye contact (with humans) increased with age and experience, so they were predisposed for this type of social contact, but the environment also shaped it in individual dogs (Persson *et al.*, 2015). A follow-up to this study (Persson *et al.*, 2018) incorporated additional breeds, including Golden Retrievers and Labrador Retrievers, as well as wolves. In addition, they divided the Labrador Retrievers into "field" and "show" to look at recent selection. They used the same behavioral measure as before (impossible task), but they also collected genetic information to evaluate two single nucleotide polymorphisms (SNPs) previously linked with these social behaviors with humans but found in their Beagle study. They found that there was variation between individuals of Labrador Retrievers and Golden Retrievers for SNP1, but no variation in wolves for this genetic marker (they were all the same). There were differences in both breeds and the wolves in SNP2. There were significant differences between field and show Labrador Retrievers for both markers. They concluded that their work shows that there are genetic breed differences in these human-directed social behaviors, and that they have identified the genomic region for these differences across breeds.

In similar fashion, Kubinyi *et al.* (2017) examined variation in the *OPRM1* gene, which is associated with social behavior and friendliness. The researchers found that Siberian Huskies and gray wolves were more similar to one another, and differed significantly from Border Collies and German Shepherd Dogs. This study had a small sample size and they caution that more work needs to be done to fully explore these exact breed relationships, but that this gene appears to affect human-directed social behavior and appears to vary by breed.

During artificial selection, it is thought that humans selected for sociability in dogs. Studying the genetic markers for this was approached by targeting a section of genome understood in humans to result in hypersociality or extreme friendliness, something called Williams–Beuren Syndrome in humans (Chapter 2).

vonHoldt *et al.* (2017) studied the variation in this gene using SNPs from 701 dogs (85 breeds) and 92 gray wolves. They concluded that variants they identified corresponded to sociality in dogs, and that these may have been associated with our domestication of dogs. But dogs aren't the only domesticate with gene variants related to their sociability. Horses, too, are sensitive to the cues of their humans, and will indicate what they want via tactile and visual gestures, even modifying their signals if they believe that their human does not know where their food has been hidden (Ringhofer and Yamamoto, 2017). This is similar to how dogs will look to their humans to help, rather than perseverating with a task. Recall that dogs experienced genetic mutations that led to their hypersocial temperament; the mutations on one gene in particular, *WBSCR17*, have been implicated in these changes. It's worth examining equine genetics to see if there's a similar mechanism for sociability in dogs and in horses. For decades, horse owners anecdotally believed that there was interbreed variance when it comes to personality traits, and now the science is verifying what many of us already suspected. One such study examined the variance of personality traits across eight horse breeds, finding significant interbreed differences between excitability and anxiousness (Lloyd *et al.*, 2008). Recognizing that we have been able to tame horses, and not their close relatives, the zebras, scientists are now examining the genetics of horses to see if they might have mutations similar to *WBSCR17* (Ratliff, 2011).

Why the long face?

One mutation that certainly sets horses apart is their capacity for emotional memory. If you've ever spent a considerable amount of time with horses, you'll notice that they are very observant and very sensitive. And if you suspect that a horse will hold your bad mood against you the next time you see her, you're absolutely right—so it's best to check that bad attitude at the barn door. Those who work with horses have long thought that their emotional states impact their relationships with their equine partners, but a 2018 study (Proops *et al.*, 2018) quantified just how much horses perceive, remember, and react to human emotions.

Researchers from the Universities of Sussex and Portsmouth in England wanted to measure horses' reactions to human facial expressions, so they showed them images of people with either a happy or angry expression. When the horses viewed the images of the angry expressions, they turned their heads to the right side and viewed the image with their left eye. Their heart rate also increased. When a horse viewed the happy expressions, they didn't change the position of their head or show left or right eye viewing preference and their heart rate remained steady. The horses then interacted with the people at a later time, when the people were stoic rather than emotional. Horses who had viewed an angry image of the person prior to meeting them tended to turn their heads and look at the person primarily with their left eye, while horses who had viewed a happy image of the person prior to the interaction again did not show a left or right eye bias. Those who work with horses often say that horses can "reflect your emotions back to you," but this new research reveals that they aren't just emotional mirrors—they're also emotionally *memorious*—they will remember the last expression that you had on your face. (So, try to make it a positive one!) Laterality indicates that horses not only *remembered* the prior image, but they knew what it *meant*, too. This is because of how the brain processes stimuli: the right hemisphere processes items that are perceived as threatening, and the left eye corresponds to the right hemisphere. Beyond remembering and understanding the meaning of the image, they also knew how to modify their behaviors during their interactions. This study marked the first time that a non-human animal was found to have the ability to remember and catalogue emotions and modify their later behaviors based upon this emotional memory.

Being able to read, remember, and respond appropriately to human emotions is an important skill for domesticated animals. Dogs, having co-evolved with humans for tens of thousands of years, have long been the kings of reading and responding to human emotions. In this study, however, horses excelled. Horses are one of humans' oldest domesticates after dogs, and they've experienced a wide range of roles throughout our shared histories, accompanying us on the farm, battlefield, sports arena, and in therapeutic settings. Emotional expression is an important part of human communication, so it should come as no surprise that having co-evolved with us for so long, horses can read our emotions, including our facial expressions, with striking accuracy, and then respond appropriately to those emotions. If someone is angry, they're perceived as a potential threat. If someone is happy, they aren't a threat. So, the next time you interact with a horse, try to end things on a good note: remember that they have an exceptional emotional memory, and that they'll modify their later behaviors based upon what you emoted the last time you saw them.

Clever Hans

Tap, tap, tap, tap … tap. Hans, a dark, stately Orlov Trotter, stopped and gazed at his trainer, Herr Wilhelm von Osten, expectantly. The clever horse had once again tapped out the correct answer. This time, von Osten had asked him an arithmetic problem, but Hans wasn't limited to answering arithmetic; his reported repertoire included a slew of subjects, including reading, spelling, fractions, addition, subtraction, multiplication and division. Hans couldn't "say" his responses, so he tapped them out with his hoof. During the early 1900s, Clever Hans became a household name, but he wasn't an equine genius; at least, not in the way people initially thought. Rather than knowing these concepts, he knew how to read people.

Developed in Russia by Count Alexei Orlov during the late 18th century, the Orlov Trotter is a horse breed that was bred for stamina and a very fast trot, a two-beat gait where alternate diagonal legs, e.g. right foreleg and left hind leg, are separated by suspension in between touching the ground.

As we learn more about the genetic mechanisms behind domesticated species, we learn more about ourselves, too. But how well do we know and read our canine companions? The easy answer is: not as well as dogs or horses read humans. Clever Hans is the classic example of horses reading humans proffered by psychology professors. Rather than knowing how to divide and subtract, Hans could add the multitude of behavioral signs he was receiving from the person who had asked them the question. As long as the asker knew the answer, Hans "knew" the answer, too, because subtle non-verbal signals alerted him to when he'd stopped at the correct answer. Hans could read the people, but they were unable to discern what Hans was doing (until 1907, at least, when psychologist Oskar Pfungst discovered Hans' actual ability).

When we began our journey together discussing the differences between dog breeds, we noted that the lineages of canines and humans split approximately 95 mya. Because of convergent evolution, though, dogs have learned to read humans, often better than we can read ourselves—but the same can't be said for humans' understanding of dogs. We're learning, though! If you've ever seen an image of a dog with whale eyes (Fig. 8.7) being hugged by a human (Coren, 2016), or leaning away with their teeth showing, or with wide eyes as they're "dog shamed" and labeled as feeling "guilty,"

Fig. 8.7. Whale eyes. Contributed by June Krisko from Pixabay. Free for use under the Pixabay content license. Available at: https://pixabay.com/photos/dog-chocolate-labrador-retriever-731979.

you'll have an idea of what's going on here. This is the Dunning–Kruger Effect (cognitive bias stating that people with low ability at a task overestimate their ability) at its worst: we believe that we "know" what our dogs are feeling or thinking, but many of us are greatly overestimating our ability to decode our dogs. Perhaps because dogs are so familiar to us, we assume that we can understand what they're feeling and communicating, but a foundation in animal behavior—and understanding a dog ethogram, in particular—can go a long way in bridging the communication/interpretation gap. But are they also "acting" how we'd act? Meaning, are we interpreting their behavior through our human lens, the way that we'd interpret human behavior? As humans, we experience the world through our own umwelts, and even when we "know better," we tend to not adjust to the species.

This is much like a Theory of Mind (ToM) issue. When one has a "ToM", they have the ability to understand that beings other than themselves can have perspectives, information, emotions, thoughts, beliefs, and desires that differ from their own. ToM is a popular psychological area of study because its presence in humans is associated with development; not all animals have one. While the topic remains controversial, researchers have found evidence that several non-human species have a ToM, though, including, but not limited to, Japanese macaques (*Macaca fuscata*) (Hayashi *et al.*, 2020), rhesus macaques (*Macaca mulatta*) (Flombaum and Santos, 2005), the large-bodied apes, including bonobos (*Pan paniscus*), chimpanzees (*Pan troglodytes*) and orangutans (*Pongo abelii*) (Krupenye *et al.*, 2016), corvids, including Western scrub jays (*Aphelocoma californica*) (Bugnyar, 2010), ravens (*Corvus corax*) (Bugnyar, 2010), and crows (*Corvus brachyrhynchos*) (Emery and Clayton, 2004), gray parrots (*Psittacus erithacus*) (Pepperberg, 2012), cetaceans, including bottle-nosed dolphins (*Tursiops truncatus*) (Pack and Herman, 2004), and with compelling research on dogs.

The challenge in ToM research in non-human animals has been in teasing apart whether or not the animal is merely making use of learned social cues or truly understands that the perspective of another is different than their own. For example, if another individual's gaze is directed to a specific location, the observer may have learned that "it is productive to explore locations that others notice/use" or intuit that "they know about something I don't because they are looking intently over there". Some studies have reported that dogs show superior performance on ToM tests compared to primates when the test involves "reading" the gaze or directional finger pointing of humans (Maginnity and Grace, 2014). Those authors suggest that dogs perform as if they have a functional ToM, whether or not they have a mentalistic understanding of ToM. In other words, the studies on dogs suggest that dogs respond to the appearance and behavior of others, and act accordingly (functional ToM), but we don't yet have the evidence to know that they understand that another's mental state is different than their own (mentalistic version of ToM) (Horowitz, 2011).

Regardless, the proficiency that dogs show in reading human cues is quite remarkable, and has been documented in numerous studies (Hare and Tomasello, 2005; Miklosi, 2008). This finding led to a number of studies comparing wolves and dogs in this ability, and debate over whether dogs "inherited" this social trait evolutionarily from wolves, learn it through experience with humans, or whether it is a result of artificial selection during domestication. While hotly debated, the evidence suggests that both genetic predispositions and the environment influences this ability, and that the genetic predispositions were likely influenced by artificial selection during domestication. For our purposes, behavioral differences among canines is of the most interest.

Work that compares dogs and wolves has found:

1. Wolf puppies do not use human communication cues to the same degree of skill as dog puppies, unless they are trained to do so (Agnetta *et al.*, 2000; Hare *et al.*, 2002; Viranyi *et al.*, 2008).
2. Dog puppies develop this ability even if they have very little exposure to humans (Hare *et al.*, 2002).

Based on the performance of wolves on these assessments, we might predict that more primitive dog breeds would respond to human gestural cues less readily than more domesticated breeds, and the literature supports that (Konno *et al.*, 2016). Whether or not reading human gestural cues is sufficient to suggest that dogs have a ToM, remains a topic of discussion (Horowitz, 2011).

It's clear that dogs aren't fond of hugs, and people often indulge themselves in hugging dogs. If we assessed ToM in humans based on how well they read dog communication, many adults would fail, so it's unsurprising that children also struggle. A recent study using questionnaires with images of dogs suggests that 8–12-year-old children only recognized "attack" and "threat" 7% and 9% of the time, respectively. Their best interpretation of dog behavior was "fear" at 34% accuracy; they recognized "joy" and "friendliness" in dogs 20% and 21% of the time, respectively (Chlopčíková and Mojžíšová, 2010).

Conclusion

There are significant social differences among dog breeds and breed groups, including communication, play, and social structure. These differences play important roles in how different dogs will communicate with one another and with their human companions. By understanding these differences—and the genetics that play an important role in behavior—we can mitigate potentially stressful situations and set our dogs up for success.

Given the almost complete change between derived dogs and wolves from their last common shared ancestor, derived dogs and wolves can't necessarily communicate with one another. It's not just that they have different dialects; they literally speak different languages. There are practical applications for this, given the differences in social structure between derived and ancient breeds, the latter again being largely genetically indistinguishable from their ancient wolf-life ancestors, and thus exhibiting those stronger tendencies for social structure with one another. In mixed breed households, you can overcome some of these translation issues by ensuring proper early socialization. Think of it as allowing derived and ancient breeds to learn one another's cultures; if you had two dogs from the same "culture," things would go smoother, but a skilled course in dog culture 101 for each representative breed can help smooth things over. It's important to keep in mind, as well, that those dog breeds that were specifically created to work alongside humans are going to hold eye contact longer, look to the human for help, and be more attuned to hand gestures than those dogs who were not bred for this purpose.

While dominance continues to be a controversial topic in dog training and behavior, the importance of dominance in dogs relates more to their standing as an ancient breed than in their interest in following human rules. Dogs that exhibit dominance aren't doing so to usurp their human's power; they're simply acting according to their behavioral predisposition (don't take it personally, but do be aware of it!). Managers at animal shelters should be cautious about integrating ancient and derived breeds in social play groups without knowledge of their individual histories. It would be a good precaution to assume they might not mix well, given their different cultures and languages.

Gregg and Little's power lunch was the catalyst behind understanding many of the mysteries of animal behavior, human and non-human alike. We're excited to see what the next important meetings will hold: our tables are set and our napkins are in our laps.

References

107th Congress, H.R. 2646 (2002) Public Law 107-171--Farm Security and Rural Investment Act of 2002. US Department of Agriculture – National Agricultural Library. Archived from the original on 7 July 2012. Available at: www.govinfo.gov/content/pkg/PLAW-107publ171/pdf/PLAW-107publ171.pdf (accessed 11 December 2023).

Agnetta, B., Hare, B. and Tomasello, M. (2000) Cues to food location that domestic dogs (*Canis familiaris*) of different ages do and do not use. *Animal Cognition* 3(2), 107–112.

Barnard, S., Passalacqua, C., Pelosi, A., Valsecchi, P. and Prato-Previde, E. (2019) Effects of breed group and development on dogs' willingness to follow a human misleading advice. *Animal Cognition* 22(5), 757–768.

Bonanni, R., Cafazzo, S., Abis, A., Barillari, E., Valsecchi, P. *et al.* (2017) Age-graded dominance hierarchies and social tolerance in packs of free-ranging dogs. *Behavioral Ecology* 28(4), 1004–1020.

Bugnyar, T. (2010) Knower–guesser differentiation in ravens: others' viewpoints matter. *Proceedings of the Royal Society B: Biological Sciences* 278(1705), 634–640.

Burghardt, G.M. (2006) *The Genesis of Play, Testing the Limits*. The MIT Press, Cambridge, MA.

Campion, T.L., Jensvold, M.L. and Larsen, G. (2011) Use of gesture sequences in free-living chimpanzees (*Pan troglodytes schweinfurthii*) in Gombe National Park, Tanzania. *American Journal of Primatology* 73(Suppl. 1), 97.

Chlopčíková, M. and Mojžíšová, A. (2010) Risk factors in the mutual relationship between children and dogs. *Journal of Nursing, Social Studies, and Public Health* 1, 102–109.

Coren, S. (2016) The data says "Don't hug the dog!" *Psychology Today*. Available at: www.psychologytoday.com/us/blog/canine-corner/201604/the-data-says-dont-hug-the-dog (accessed 30 October 2023).

Craig, W. (1918) Appetites and aversions as constituents of instincts. *The Biological Bulletin* 34(2), 91–107.

Dale, R., Range, F., Stott, L., Kotrschal, K. and Marshall-Pescini, S. (2017) The influence of social relationship on food tolerance in wolves and dogs. *Behavioral Ecology and Sociobiology* 71(7), 1–14.

Darwin, C. (1871) *The Descent of Man, and Selection in Relation to Sex*. J. Murray, London.

Dobos, P. and Pongrácz, P. (2023) Would you detour with me? Association between functional breed selection and social learning in dogs sheds light on elements of dog–human cooperation. *Animals* 13(12), 12.

Emery, N.J. and Clayton, N.S. (2004) The mentality of crows: convergent evolution of intelligence in corvids and apes. *Science* 306(5703), 1903–1907.

Flombaum, J.I. and Santos, L.R. (2005) Rhesus monkeys attribute perceptions to others. *Current Biology* 15(5), 447–452.

Fournier, F. and Festa-Bianchet, M. (1995) Social dominance in adult female mountain goats. *Animal Behaviour* 49(6), 1449–1459.

Fouts, R.S. (1972) Use of guidance in teaching sign language to a chimpanzee (*Pan troglodytes*). *Journal of Comparative and Physiological Psychology* 80(3), 515–522.

Gardner, R.A. and Gardner, B.T. (1969) Teaching sign language to a chimpanzee. *Science* 165(3894), 664–672.

Groos, K. (1898) *The Play of Animals*. Translated by Elizabeth L. Baldwin. Appleton, New York.

Hall, S.L. and Bradshaw, J.W.S. (1998) The influence of hunger on object play by adult domestic cats. *Applied Animal Behaviour Science* 58(1–2), 143–150.

Hare, B. and Tomasello, M. (2005) Human-like social skills in dogs? *Trends in Cognitive Sciences* 9(9), 439–444.

Hare, B., Brown, M., Williamson, C. and Tomasello, M. (2002) The domestication of social cognition in dogs. *Science* 298(5598), 1634–1636.

Hayashi, T., Akikawa, R., Kawasaki, K., Egawa, J., Minamimoto, T. *et al.* (2020) Causal role of the macaque medial prefrontal cortex in false belief attribution. *Cell Reports* 30, 4433–4444.

Horowitz, A. (2011) Theory of mind in dogs? Examining method and concept. *Learning & Behavior* 39(4), 314–317.

Jakovcevic, A., Elgier, A.M., Mustaca, A.E. and Bentosela, M. (2010) Breed differences in dogs' (*Canis familiaris*) gaze to the human face. *Behavioural Processes* 84(2), 602–607.

Kellogg, W.N. and Kellogg, L.A. (1933) *The Ape and The Child: A Comparative Study of the Environmental Influence Upon Early Behavior*. Hafner, New York.

Kerswell, K.J., Butler, K.L., Bennett, P. and Hemsworth, P.H. (2010) The relationships between morphological features and social signalling behaviours in juvenile dogs: the effect of early experience with dogs of different morphotypes. *Behavioural Processes* 85(1), 1–7.

Kis, A., Ciobica, A. and Topál, J. (2017) The effect of oxytocin on human-directed social behaviour in dogs (*Canis familiaris*). *Hormones and Behavior* 94, 40–52.

Klinghammer, E. (1979) *The Behavior and Ecology of Wolves*. Garland, New York.

Konno, A., Romero, T., Inoue-Murayama, M., Saito, A. and Hasegawa, T. (2016) Dog breed differences in visual communication with humans. *PLoS One* 11(10), e0164760.

Kovács, K., Kis, A., Pogány, Á., Koller, D. and Topál, J. (2016) Differential effects of oxytocin on social sensitivity in two distinct breeds of dogs (*Canis familiaris*). *Psychoneuroendocrinology* 74, 212–220.

Krupenye, C., Kano, F., Hirata, S., Call, J. and Tomasello, M. (2016) Great apes anticipate that other individuals will act according to false beliefs. *Science* 354(6308), 110–114.

Kubinyi, E., Bence, M., Koller, D., Wan, M., Pergel, E. *et al.* (2017) Oxytocin and opioid receptor gene polymorphisms associated with greeting behavior in dogs. *Frontiers in Psychology* 8, 1520.

Linnaeus, C. (1792) *The Animal Kingdom; or, Zoological System of the Celebrated Sir Charles Linnaeus. Class I. Mammalia and Class II. Birds*. Translated and revised by R. Kerr. J. Murray, London. Being a Translation of That Part of the Systema Naturae, as Lately Published with Great Improvements by Professor Gmelin, Together with Numerous Additions from More Recent Zoological Writers and Illustrated with Copperplates.

Lloyd, A.S., Martin, J.E., Bornett-Gauci, H.L.I. and Wilkinson, R.G. (2008) Horse personality: variation between breeds. *Applied Animal Behaviour Science* 112(3–4), 369–383.

Lorenz, K. (1949) *So kam der Mensch auf den Hund [Man Meets Dog]*. Verlag G. Borotha-Schoeler, Vienna.

Maginnity, M.E. and Grace, R.C. (2014) Visual perspective taking by dogs (*Canis familiaris*) in a Guesser–Knower task: evidence for a canine theory of mind? *Animal Cognition* 17(6), 1375–1392.

Mech, L.D. (1970) *The Wolf: The Ecology and Behavior of an Endangered Species*. Natural History Press, Garden City, NY.

Mech, L.D. (1999) Alpha status, dominance, and division of labor in wolf packs. *Canadian Journal of Zoology* 77(8), 1196–1203.

Mech, L.D. and Boitani, L. (2003) *Wolves: Behavior, Ecology, and Conservation*. The University of Chicago Press, Chicago, IL.

Mehrkam, L.R., Hall, N.J., Haitz, C. and Wynne, C.D.L. (2017) The influence of breed and environmental factors on social and solitary play in dogs (*Canis lupus familiaris*). *Learning & Behavior* 45(4), 367–377.

Miklosi, A. (2008) *Dog Behavior, Evolution & Cognition*. Oxford University Press, New York.

National Institutes of Health (2005) Researchers publish dog genome sequence. Analysis sheds light on human disease; Differences among canine breeds. *NIH News*. Available at: www.genome.gov/17515860/2005-release-researchers-publish-dog-genome-sequence (accessed 30 October 2023).

Pack, A.A. and Herman, L.M. (2004) Bottlenosed dolphins (*Tursiops truncatus*) comprehend the referent of both static and dynamic human gazing and pointing in an object-choice task. *Journal of Comparative Psychology* 118(2), 160–171.

Parker, H.G., Dreger, D.L., Rimbault, M., Davis, B.W., Mullen, A.B. *et al.* (2017) Genomic analyses reveal the influence of geographic origin, migration, and hybridization on modern dog breed development. *Cell Reports* 19(4), 697–708.

Pepperberg, I. (2012) Theory of mind. In: Tallerman, M. and Gibson, K.R. (eds) *The Oxford Handbook of Language Evolution*. Oxford University Press, New York, p. 114.

Persson, M.E., Roth, L.S.V., Johnsson, M., Wright, D. and Jensen, P. (2015) Human-directed social behaviour in dogs shows significant heritability. *Genes, Brain, and Behavior* 14(4), 337–344.

Persson, M.E, Sundman, A.-S., Halldén, L.-L., Trottier, A.J. and Jensen, P. (2018) Sociality genes are associated with human-directed social behaviour in golden and Labrador retriever dogs. *PeerJ* 6, e5889.

Proops, L., Grounds, K., Smith, A.V. and McComb, K. (2018) Animals remember previous facial expressions that specific humans have exhibited. *Current Biology* 28(9), 1428–1432.

Pusey, A.E. and Packer, C. (1997) The ecology of relationships. In: Krebs, J.R. and Davies, J.R. (eds) *Behavioural Ecology: An Evolutionary Approach*. Blackwell Science, Oxford, pp. 254–283.

Ratliff, E. (2011) Taming the wild. *National Geographic*. Available at: www.nationalgeographic.com/magazine/article/animal-domestication (accessed 30 October 2023).

Ringhofer, M. and Yamamoto, S. (2017) Domestic horses send signals to humans when they are faced with an unsolvable task. *Animal Cognition* 20(3), 397–405.

Rumbaugh, D.M., Gill, T.V., Brown, J.V., von Glasersfeld, E.C., Pisani, P. *et al.* (1973) A computer-controlled language training system for investigating the language skills of young apes. *Behavior Research Methods & Instrumentation* 5(5), 385–392.

Scott, J.P. and Fuller, J.L. (1965) *Genetics and the Social Behavior of the Dog*. The University of Chicago Press, Chicago, IL.

Spinka, M., Newberry, R.C. and Bekoff, M. (2001) Mammalian play: training for the unexpected. *The Quarterly Review of Biology* 76(2), 141–168.

The Nobel Prize (1973) The Nobel Prize in Physiology or Medicine 1973. *Nobel Prize.org*. Available at: www.nobelprize.org/prizes/medicine/1973/summary (accessed 30 October 2023).

Udell, M.A.R., Ewald, M., Dorey, N.R. and Wynne, C.D.L. (2014) Exploring breed differences in dogs (*Canis familiaris*): does exaggeration or inhibition of predatory response predict performance on human-guided tasks? *Animal Behaviour* 89, 99–105.

Vandenberg, B. (1978) Play and development from an ethological perspective. *American Psychologist* 33(8), 724–738.

Viranyi, Z., Gácsi, M., Kubinyi, E., Topál, J., Belényi, B. *et al.* (2008) Comprehension of human pointing gestures in young human-reared wolves (*Canis lupus*) and dogs (*Canis familiaris*). *Animal Cognition* 11(3), 373–387.

vonHoldt, B.M., Pollinger, J.P., Lohmueller, K.E., Han, E., Parker, H.G. *et al.* (2010) Genome-wide SNP and haplotype analyses reveal a rich history underlying dog domestication. *Nature* 464(7290), 898–902.

vonHoldt, B.M., Shuldiner, E., Koch, I.J., Kartzinel, R.Y., Hogan, A. *et al.* (2017) Structural variants in genes associated with human Williams–Beuren syndrome underlie stereotypical hypersociability in domestic dogs. *Science Advances* 3(7), e1700398.

Wobber, V., Hare, B., Koler-Matznick, J., Wrangham, R. and Tomasello, M. (2009) Breed differences in domestic dogs' (*Canis familiaris*) comprehension of human communicative signals. *Interaction Studies* 10(2), 206–224.

Zimen, E. (1971) Wölfe und Königspudel: *Vergleichende Verhaltensbeobachtungen [Wolves and King Poodle: Comparative Behavioral Observations]*. Piper, Munich, Germany.

Zimen, E. (1981) *The Wolf: A Species in Danger [Der Wolf: Mythos und Verhalten]*. Delacorte Press, New York.

9 Aggression and Breed Differences

Abstract

Chapter 9 reviews the known science of aggression and breed differences, including the issues that complicate the interpretation of dog bite data, and thus our understanding of dog aggression by breed. There is an examination of the heritability of certain personality traits, including dominant and aggressive behavior and it addresses the fact that there are many types of aggression, including dominant, territorial, possessive, protection-of-litter-, pain-, fear-, and play-motivated, predation, redirected, intraspecific, idiopathic, physiopathological, and learned aggression. Case studies are used to examine "loaded gun" scenarios with an aggression-prone breed, a dog breed that's typically friendly, but showing aggression due to an underlying health issue, and a breed that has a prior history for specific types of aggression. Genetic predispositions in some breeds and/or poor environments can increase the risk of aggressive behavior, as can a lack of human understanding of the origins of the many types of aggression.

"Bad" to the Bone?

Patty Rocha's sneakers slapped against the pavement as she exhaustedly neared the crest of her hill. Mile two. Her Doberman, Gordon, padded along beside her, matching her diminishing strides. At the top of the incline, Patty and Gordon finally stopped. The former was winded and clumsily searched her training fanny pack for a dog treat. Gordon had eyes only for Patty. He excitedly ate his treats, oblivious to the jogger who was beginning his own descent down their hill. But as the jogger approached, Gordon realized how close this stranger was. He barked and lunged. Patty, startled, felt the wrapped leash tighten around her wrist. Her big dog nearly pulled Patty off of her feet, his jaws snapping menacingly.

"Gordon! Stop!" she cried out loudly. Her driveway was in sight; would her husband hear her? "Don! Don!" she screamed, as Gordon continued to lunge at the jogger.

"Call your dog off!" the man yelled, running to the opposite side of the street. Gordon, dragging Patty behind, continued in pursuit. Patty's husband, Don, heard the commotion and emerged from their driveway to investigate. There he saw Gordon, tethered like a kite, pulling as Patty helplessly hung on.

"Gordon!" Don yelled, running toward the dog, bat brandished above his head. He brought the bat down on the dog's side with a sickening thud. Gordon yelped and ducked; Don yanked the dog back harshly. Gordon, tail tucked and eyes averted, continued to cry. The jogger, shaken, cursed at all three of them as he continued down the road.

Gordon had no idea what had transpired, or why he was being reprimanded. Patty was equally mystified over her dog's behavior and told her husband that they needed to get professional behavioral help for their dog. It was during a consultation to help them with another of their animals that Don reluctantly admitted that he'd been "training" Gordon to be aggressive. Don had been withholding food and encouraging Gordon to aggress toward strangers, feeding him once he'd performed the desired behaviors. Don's excuse? He wanted to keep Patty "safe," but Don Rocha was neither a man of compassion (Gordon wasn't their only questionably "trained" animal) nor one of much foresight. Between Don's mistreatment and Gordon's breeding to be a guard dog, this was a recipe for disaster. The situation Don had created was far from "safe," but it was Gordon who would be penalized for being aggressive.

Dobermans are one of several breeds that have been discriminated against because they're considered to be an aggressive breed. While Don and Patty's neighbors might disagree, Gordon *wasn't* an

DOI: 10.1079/9781800624566.0009

inherently mean or aggressive dog. Pairing his propensity to protect with what looked like long-term abuse, however, Gordon's worst side was showing every time he accompanied Patty. Under Don's unskilled tutelage, Gordon wasn't becoming a better protector of Patty; he was becoming increasingly aggressive and unpredictable. They were setting Gordon up to fail—and they needed to work long-term with a real trainer that had a solid foundation in animal behavior.

Let's Talk About Aggression

Aggression. Some people don't even want to use that word, especially when it pertains to dogs. While we'd like to think that aggression is an "uncommon" trait among our closest companions, it's perfectly natural for any animal to feel or exhibit aggression from time to time—and there are biological, evolutionary, and emotional reasons behind aggressive behaviors. We have all felt aggression at one point in our lives. It's a trait that could have adapted for certain species during certain contexts; in some cases, being aggressive was the most successful strategy. Aggression has been a part of human behavior since our earliest days; ancient hominid fossils have been discovered with rib fractures that appear to have come from human weapons (Trinkaus and Zimmerman, 1982). Sometimes, weapon fragments are found with the skeletal remains (Buss and Shackelford, 1997). The more we examine the remains of ancient hominids, the more evidence accumulates that aggression was just a part of everyday life. Warfare, which can be defined as "organized aggression between autonomous political units" (Thorpe, 2003), has shaped the course of human history since our earliest days. While aggression is often frowned upon in modern society (e.g. road rage or "going postal"), evolutionarily, it makes sense. The hormones that influence aggression include serotonin and testosterone, and these same hormones can help drive an individual to be more successful, whether they're a human athlete or a canine in the wild. Adaptive problems to which aggression may have evolved as a solution for early humans include co-opting the resources of others, defending against attack, inflicting costs on intrasexual rivals, negotiating status and power hierarchies, deterring rivals from future aggression, deterring long-term mates from sexual infidelity, and reducing resources spent on unrelated children (Buss and Shackelford, 1997). Early on in the domestication process, canines that demonstrated friendliness and lower levels of reactivity were selected to breed with one another and share our homes, but aggression wasn't "bred out" of dogs; many dogs had early roles, such as hunting, herding, and guarding, in which aggression gave them an advantage. While aggression is often considered to be one thing, there isn't just "one" type of it; there are numerous context-specific instances of aggression.

Definitions and Types of Aggression

First, let's define aggression as it pertains to dogs and then clarify the different types. Aggression in dogs is sometimes categorized based upon the *target* of the aggression: owner-directed aggression, stranger-directed aggression, or dog-directed aggression (Mehrkam and Wynne, 2014). It's also described based on the root cause, or stimulus of the aggression (Borchelt, 1983; Blackshaw, 1991). It's important to differentiate between the different kinds of aggression to determine whether they're "true" aggression or anxiety related. Because behaviors from different social contexts can often *appear* to be the same on the surface (e.g. a nervous smile with humans or a "fear grimace"—the face you make if you burn yourself) it's important to identify true aggression versus fear reactivity, as the two have very different causes and treatments.

Definitions based on the stimulus

- *Fear or anxiety aggression* is the most common type seen in clinical cases. This type of aggression evokes submissive behaviors in the dog such as a lowered or tucked tail, lowered head, lowered ears, avoidance of eye contact, etc., and typically occurs in response to stimuli such as the presence of strangers, children, or an unusual-looking person. The dogs may initially bark, but not bite or growl unless approached, particularly if the stimulus is fast or unpredictable.
- *Pain-related aggression* is surprisingly common. It requires a careful physical exam to detect pain that may be well hidden by the canine, and careful observation to detect sensitivity or body issues that the owners may not recognize. This is often associated with an abrupt and inexplicable onset of aggression in an otherwise non-aggressive dog.
- *Redirected aggression* describes aggression typically focused on younger, weaker individuals, whether animal or human. This may occur because

of trigger stacking, when the dog experiences multiple scary or evoking stimuli simultaneously or sequentially (close in time), and cannot strike out at the source.

- *Resource guarding or protective aggression* can be diagnosed because it typically occurs on the property or in the owner's vehicle, or in the presence of the owner, and is not observed elsewhere. A resource can include, but is not limited to, food, toys, bedding, or the members or perceived members of a dog's family.
- *Dominance aggression* is a rare form of aggression presented in clinical cases. In these situations, the dog typically reacts to the owner rather than to an outsider, and it is typically displayed over resources, and generally occurs when the owner does not demonstrate clear rules and boundaries. This type of aggression, while rare, is more likely to occur in some of the more primitive or "ancient" breeds of dogs that have undergone less artificial selection than the more modern breeds have.
- *Predatory aggression* is a rare form of aggression, and typically presents when a large dog goes after a very small dog or a cat.
- *Male–male aggression* is another rare form of aggression, and typically occurs in un-neutered dogs.
- *Other aggression*, including extremely rare situations of female–female aggression or maternal aggression, driven by hormones.

The problems with assessing aggression by breed

While breeds and breed groups can be used as a point of departure for aggression and aggressive tendencies, recall that individuals vary. We can make assertions like "dominance hierarchies, and thus dominance-related aggression, are more likely with ancient breeds," but Cattle Dogs have dominance hierarchies, as well. And while this *can* lead to aggression, it doesn't guarantee it.

Some breeds are more prone to resource guarding than others, but there are both genetic and learned variations of resource guarding. (A behavior that's learned is far easier to deal with than something with a genetic predisposition.)

Honeymoon Hounds

Skyler and Debora fell in love with the "little" beach dog that they met on their honeymoon in Puerto Rico. During their 2 weeks on the beach, they celebrated their fledgling marriage and their new friend. When it was time to go home, they couldn't leave their little dog—who they'd appropriately dubbed "Sunny," behind. But they didn't realize what they were getting into.

Once they returned to the US, Sunny joined their household with their other dog, Lucy, but all was not copacetic. Sunny was apprehensive, fearful, and exhibited behaviors that appeared to be aggressive. She was in an entirely new environment—one where she could no longer roam freely, forage, and make her own choices. She felt out of sorts. When it was breakfast or dinner time, Sunny savagely "defended" and "protected" her food bowl, even though their Labrador/Shepherd mix, Lucy, showed no interest in her food. Lucy had her own food, and she'd never known otherwise. She'd never had to fight for or protect her food. But Sunny, who had lived on the beach her entire life, only knew a life where she had to scrap for every bite ... and it showed. You hear stories about the couple that fell in love with a beach or pariah dog (Fig. 9.1) on their honeymoon, but then when they got the dog home, the dog guards its resources. For many owners, this is a novel behavior that's above their experience level—and beyond their ability to cope with. When left unchecked, these behaviors can quickly escalate.

Unreported Bites

A 6-year-old was being swung around in circles by her dad when she was bitten in the leg by a family friend's dog. The dog, a Schnauzer who wasn't particularly friendly, sat and watched as they rushed her into the house, cleaned the wound and wrapped it up with bandages. No stitches were required, but the bite left a scar. Because it was a dog the family knew, and one that belonged to friends, the bite went unreported. It was explained as "play." But this definitely wasn't a playful encounter—the dog wasn't accustomed to being around young children, and he was likely anxious of the screaming as the child swung around. He came running out, bit, and then retreated. Looking back on this incident, it's easy to see how so many similar bites likely go unreported. The Schnauzer isn't a breed that "typically" comes to mind when people are thinking about aggressive dogs, but that doesn't mean they don't exhibit aggression.

Fig. 9.1. Pariah dogs in a city park in Yelahanka, Bengaluru, India. They are alert to the presence of an unfamiliar pariah dog from outside the park. This file is licensed under the Creative Commons Attribution-Share Alike 4.0 International license and provided by Wikipedia. Available at: https://commons.wikimedia.org/wiki/File:Pariah_dogs_in_a_city_park_in_Yelahanka,_Bengaluru,_India._They_are_alert_to_the_presence_of_an_unfamiliar_pariah_dog_from_outside_the_park..jpg.

Unreported bites—and bites from dogs that don't receive as much recognition for being "aggressive"—are far more common than most people realize.

Studying canine aggression can be problematic for numerous reasons. A review by researchers Mehrkam and Wynne (2014) examined the behavioral differences among domestic dog breeds. They reported that there was low reliability between different methodologies for studying aggression, and suggested that perception and reporting bias also contributed to the challenge of studying patterns of breed-specific aggression. With so many different types of aggression, and with behavioral contexts often confused with one another, it's easy to see how studies would be inconsistent. However, Mehrkam and Wynne (2014) also noted that Golden Retrievers and Labrador Retrievers are reported as consistently low in aggression. This is supported by an owner survey conducted in Finland that included 14 purebreds and a "mixed breed" category. This survey found that Labrador Retrievers were significantly less aggressive than all other purebreds and mixed breeds in the study. Golden Retrievers were a not part of this study, but we suspect, given all other evidence, that similar results would have been found for this breed, as well. Miniature Schnauzers in this survey had the highest reported levels of stranger-directed aggression, while the lowest levels of stranger-directed aggression belonged to Labrador Retrievers, Smooth Collies, and Staffordshire Bull Terriers (Salonen *et al.*, 2020).

There are several issues that complicate the interpretation of dog bite data, and thus our understanding of dog aggression by breed. One is that bite reporting is biased towards more severe bites that require more medical intervention, generally from larger and more powerful dogs, and this type of reporting underestimates bites from smaller dogs. These types of sources generally can't explain whether certain breeds are reported more often because they bite more often, or because the breed is popular, and thus represents a higher proportion of the population of dogs. In other words, does breed X bite more than other breeds, or are there more of these dogs out there, and thus by chance more of them happen to bite? Are these breeds incorrectly over-represented by visual identification? Lastly, the tendency to bite in one context is not necessarily predictive of other contexts (Casey *et al.*, 2014).

Approximately half of the population in the US receives dog bites as children (Beck and Jones, 1985), and the Centers for Disease Control and Prevention (CDC) states that there are 4.7 million dog bites annually in the US, approximately 16 of which are fatal (Centers for Disease Control and Prevention, 2003). Pit Bulls are generally singled out as the "breed" responsible for most of these bites, but statistics pertaining to the breed of the biter are a bit more difficult to obtain, given the variability in correctly identifying a dog's breed. The American Veterinary Medical Association (AVMA) cautions that it's "not a dog's breed that determines whether it will bite, but rather the dog's individual history and behavior" (AMVA, 2023a). Depending upon who you ask, what their bias is, and how accurate their source is, the "dog breeds that bite the most" vary from Chihuahuas to Chow Chows, American Pit Bull Terriers to Pomeranians, and Rottweilers to Russell Terriers.

The Damage of Breed Stereotypes

In July 2019, the *International Business Times* ran a story entitled, "Man suffocates neighbor's pit bull to death in front of owner." The Pit Bull, named Rex, was attacked by two other dogs, a mixed breed (Golden Retriever/Poodle cross, also referred to as a goldendoodle), and a Shih Tzu. The article, written by Vaishnavi Vaidyanathan, stated (Vaidyanath, 2019):

> ... Dominic Primerano was walking his dog, a Pit Bull named Rex, around his apartment complex when two unleashed canines, a Shih Tzu and a goldendoodle, [started] attacking his pet. While the Shih Tzu's owner took the dog away, the other canine continued to attack Rex. The Doodle also bit the man on the face and hands as he tried to save his dog.
>
> As Primerano tried to save Rex, Huynh Toquoc, a neighbor, came out and held Rex by his snout, before suffocating the canine. Toquoc was charged with killing the dog and he was cited with an appearance ticket.
>
> "I tried pulling the guy off my dog, but he was locked on to my dog and my dog just died right in front of me," Primerano said.

While Rex was the innocent victim in this horrific incident, the *International Business Times* compounded this tragedy by using a stock image of an aggressive-looking Pit Bull as the main article's main photo. Goldendoodles and Shih Tzus don't have a reputation for being aggressive, but Pit Bulls do ... and images like these sell papers. Never mind that it was sensational or inaccurate; a cursory glance would lead one to believe that Rex was somehow at fault, simply for being a Pit Bull.

The friendly Labrador. The happy Golden Retriever. The hyper Chihuahua. The aggressive Doberman, Rottweiler, German Shepherd, and Pit Bull. Breed stereotypes are more harmful than helpful, and often paint a picture that's based upon anecdotes, and not evidence. (The plural of anecdotes is not evidence). That's not to say there isn't "something" to these stereotypes. Recall Chapter 5, where we discussed how different breeds are. There's a genetic basis to breed differences, and thus, behavioral differences, but these can only be used as a point of departure, not as a certainty.

There was an unfortunate incident in March 2019 where a small child's hand was bitten off by the neighbor's dogs, but the dog breed was mentioned almost nowhere. Had the incident involved a Pit Bull, instead of Huskies, it would have been an entirely different scenario. This is because we have stereotypes about certain dog breeds. Breed X is "aggressive," while Breed Y is "friendly;" and those stereotypes stick, to the detriment of many dogs. Dogs labeled under the "Pit Bull" umbrella category are stereotypically seen as "aggressive" across all behavioral contexts, and without much scientific evidence to support this. These stereotypes have become so pervasive that most of the heavy hitters, such as the AVMA and American Society for the Prevention of Cruelty to Animals (ASPCA), are in agreement that they are harmful and inaccurate. The topic of "Pit Bulls" in particular has become so contentious that the ASPCA (2023b) has a position statement on them. While it's rather lengthy, we are presenting the statement in its entirety, as recommended by the ASPCA (emphases are ours):

> Dog breeds are characterized by certain physical and behavioral traits. Each breed was developed to perform a specific job, whether that job is hunting

rabbits, retrieving downed birds, herding livestock, or sitting on people's laps. When developing a breed, breeders selected only those dogs that performed their job best to produce the next generation.

Physical abilities and behavior are both important facets of any breed. A well-bred dog should have both the physical attributes necessary to perform its job and the behavioral tendencies needed to learn it. It's not surprising that individuals of a specific breed tend to look and behave somewhat similarly. Pointers are more likely than Poodles to point, and sheepdogs are more likely than lapdogs to herd. However, while a dog's genetics may *predispose* it to perform certain behaviors, *tremendous behavioral variation exists among individuals of the same breed or breed type.* It's also important to note that some dog breeds are now bred for entirely different jobs than those for which they were originally developed. For example, certain strains of Golden Retrievers are now being bred as service dogs, a far cry from their original job of retrieving downed birds.

Today's Pit Bull is a descendant of the original English bull-baiting dog—a dog that was bred to bite and hold bulls, bears, and other large animals around the face and head [Fig. 9.2]. When baiting large animals was outlawed in the 1800s, people turned instead to fighting their dogs against each other. These larger, slower bull-baiting dogs were crossed with smaller, quicker Terriers to produce a more agile and athletic dog for fighting other dogs.

Some pit bulls were selected and bred for their fighting ability. That means that they may be more likely than other breeds to fight with dogs. It doesn't mean that they can't be around other dogs or that they're unpredictably aggressive. Other pit bulls were specifically bred for work and companionship. These dogs have long been popular family pets, noted for their gentleness, affection and loyalty. *And even those pit bulls bred to fight other animals were not prone to aggressiveness toward people. Dogs used for fighting needed to be routinely handled by people; therefore, aggression toward people was not tolerated. Any dog that behaved aggressively toward a person was culled, or killed, to avoid passing on such an undesirable trait.* Research on pet dogs confirms that dog aggressive dogs are no more likely to direct aggression toward people than dogs that aren't aggressive to other dogs.

Fig. 9.2. Bull baiting. "Bull broke lose." From Leighton (1911). This image was originally posted to Flickr by Internet Archive Book Images at https://flickr.com/photos/126377022@N07/14576985288. It was reviewed on 15 September 2015 by FlickreviewR and was confirmed to be licensed under the terms of the No known copyright restrictions. Available at: https://commons.wikimedia.org/wiki/File:The_new_book_of_the_dog_-_a_comprehensive_natural_history_of_British_dogs_and_their_foreign_relatives,_with_chapters_on_law,_breeding,_kennel_management,_and_veterinary_treatment_(1911)_(14576985288).jpg.

It is likely that the vast majority of Pit Bull type dogs in our communities today are the result of random breeding—two dogs being mated without regard to the behavioral traits being passed on to their offspring. The result of random breeding is a population of dogs with a wide range of behavioral predispositions. For this reason, it is important to evaluate and treat each dog, no matter its breed, as an individual.

While a dog's genetics may predispose it to behave in certain ways, genetics do not exist in a vacuum. Rather, behavior develops through a complex interaction between environment and genetics. *This is an especially important consideration when we look at an individual dog versus a breed.* Many diverse and sometimes subtle factors influence the development of behavior, including, but not limited to, early nutrition, stress levels experienced by the mother during pregnancy, and even temperature in the womb. [Recall our discussion of epigenetics.] And when it comes to influencing the behavior of an individual dog, factors such as housing conditions and the history of social interactions play pivotal roles in behavioral development. The factors that feed into the expression of behavior are so inextricably intertwined that it's usually impossible to point to any one specific influence that accounts for a dog becoming aggressive. This is why there is such variation in behavior between individual dogs, even when they are of the same breed and bred for the same purpose. Because of the impact of experience, the Pit Bull specifically bred for generations to be aggressive may not fight with dogs and the Labrador Retriever bred to be a service dog may be aggressive toward people.

Early positive experiences, most notably socialization, are considered key in preventing aggressive tendencies in dogs. Puppies that learn how to interact, play, and communicate with both people and members of their own and other species are less likely to show aggressive behavior as adults. Given the powerful impact of socialization, it's no surprise that dogs that are chained outside and isolated from positive human interaction are more likely to bite people than dogs that are integrated into our homes. Unfortunately, Pit Bull type dogs that find themselves in these conditions may be at greater risk for developing aggressive behavior. But because these factors are ones that can be controlled by better educated owners, it is possible to reduce these risks, not just in pit bulls but in dogs of all breeds.

The reality is that dogs of many breeds can be selectively bred or trained to develop aggressive traits. Therefore, the responsible ownership of any dog requires a commitment to proper socialization, humane training and conscientious supervision. Despite our best efforts, there will always be dogs of various breeds that are simply too dangerous to live safely in society. We can effectively address the danger posed by these dogs by supporting the passage and vigorous enforcement of laws that focus, not on breed, but on people's responsibility for their dogs' behavior, including measures that hold owners of all breeds accountable for properly housing, supervising and controlling their dogs. Breed neutral "dangerous dog" laws, "leash laws" that prohibit dogs from running loose off their owners' property, and "anti-chaining" laws can control the behavior of individual dogs and individual owners and thereby help reduce the risk of harm to people and other animals.

Laws that ban particular breeds of dogs do not achieve these aims and instead create the illusion, but not the reality, of enhanced public safety. Notably, there are no statewide laws that discriminate based on dog breed, and 18 states have taken the proactive step of expressly banning laws that single out particular breeds for disparate legal treatment. Even the White House has weighed in against laws that target specific breeds. In a statement issued in 2013, President Obama said, "[w]e don't support breed-specific legislation—research shows that bans on certain types of dogs are largely ineffective and often a waste of public resources. And the simple fact is that dogs of any breed can become dangerous when they're intentionally or unintentionally raised to be aggressive."

All dogs, including pit bulls, are individuals. Treating them as such, providing them with the care, training and supervision they require, and judging them by their actions and not by their DNA or their physical appearance is the best way to ensure that dogs and people can continue to share safe and happy lives together.

[Please note: This position statement is intended to be considered in its entirety and excerpting is not recommended.]

One of the "solutions" offered by those who fear aggressive dogs is breed-specific legislation (BSL). This legislation enforces blanket policies against dogs that are considered to be "dangerous." Depending upon the time or jurisdiction, a number of breeds have been targeted for BSL, including, but not limited to, German Shepherds, Rottweilers, Doberman Pinschers, Chow Chows, Dogo Argentinos, and those breeds falling under (or appearing to fall under) the "Pit Bull" umbrella. Pit Bull-type dogs have increasingly been the target of BSL, without considering genetics, behavior, or temperament, but rather their physical appearance. There are BSL policies throughout the US and Canada, but the AVMA and the US CDC are opposed to these policies, as they don't do what they claim to do: prevent dog bites.

The AVMA's Policy on Dangerous Animal Legislation states: "The AVMA supports dangerous animal legislation by state, county, or municipal governments provided that legislation does not refer to specific breeds or classes of animals. This legislation should be directed at fostering safety and protection of the general public from animals classified as dangerous." (AMVA, 2023b).

But breed isn't actually a good predictor of whether a dog will bite or not. A 2013 study found that "Most DBRFs [dog bite-related fatalities] were characterized by coincident, preventable factors; breed was not one of these. Study results supported previous recommendations for multifactorial approaches, instead of single-factor solutions such as BSL, for dog bite prevention" (Patronek *et al.*, 2013). Government agencies are in agreement with this. According to the CDC, *any* dog can bite—and one in five people who are bitten by a dog will require medical attention for their injuries.

Not only is breed-based legislation ineffective, but properly identifying dogs that fall under the "Pit Bull" umbrella is difficult to do, even for seasoned professionals. A 4-year shelter study conducted by Maddie's Fund (Maddie's Fund, 2012) found that out of 120 dogs, 55 were identified as "Pit Bulls" by the shelter staff, but only 25 belonged to the "Pit Bull" umbrella when their DNA was analyzed. This makes sense to those who have long been saying that Pit Bulls are over-represented because of inaccurate identifications.

Behavior-Based Legislation

Effective January 1, 2020, Washington State's House Bill 1026 added a new chapter to 16.08 Revised Code of Washington (RCW) that required local jurisdictions to focus not on a dog's breed, but on its behavior as a reason for banning it. The bill stated, in part:

> Sec. 1 (1) A number of local jurisdictions have enacted ordinances prohibiting or placing additional restrictions on specific breeds of dogs. While the legislature recognizes that local jurisdictions have a valid public safety interest in protecting citizens from dog attacks, the legislature finds that a dog's breed is not inherently indicative of whether or not a dog is dangerous and that the criteria for determining whether or not a dog is dangerous or potentially dangerous should be focused on the dog's behavior.
>
> (2) The legislature further finds that breed-specific ordinances fail to address the factors that cause dogs to become aggressive and place an undue hardship on responsible dog owners who provide proper socialization and training. The legislature intends to encourage local jurisdictions to more effectively and fairly control dangerous dogs and enhance public safety by focusing on dogs' behavior rather than their breeds.

This bill enabled dog owners in Washington State to keep their dogs, irrespective of their breed, if the dog is able to pass a behavior assessment such as the American Kennel Club's Canine Good Citizen (CGC) program (AKC, 2022). The test can help evaluate whether a dog may behave aggressively across a variety of contexts, including interacting with unfamiliar dogs and unfamiliar humans. It states, "If an Evaluator observes any signs of aggression (biting, snapping, growling, attempting to attack) the test should not be continued."

Local jurisdictions in Washington State that still have breed bans will be able to continue to enforce them, they will have to exempt those dogs that pass the CGC or its equivalent. Passing the test provides a 2-year exemption.

> Sec. 2 (2) This section does not apply to the act of documenting either a dog's breed or its physical appearance, or both, solely for identification purposes when declaring a dog dangerous or potentially dangerous.

The law pertains to domesticated dogs, and excludes wolves, coyotes, wolf–dog hybrids, and coyote–dog hybrids.

For those of us who have been saying that dogs should be judged by their actions, and not by their appearance, this is an important step. BSL is slowly losing its support ... In 2018, a 30-year breed ban on Pit Bulls in rural Yakima, Washington, was lifted. Incidents involving dangerous dogs did not increase.

Bad Science

Where, exactly, did the Pit Bull's aggressive reputation originate from? Well, Baack *et al.* (1989) first made the extraordinary claim that Pit Bulls were "different" from other dogs, because they produce a bite force of up to 1800 pounds per square inch (psi). Now, Baack's extraordinary claim was backed up with a reference: a paper from 1983 (Boenning *et al.*, 1983). The problem? The 1983 paper made no claims about the bite force of a Pit Bull ... because there's no data for this. Unfortunately, though, other papers followed suit. A 2006 report (Akhtar *et al.*, 2006) repeated this inaccurate psi, citing a 1997 paper (Presutti, 1997) that, again, didn't include this statistic. So rather than fact-checking this source, media outlets, scientists who were content to use second- or third-hand references, and anti-Pit Bull crusaders took this number and ran with it. For decades. (The only place you can find that 1989 paper available now is on an anti-Pit Bull website; much like Dr Andrew Wakefield, whose dangerous and incorrect assertions that vaccines caused autism, Baack's debunked paper exists now only to serve conspiracy theorists and the scientifically illiterate). Recall that "Pit Bull" is an umbrella term used for a dog that may or may not have genetic heritage linking back to one of several breeds, including American Staffordshire Terriers, *or*, they may have the "appearance" of a Pit Bull. You can't genotype based on appearance, though, and thanks to this 1989 paper and its undeniably bad science, Pit Bulls earned a bad (and undeserved) rap. To put this extraordinary claim into perspective, a hyena has a bite force of 1100 psi, a grizzly bear, 1160 psi, a bull shark, 1350 psi, a hippopotamus, 1800 psi, and coming in at the very top, a Nile crocodile at 5000 psi (Spanner, 2023). (There have been some reports of the Orca having a bite force of up to 19,000 psi, so while it's worth mentioning, this number has not been fully substantiated). So, what is the maximum bite force of any dog, much less a Pit Bull? We aren't sure, and it would be ethically and scientifically irresponsible of us to pretend that we were, but it's probably not more than a bull shark and equal to a hippopotamus!

A 1995 study attempted to find the range of bite forces of 22 dogs who ranged in size from 15 to 121 pounds. The bite force ranged from 13 to 1394 Newtons. While the study didn't include Pit Bulls, it did include Rottweilers, which had the largest bite force in this study (Lindner *et al.*, 1995). Unfortunately, because the authors of this paper created their own tool to measure the bite force of the dogs, and because psi (a unit of pressure) and Newtons (a unit of force) aren't comparable, we can't compare them (believe us, we've tried). This definitely looks like an interesting area for future researchers to delve into, as we continue to try to undo so much of the bad science that has plagued so many breeds.

Born to Be Mild

Individuals will vary, but there's still much to be said about what one may have been bred to do. There will be a range of motivation and skill levels, but herding dogs typically herd other animals. Bird hunting dogs have a "soft" mouth to pick up bodies without damaging them. Fighting dogs will fight. But there are always exceptions to the rule—and environmental factors play a big part. One's lineage isn't predestined, but it's the blueprint for their likely behavior: they are predisposed to act in a particular way. Take, for example, the trait of impulsivity. Impulsiveness is on a continuum, from high to low. In psychological terms, when one has high impulsivity, they tend to act with little or no prior forethought and without weighing the consequences of their actions. Think of a child who can't stop themselves from eating a cookie candy that's been left out, or a dog from sneaking cat food or chasing a squirrel. Many, when presented with a salient cue or a high-value item, are unable to inhibit their response.

Inhibitory control is a part of one's development, with more inhibition exhibited after the adolescent period, but some will still struggle with this, regardless of their age. Inhibition has been clinically studied with children for decades, and often with hilarious results. Studies conducted with 3- and 4-year-old children reveal difficulty with inhibiting an impulse and focusing on a competing goal, while by age 6 or 7 years, it's far easier for children to do so (Kirkham *et al.*, 2003). The "Stanford Marshmallow Studies" examined whether children could delay their gratification, either receiving a larger food reward for waiting for a certain duration of time, or fewer if they chose to eat them immediately. Led by Stanford University professors Walter Mischel and Ebbe B. Ebbeson, the original experimental design involved 16 boys and 16 girls, with an average age of 4 years, 6 months, at Stanford University's King Nursery School. An experimenter presented the children with a small food reward (a marshmallow, Oreo cookie, or pretzel stick) if they chose to eat it immediately, or a larger reward if they waited for approximately 15 minutes. After explaining the terms and putting out the food reward, the experimenter left the testing room and observed the children's reactions. Of the children who tried to delay gratification (often by covering their eyes or trying to distract themselves), one-third were able to do so long enough to get the second food reward. A small minority of the children were unable to wait at all, gobbling the food reward the moment the experimenter left the room (Mischel and Ebbesen, 1970). The most "successful" strategy for receiving a larger food reward was distracting oneself in the absence of actually having strong inhibitory control. You might notice, if you ask your dog to wait for a treat, that they'll be drooling; they'll also often look away from it, or toward you, as hyper-focusing on it almost makes it too difficult for them to stand.

Studies examining brain maturity and inhibitory control have shown that the ability to inhibit one's response more quickly improves with age. A study with youths aged 8–20 years revealed not only more inhibitory control for the older subjects in the study, but corresponding changes in certain regions of the brain, as well. Functional magnetic resonance imaging (fMRI) revealed a positive correlation between age and activation in the left inferior frontal gyrus and a negative correlation between age and activation in the left middle/superior frontal gyri (Tamm *et al.*, 2002). While neuroimaging studies have shown that the right inferior frontal gyrus is associated with inhibitory control (Hampshire *et al.*, 2010), a growing body of research shows that the left inferior frontal gyrus is associated with inhibition, as well (Swick *et al.*, 2008).

Inhibitory impairment often coincides with aggressive behavior among preschool children (Raaijmakers *et al.*, 2008). Those with less inhibitory control have higher rates of aggression than their peers who exhibit more inhibitory control. With dogs, high impulsivity has been associated with certain groups or breeds of dogs. An examination of 1161 dogs assessed impulsivity in Border Collies (who specialize in herding work) and Labrador Retrievers (who specialize in gun work) (Fadel *et al.*, 2016). The study found that working dogs that specialize in herding have high impulsivity, but dogs of the same breed that were bred for show have comparatively lower rates of impulsivity. This is called "trait-level" impulsivity. While it's still in the infancy of its research, new tools are being created to measure impulse control rates. The Dog Impulsivity Assessment Scale (DIAS), developed in 2011, is an excellent point of departure for this line of research, and covers 18 dimensions of impulsivity (Wright *et al.*, 2011):

1. My dog shows extreme physical signs when excited (drooling, panting, raising hackles, urinating, licking lips, widening of eyes)
2. When my dog gets very excited, it can lead to fixed repetitive behavior (an action that is repeated in the same way over and over again), such as tail chasing or spinning around in circles
3. I would consider my dog to be very impulsive (has sudden, strong urges to act; acts without forethought; acts without considering effects of actions)
4. My dog doesn't like to be approached or hugged
5. My dog becomes aggressive (growls, snarls, snaps, or bites) when excited
6. My dog *appears* to be "sorry" after it has done something wrong
7. My dog does not think before it acts (they would steal food without first looking to see if someone is watching)
8. My dog can be very persistent (they will continue to do something even if they know they will get punished or told off)
9. My dog may become aggressive (growl, snarl, snap, or bite) if frustrated with something
10. My dog is easy to train
11. My dog is not keen to go into new situations
12. My dog takes a long time to lose interest in new things
13. My dog calms down very quickly after being excited
14. My dog appears to have a lot of control over how it responds
15. My dog is very interested in new things and new places
16. My dog reacts very quickly
17. My dog is not very patient (they get agitated waiting for their food, or waiting to go out for a walk)
18. My dog seems to get excited for no reason

Test-takers can grade their dogs on a scale, with "strongly agree" on one end of the continuum and "strongly disagree" on the other end, with an additional category of "don't know/not applicable." The instructions on the DIAS state: "For each of the [18 statements] please place a cross in the box that most accurately describes your level of agreement: the answer should reflect the general personality of the dog, so for example if a statement applies to your dog in some situations but not others, please make a judgment as to how much you agree."

The tests were scored as follows for numbers 1–9 and 15–18: 5=strongly agree, 4=partly agree,

2=partly disagree, 1=strongly disagree. For numbers 10–14, the scores were reversed: 4 becomes 2; 1 becomes 5. The scores are divided by the number of questions answered, with Overall Questionnaire Score as follows: higher score = higher impulsivity (Wright *et al.*, 2011):

- *Factor 1 (Behavioral Regulation)*: higher score = lower behavioral regulation (i.e. higher impulsivity)
- *Factor 2 (Aggression and Response to Novelty)*: higher score = higher aggression/negative responses to novelty
- *Factor 3 (Responsiveness)*: higher score = higher responsiveness

Even when a dog has been trained and they know that they aren't supposed to do something, it's hard to inhibit their response prior to 2 years of age due to a lack of myelinization in the brain. Myelination is the process of coating the axon of each neuron with a fatty covering called myelin. This coating protects the neuron and helps it conduct signals more efficiently. Myelination begins in the brainstem and cerebellum before birth, but isn't completed in the frontal cortex until late adolescence (Arain *et al.*, 2013). Dogs, much like the children in the impulse control studies, will fail at these tasks prior to a certain age (typically around 1.5–2 years).

With this in mind, let's take a closer look at trait-level characteristics. "Pit Bull" is an umbrella category for several different "bully" breeds, many of which have been bred for different purposes. The data on purpose-bred Pit Bulls is difficult to find, however, as breeding dogs to fight other dogs is unethical and illegal. A backyard breeding operation that selects individuals with the highest rates of dog-related aggression to produce puppies with similarly high rates of dog-related aggression likely isn't going to be the most reliable source of information on behavior or genetics. As of 2008, dog fighting is a felony in every US state and territory. Being a spectator at a dog fighting event is illegal in all states excepting Hawaii and Montana, and possessing a dog who was bred for the purpose of fighting other dogs is also a felony. Thus, while we know that some Pit Bulls are bred for show or companionship and others for fighting, other than instances where dogs have been rescued from dog fighting operations, we're only able to comment on these trait-level differences anecdotally.

Of those instances where the dogs were rescued, the dogs' aggression with one another would then need to be systematically measured. Fortunately, the ASPCA has done exactly that. During the 2012 American College of Veterinary Behaviorists/ American Veterinary Society of Behavior conference, researchers Pamela Reid and Kristen Collins of the ASPCA's anti-cruelty behavior team reported on 292 adult Pit Bulls who had been rescued from dog fighting rings during 2009 (Reid and Collins, 2012). These dogs had been specifically bred to exhibit dog-directed aggression and had been trained for dog fighting. Beginning 15 days after the dogs were seized, the researchers evaluated them, focusing on dog–dog aggression. The researchers used four different conditions, including a dog of the same sex, a dog of the opposite sex, a stuffed dog, and a benign control object. The researchers classified the dogs' reactions to the conditions as Friendly, Fearful, Neutral, Pushy, Aggressive, or Sexual. The "Aggressive" classification required behaviors such as biting, showing teeth, snarling, or growling. The dogs that exhibited dog–dog aggression did so to both the real dogs and to the stuffed dog. Most of the dogs had a neutral response in the control condition. While rates of aggression were lower than the researchers had initially hypothesized, the dogs had a significantly higher rate of aggression (27.6%) than what been found among non-dog-fighting shelter dogs (14.3%).

Promisingly, rates of dog–dog aggression among puppies that were rescued from this dog fighting operation were even lower. A separate study (Collins *et al.*, 2012) examining 34 Pit Bull puppies seized from this operation found few incidents of aggressive behaviors; none of the puppies exhibited aggression toward a fake puppy and only three puppies exhibited aggression toward another puppy. The three puppies that exhibited aggressive behaviors did so across five interactions, but there was no physical contact during these incidents. The behaviors (including snarling, growling, and snapping) were deemed to be ritualized, rather than expressions of true aggression. We don't know if this is because the dog fighters aren't selecting for dog–dog aggression enough, or if we just haven't been selecting for this trait for long enough, but it gives hope for puppies that are rescued from these situations.

What we do know, given data from other lines of dogs, is that dogs who have been bred for a certain purpose will generally act differently from dogs who were bred for another purpose. Recall the rates of impulsiveness for herding dogs that were bred for show versus work; while these are the same breed, they have differences in their genetics and in their behavior. Dogs who have been bred to fight other dogs for generations were selected for their dog-related aggression, and have a higher propensity to fight with other dogs, as will their offspring. But even when you select for a trait, that doesn't guarantee that that trait will be seen in following generations. We can examine this phenomenon of the heritability of trait-level characteristics and individual variation with case studies, where you can see trends in behaviors. Because this is a controversial topic, we want to be abundantly clear here: these are *observed trends* in case studies, based upon the dog's *reported* breed or breed group.

The idea of lines or different (genetic) forms of a "breed" is most often seen in clinical cases. While behavior clinicians are called in on a lot of cases of aggression in "Pit Bulls" (there are several closely related breeds that are included in this umbrella term), they also see cases of aggression in other breeds while the pittie in the same household rests its ginormous head gently in their lap. There seem to be two kinds of Pit Bulls and, we suspect, the difference comes down to inhibition. We know that inhibition, or the ability to stop and control your behavior, is highly influenced by genetics, and is controlled by feedback pathways in the brain which don't completely develop until two years of age in dogs. It makes sense that people breeding Pit Bulls for fighting other dogs would be breeding for dogs that lack inhibition: if a dog hesitates to attack, these breeders don't want them. On the other hand, people breeding Pit Bull-types for other reasons, such as being a family dog or a show dog, would most certainly want the highest level of inhibition. Being largely genetically controlled, this trait would be quick and easy to manipulate as it reacted to human selection.

This hypothesis is based upon behavioral observations, which are based upon the reported breed or breed group of the dogs in case studies. And again, even when one breeds for a certain trait, that doesn't guarantee that the trait will be seen in the following generation. For clinical use, a tool to distinguish between these two lines of Pit Bulls is an assessment for inhibition. For example, you could assess their ability to learn a new task, since learning requires well-developed inhibition ("Don't do what I want, do what I am supposed to do to get the reward"). Additionally, if there has been an aggressive incident, bite severity reflects inhibition: did the dog "pull its punch," or did it not show any inhibition? The idea of trait-level characteristics, genetic selection, and the development of "lines" within a breed is an important one for understanding these sharp differences in what superficially may appear to be similar animals.

The Champions

The same year that four coordinated terrorist attacks conducted by al-Qaeda ended the lives of almost 3000 people in the US, a rookie football player quietly purchased a large parcel of land in Surry County, Virginia, and acquired a dog breeding license. It was at the kennel on this property, dubbed "Bad Newz" by its three associates, Michael Vick, Purnell Peace, and Quanis Phillips, that countless dogs were tortured and forced to fight one another. Bad Newz Kennels bred Pit Bulls on-site and used torture and starvation to attempt to bring out higher rates of aggression among the dogs; those dogs that didn't perform to their "standards" were electrocuted, shot, drowned, or hung. In 2007 alone, the trio executed eight dogs. For years, dogs suffered at this site, until the dog fighting operation was discovered in April of 2007. On April 25, 2007, Michael Vick's cousin, Davon Boddie, was arrested on drug-related charges and provided the Bad Newz Kennels property as his address. The authorities found probable cause to search the property for dog fighting; during that time, they discovered 54 scarred, injured, malnourished dogs. Most were Pit Bulls. Fighting paraphernalia, including performance-enhancing drugs and a "break" stick to pry open dogs' mouths, was also found on-site.

One would think, given the extreme conditions that these dogs were living in, and the stereotypes about Pit Bulls' "aggressive" natures, that these dogs would all need to be euthanized. PETA and the Humane Society of the United States (HSUS) recommended euthanasia. But even with everything stacked against them, these weren't "aggressive" animals. Of the 54 dogs, 51 were classified under the umbrella term of "Pit Bulls," with no breed descriptor provided for the other three dogs. Only one of these dogs was so physically and emotionally traumatized that euthanasia was the only humane alternative. Two other dogs died shortly after they were taken from Bad Newz Kennels, but the remaining 48 overcame their cruel beginnings.

Many became beloved pets, therapy animals, and ambassadors for Pit Bulls everywhere; some passed the Canine Good Citizen test, indicating that they were well-mannered canine citizens. The story of five of these dogs was told in Filmmaker Darcy Dennett's documentary, *The Champions*. An organization called BADRAP took in ten of the dogs and Best Friends Animal Society took in 22. The majority were successfully placed into adoptive homes.

While most of these dogs were able to overcome the abuse they experienced at the hands of Vick and his associates, animal behavior experts want people to be aware of just how much their rehabilitation entailed. Celeste Walsen, Executive Director of the Courthouse Dogs Foundation of Washington State, points out that this rehabilitation came at great temporal and financial cost ... As a veterinarian and an experienced handler of service dogs, Walsen's opinion carries a lot of weight. While these dogs were able to be rehabilitated, it wasn't "easy."

So *can* you predict an individual dog's behavior, based solely on breed? Breed is a good point of departure, but it's correlational, not predictive; consider again the other potential factors that go into determining whether an individual will act aggressively in a particular context. Not only are there myriad factors at play; it's also difficult to find good data on whether an individual will be aggressive, even when we do research this topic. Amat *et al.* (2009) compared 19 English Cocker Spaniels to 20 other dogs (both mixed and purebred) all of whom came to a veterinary clinic with reports of aggression toward owners, strangers, or other dogs. The researchers found that English Cocker Spaniels had lower levels of serum serotonin compared to other breeds. Serotonin is a neurotransmitter that helps regulate social behavior, mood, memory, sleep, appetite, digestion, and sexual desire; low serotonin levels are correlated with depression. But is this a significant finding, or merely a genetic artifact from somewhere in their lineage? Unfortunately, it could be that English Cocker Spaniels have lower levels of serum serotonin even if they aren't aggressive, so in our opinion, a control group of Cocker Spaniels and other breeds that were not aggressive should have been included in the study.

Evidence for Breed Differences in Aggression—Based on Questionnaires

Recall our discussion of temperament and personality in Chapter 7 and Hart and Hart's study that examined the behavioral traits of 56 dog breeds as determined by dog experts (Hart and Hart, 1985). In that study, the three factors that explained most of the variation between breed groups was reactivity, aggressiveness, and trainability. They argued that this information should be used in combination with an evaluation of the dam and sire and the environmental conditions in which individuals are raised. Notably, they reported that breeds with very high aggression, high reactivity, and medium trainability included a list of Terriers (the Cairn, Scottish, Fox, Airedale, Highland, and Silky Terriers, plus the Chihuahua, Schnauzer, and Dachshund). German Shepherds, Akitas, Dobermans, and Rottweilers were also reported as having very high aggression by the experts, but with the benefits of very low reactivity and very high trainability. Breeds with "very low" aggression were mostly hounds (the Basset Hound, Bloodhound, and Elkhound) as well as English Bulldog and English Sheepdog. Breeds rated as "low" aggression included Retrievers (the Chesapeake Bay, Golden, and Labrador), as well as the Australian Shepherd, Brittany Spaniel, Collie, Keeshond, Newfoundland, and Shorthaired Pointer.

Duffy *et al.* (2008) examined breed differences in aggression using the Canine Behavioral Assessment & Research Questionnaire (C-BARQ) to survey owners about their purebred dogs. Samples were acquired through breed club recruitment (n=1521, representing 11 different breeds) and via online distribution of the survey (n=3791, representing 33 different breeds). They found that bites and bite attempts towards *both owners and strangers* were reported at a significantly higher percentage for Dachshunds, Chihuahuas, and Jack Russell Terriers. Similarly, Australian Cattle Dogs were reported at higher frequencies, but only for *stranger-directed* bites and bite attempts, while American Cocker Spaniels and Beagles were reported at higher frequencies towards *owners*. Akitas, Jack Russell Terriers, and Pit Bull Terriers were reported more frequently for bites and bite attempts towards other dogs (dog-directed aggression). The least aggressive breeds (including owner, stranger, and dog aggression) were Golden Retrievers, Labrador Retrievers, Bernese Mountain Dogs, Brittany Spaniels, Greyhounds, and Whippets.

Different lines of one dog breed can have differences in their temperament, as well. With English Springer Spaniels, those that were bred for conformation showed more aggression to humans than those that had been bred for the field. They showed more stranger-, owner-, and dog-directed aggression, as well. With Labrador Retrievers, however, the opposite pattern was observed, with field-bred dogs showing more aggression than conformation-bred ones (Duffy *et al.*, 2008).

Belyaev and Sorokina's Bad Boys

Do you remember the work of Belyaev and Sorokina, and their silver foxes who were bred to be friendlier, but changed in both appearance and behavior, discussed in Chapter 2? Yes, there was the well-known line of friendlier foxes who were more "dog-like," but did you know that there was a second, "not-so-friendly," line? The less-discussed canines from Belyaev and Sorokina's Siberian research consisted of individuals who retained their wild appearance and acted far more aggressively toward humans than the friendly foxes or the founding population. Beginning in 1959, the well-known line of foxes was selected to be friendly, and beginning in 1970, another line was selected to be aggressive (Wang *et al.*, 2018). Given the ease with which these lines of foxes emerged (one very friendly toward humans, the other not), it's easy to see how this could follow with other species of canines.

It's possible that there are now two different lineages of Pit Bulls: lines that were bred for show, and lines that were bred for fighting. While this is still anecdotal, we are seeing certain trends that warrant further investigation, with some dogs that fall under the "Pit Bull" umbrella demonstrating aggression toward other dogs, and some showing none at all. But when does this occur?

Typically, those Pit Bulls that exhibit dog–dog aggression are of a certain age and background. They have many factors in common: they come from shelters in Southern California, they're approximately 1.5–2 years of age, and they're super persistent. They haven't been trained, and especially, they haven't been obedience trained, and they have limited inhibition. The reason behind this? They may have been purpose-bred to fight other dogs.

These dogs were then abandoned or left on the streets of Los Angeles for one reason or another. Because they weren't obedience trained and lacked adequate socialization, they didn't learn inhibition. These are the dogs that can have the potential to become dangerous—it's the combination of breeding, lack of training, and lack of socialization. Some dogs are bred for lack of inhibition, but those that have "a little bit too much" inhibition will get tossed out. If stitches are needed, you have a lack of inhibition. According to Seattle dog trainer Julie Forbes, aggression can be breed- and purpose-specific.

"When working with aggression issues, I think about the breeds that have been specifically bred for guarding. These include the Shar-Pei, Chow, and Akita; the Asian-originating guarding breed category can be particularly tough with aggression. This trait—to guard, and to be aggressive while doing so—is bred into them. The individual genetics here are *so* important—were the individual parents aggressive, or were they not? It's so important to know the behavior of those lines."

In Julie's experience, Scenthounds can also be problematic when it comes to aggressive tendencies.

"Scenthounds can have a tendency for resource guarding and aggression toward people," she explained, "and they can be really tough to work out of it ... They can be stubborn and not motivated to another way of being. Dogs in the Terrier Group, too, can be challenging, but they're more straightforward with their retraining. And of the Toy dogs, the Lhasa Apso can be particularly aggressive toward people, but that breed isn't as common."

So, what other trends do we find with breeds, breed groups, and aggression? Flint (2017) also used C-BARQ to assess risk factors for stranger-directed aggression in 14,310 dogs. Sporting breeds, based on the AKC 2016 criteria, were found to be the least likely to show severe stranger aggression. Hsu and Sun (2010) also used the C-BARQ survey to examine aggressive behaviors among pet dogs in Taiwan. The researchers surveyed 852 Taiwanese dog owners to examine owner, dog and environmental predictors associated with aggressive behavior. They categorized the dogs in the study into one of 15 breeds, 13 of which were pure breeds, with a mixed breed group and an "other breeds" group (see below for list in results). The "other breeds" was created as a mixed group of purebreds that were represented only by a few individuals, so they grouped them together. Based on owner report, they found that breed was a significant predictor of aggressive behavior directed to the owner, strangers and/or other dogs. Specifically, Golden Retrievers scored significantly lower in aggressiveness towards all three categories (stranger, owner and dog–dog), while other breeds varied between these categories. For *stranger-directed aggression*, these breeds were most-to-least aggressive, in this order: Dachshund, Chihuahua,

Miniature Schnauzer, Mixed Breed, Pomeranian, Maltese, Yorkshire Terrier, Other Breeds, Shih Tzu, Toy Poodle, Beagle, Labrador Retriever, Siberian Husky, Shiba Inu, Golden Retriever. For *owner-directed aggression*, these breeds were most-to-least aggressive in this order: Shih Tzu, Pomeranian, Yorkshire Terrier, Maltese, Mixed Breed, Shiba Inu, Other Breeds, Beagle, Dachshund, Miniature Schnauzer, Chihuahua, Toy Poodle, Siberian Husky, Labrador Retriever, Golden Retriever. For dog–dog aggression, these breeds were most-to-least aggressive, in this order: Shiba Inu, Chihuahua, Miniature Schnauzer, Beagle, Mixed Breeds, Other Breeds, Siberian Husky, Dachshund, Pomeranian, Labrador Retriever, Yorkshire Terrier, Maltese, Shih Tzu, Toy Poodle, Golden Retriever. Interestingly, the other factor that was significant in predicting aggression across all three categories was whether the dog owners used physical punishment on their dogs. However, physical punishment was only predictive of owner-directed aggression when the dogs were between 5 and 10 years of age.

It's unsurprising that dogs who had been aggressed upon by their owners with physical punishment would, in turn, aggress toward their owners. But being the recipient of aggression isn't the only risk factor for dogs exhibiting aggression. González-Martínez *et al.* (2011) performed a cross-sectional examination at a veterinary teaching hospital in Spain to identify risk factors for behavioral problems. Client pets were at the clinic for annual health and minor surgery procedures and their owners agreed to participate in the survey. A total of 232 pets were evaluated for several potential risk factors, including age, sex, and breed. Breeds that were considered to be "potentially dangerous" included the following characteristics: (i) strong musculature, athletic configuration, agility, vigorous, and resistant; (ii) strong character and great bravery; (iii) thoracic perimeter between 60 and 80 cm, height to the cross between 50 and 70 cm, and weight 20 kg; (iv) voluminous, cuboid, and robust head with wide and big skull and muscular convex cheeks, big and strong jaws, and robust, wide, and deep mouth; (v) wide, muscular, and short neck; (vi) solid, wide, big, and deep chest, arched ribs, and short and strong loins; and (vii) parallel, right, and robust forequarters and very muscular hindquarters, with relatively long legs forming a moderate angle. Some of these criteria are not supported by scientific data, but are of compulsory use for the classification of dogs according to the Spanish laws. The breeds addressed in the study include the Pit Bull Terrier, Staffordshire Bull Terrier, American Staffordshire Terrier, Rottweiler, Argentinean Mastiff, Brazilian Mastiff, Tosa Inu, and Akita Inu and crosses. Notably, the researchers did not find that these breeds were more likely to display aggressiveness than the other breeds in the study. *This is an important finding because they sampled in an unbiased manner.*

Genetic Evidence for Breed Differences in Aggression

Takeuchi *et al.* (2009) analyzed the associations between behavioral traits and genetic polymorphisms in 77 Shiba Inus in Japan. The owners of the dogs were recruited at veterinary clinics and through a listing in a magazine. The Shiba Inu was selected for this study due to its popularity, and thus its ease of acquiring a large enough sample size, and because it was a derived breed with a relatively short timespan for being artificially bred. The researchers found that one particular polymorphism (c471T>C) on the gene *SLC1A2* was significantly associated with stranger aggression in these dogs. *Importantly, the association was robust even when they analyzed the hospital and magazine recruitment samples separately.*

Våge *et al.* (2010) examined the genetics of human-directed aggression on 50 English Cocker Spaniels who were classified as "aggressive" and 81 English Cocker Spaniels who had been classified as "non-aggressive". Specifically, the researchers targeted 16 dopamine and serotonin-related genes and analyzed them for single nucleotide polymorphisms (SNPs). They found associations between four of the 16 targeted genes, including haplotypes for both low and high levels of aggression. Despite significant associations between haplotypes and aggressive behavior, there was no single haplotype that predicted the presence or absence of aggression, and not all of the variation in behavior was explained with these markers. Instead, the study supports a more complex model of genetics and the relationship between genetics and environment to explain the risk of aggression in this breed.

Zapata *et al.* (2016) used the dog genome projects in Europe and the US to evaluate fear and aggression in dog breeds. The researchers found that (i) known *IGF1* and *HMGA2* loci variants for small body size are associated with separation anxiety, touch sensitivity, owner-directed aggression, and dog rivalry; and (ii) two loci, between *GNAT3* and *CD36* on chromosome 18, and near *IGSF1* on the X chromosome, are associated with several traits, including touch sensitivity, non-social fear, and fear and aggression that are directed toward unfamiliar dogs and humans. In simpler terms, two known loci that were associated with small body size were also associated with separation anxiety, touch sensitivity, owner-directed aggression, and dog rivalry, and two known loci were additionally associated with touch sensitivity, non-social fear, and fear and aggression directed toward unfamiliar dogs and humans. All four loci were originally identified as genes associated with morphology in dogs and were under high levels of artificial selection in the formation of distinct breeds. They suggest that these particular loci are also associated with behavioral variation.

A recent study by Dutrow *et al.* (2022) sought to determine the genetic drivers of breed differences in behavior. To analyze this, the researchers used over 4000 DNA samples from domestic, semi-feral, and wild canids, as well as over 46,000 behavioral surveys from dogs. Their findings on aggression show that Terriers were more commonly associated with dog–dog aggression and predatory behavior compared to other lineages including the Pointer-Spaniel, Herder, Sled Dog, Scenthound, Retriever, Asian Spitz, and Sighthound.

It's clear that genetic mechanisms exist for breed and individual differences in a wide range of traits, including levels of aggression. So, not only is there good evidence that genes influence aggressive behaviors, but we are beginning to find specific mechanisms for how they do so. In some cases, the relationship between the genes and the behavior is relatively simple (e.g. the Shiba Inu's genetic predisposition for stranger danger and the association of small body size in breeds and owner-directed aggression). But, in other cases it is much more complex, and involves more regions of the genome as well as interactions with the environment (e.g. English Cocker Spaniels). These kinds of studies are the scientific inquiry that is needed to understand these relationships, which appear to vary both among and within breeds.

Breed Insights on Clinical Cases of Aggression

In the case of dog–human aggression, almost all cases that we see as clinicians are dog–stranger aggression—usually fear. So, our ears perk up when we receive a report of dog–owner aggression. This is *incredibly* rare and almost always involves an ancient breed, Boxer, or Australian Cattle Dog. It usually ends up being a case of a lack of structure on the part of the owner. By "structure," we mean that the owner has done little or nothing to make rules or set up boundaries for their dog. Often the issue develops during a crisis (in the dog's world): a change in home environment, introduction of a new person, contractor traffic next door, or any other change that, in the dog's mind, needs "dealing with." A strong, "I got this, I am in charge, I have it under control" attitude by the owner fills the gap; without this, when these signals are ignored, often it seems like the dog feels a need to fill the "power vacuum" at the top, and begins expressing social control signals directed to the owner. This is similar to when children without rules or boundaries will run roughshod over parents.

In a not-uncommon complication to this scenario, the dog will "redirect" its behavior. It will refuse to challenge and socially control the owner (who is larger and louder than they are), and will instead exhibit a common behavior phenomenon wherein the drive to express the social control is redirected to a less-threatening target, like a physically smaller spouse, children in the household, or even another pet. Diagnosis of these situations is complicated and often requires a behavior expert with experience in such cases.

A final practical aspect to dog–human social structure issues is that these ancient-breed dogs often "test the (social) waters" at around 1.5–2 years of age. They'll suddenly start exhibiting social dominance signals, numerous body language signals like lateral blocking and racing through doorways, and in frustration at being (unknowingly) ignored, can even lead to nipping and biting in attempting to get the owners to respond to the social signals. Again, if the owners exhibit subtle but strong relationship signals (feeding by hand, passage first through doors, positive-reinforcement training), the issue either never appears or is quickly resolved.

For anxiety or fear-based aggression, you tend to see more anxiety in more derived breeds, as well as those breeds that have had extensive recent inbreeding. You don't really tend to see anxiety, and thus anxiety-related aggression, in show dogs or breeds that aren't puppy milled, like some of the bigger dogs. You do see it a lot in breeds like the Pekingese, Pomeranians, Bichon Frise, Poodles, Papillons,

Yorkies, and Corgis—small, cute breeds that breeders can sell for maximum profit. You might see a few puppy-milled Huskies or Bernese Mountain Dogs, but it's not nearly as common.

Leroy

A young, golden dog with a blocky head and wide-set shoulders walked slowly down the road, his paw pads raw. His muzzle pushed up against one garbage can, and then another, but they were empty. He continued on, head and tail low. His jaw was slack and he panted as he continued to search. Someone began to speak, and the dog raised his head, his ears raising up. His long tail began to wag, slowly at first, and then with more speed as someone squatted down near him, coaxing him gently. In 2009, Animal Aid and Rescue Foundation (AARF), a non-profit based out of Washington State, found Leroy the Pit Bull walking the streets of the Georgetown neighborhood of Seattle (Fig. 9.3). No one knew who he was, where he'd been, how he'd gotten there, or how long he'd been on the streets. AARF founder Heather Enajibi loved this tawny dog with the golden smile and knew she had to help him. After an assessment revealed that Leroy had a strong prey drive, though, the AARF team knew that it would be more difficult to place him in an adoptive or foster home. With limited options, they reached out to Olympic Animal Sanctuary (OAS) in Forks, Washington. OAS was deemed to be Leroy's best option, as its founder, Steve Markwell, worked specifically with dogs that had behavioral issues, including dog–dog aggression, like Leroy had displayed.

Initially, OAS looked like an ideal place for dogs who would otherwise likely be euthanized, but this "sanctuary" turned out to be a warehouse for dogs, rather than a haven. By 2012, OAS housed more than 120 dogs within its 4000 square feet. Dogs were placed in crates, one atop another, three or four cages high. It was dark, smelled of urine and feces, and the barking was incessant. Opportunities to go outside were limited, but when Leroy did get to play in the grass, he was always smiling, his tail high and wagging. As the months turned to years, Leroy and many other dogs began shutting down. He avoided eye contact as volunteers walked by. His tail no longer wagged.

When Heather went to do a welfare check on Leroy in 2012, the once friendly, gregarious dog was unrecognizable. Rather than focusing on human attention, as he did when she first met him, he was instead hyper-focused on a rock that he'd found outside and carried back into his crate. Heather was appalled—and she asked Markwell to release Leroy back to AARF. He denied her request, and a court battle ensued, with Enajibi filing suit on November 19. Enajibi claimed that Markwell had failed to provide Leroy with "adequate and humane care." In December, a judge agreed, and Leroy was returned to AARF.

Fig. 9.3. Leroy. Photograph courtesy of Marika Moffitt of SoulDog Creative.

But how much dog-directed aggression would Leroy have now? Leroy was assessed by a PhD-level Applied Animal Behaviorist, who determined that the dog had no aggression toward humans, but that he *might* have post-traumatic stress from his ordeal at OAS. This made sense, given his prior assessment of a high prey drive paired with many years warehoused away with limited social contact.

Leroy was placed in a foster boarding facility with a caregiver named John Panchot, who provided Leroy with socialization and ample opportunities to play outside. After one failed adoption, Leroy found a family that could take him for the remainder of his days. A woman named Deanna Goertz and her husband agreed to foster Leroy—and that foster turned into forever. The Goertzes knew about Leroy's history and that he'd previously been classified as an "only" dog due to his prior aggression, but they wondered if it was possible to help him overcome this.

Leroy began training with Cheryl Frantz at the Zoom Room, a facility in North Seattle, Washington. Frantz began her career as a nurse, where she was a clinical liaison for a genetics laboratory, and she understood the concepts of compassionate care and consistency to achieve the best results. She applied this to her work with Leroy. But even more importantly, she took the time to connect with him—and *show* him how to connect with other canines in a positive context.

"The most important thing that I did with Leroy was teach him to play," she explained.

"When he first came in, we just took our time getting to know him and watched his reactions. We also taught his owner, Dee, how to read his signals, and her family, to determine what stresses him out and identify the behavior that constitutes a relaxed dog. I had been told that he was aggressive, that he had injured other dogs and even tried to kill them … What I saw from Leroy was fear reactivity, and not aggression. When he got stressed out, he backed up, so I allowed him to do so, and I gave him positive reinforcement. Both dogs and humans need these tools. We expect our dogs to learn our language, but we need to learn theirs, too."

Frantz started with a fake dog (Fig. 9.4), as she does with all of her clients, to determine Leroy's baseline behaviors. "We introduce all of the dogs to a fake dog, and the owner will look at the dog to see how their dog is reacting. We start with fake

Fig. 9.4. Samples of "fake" dogs for use in assessment of aggression. Photograph provided courtesy of Dr Carly Loyer.

dogs so they can learn to be calm when another dog isn't reacting," she said. "It works 90% of the time, and sometimes they wonder why they aren't moving. In my experience, around 98% of the dogs who are reactive to the fake dogs will also be reactive to a real dog. We're setting them up to be successful at their first step."

If a dog was barking and lunging at Leroy, he likely would have reacted in kind. "We go in graduated steps," Frantz said, "and if the dog isn't responding or doing well, we back up a step."

Leroy did growl and bark a little at the fake dog, but not as much as Frantz had anticipated. "And then we worked on counter-conditioning (this involves altering how one, in this case, one's dog, reacts to or feels about a given stimulus) and reading Leroy's signals. One of Leroy's big ones was that if he got uncomfortable, he would turn and want to walk away. So, we allowed him to do that."

The next step for Leroy was being in a room with one of Franz's dogs. "We didn't have them meet on leash. Most dogs don't meet well on leash. We discourage meeting on leash. We use a barrier. We want it to be one sniff, two sniff, three sniff, and then go." When Leroy passed this step, his next step was being one on one with several dogs in the room. "One of my dogs barks on cue, we call it 'singing,'" Frantz explained. "We had her 'sing,' and then we watched Leroy's reaction. And then we brought him out into a group setting, where he could hear the dogs, but not see them. And he rocked that step, too. So, then he was in with all of the other dogs."

Leroy was given a new set of skills to deal with situations that might be stressful. Leroy was once kenneled in close quarters with dozens of other dogs, unable to escape the tight confines. "When he wants to get away, we let him now," she said. "We've given him a cue—'leave it'—which means that he needs to come back to Dee instead of going toward something."

Over the course of 4 months, coming in once every week or two, Leroy passed each step, until he had passed basic Obedience 1 and graduated to Obedience 2. Frantz gave Leroy's family homework in between each session so they had new skills to build upon. Amazingly, at the end of the training, Leroy could interact with other dogs, but everyone took it very slow. (Leroy lived out his days with the Goertzes, happy and loved to his final moments. He passed away in August of 2021, surrounded by his family and friends.)

"We have to set them up to be successful," Frantz said. "If we'd tried the group class right away, it would be too distracting for him. We have to go at the dog's pace."

Frantz is hesitant to use the word "aggression" and rarely sees true cases of aggression at her facility. "I have worked with a lot of dogs that come into the Zoom Room because their owners report that they have dog–dog aggression, but actually, 90% of it is fear reactivity," she said.

We include these stories because they are exceptions to the norm, rather than the norm. A dog with true aggression, rather than mis-labeled aggression that's actually fear reactivity, or aggression that's based on a transient physical condition, can be difficult to treat.

Conclusion

Aggression is a particularly important topic to explore, for behavioral scientists and for pet owners alike. But what's also particularly important is *how* we are defining aggression and *what* the catalyst is for this aggressive behavior. We also need to consider the behavioral context and the players involved. Was this aggression that was based on fear or anxiety? Was it triggered by fear, or did it result from redirection? Was it an artifact of resource guarding or protection? Was it a rare form of dominance aggression? Was it predatory aggression, male–male aggression, or another extremely rare form of aggression, like maternal or female–female aggression? These are all critical questions to ask from the onset because they change the treatment and the prognosis for aggressive behaviors. There is evidence for a genetic predisposition for specific types of aggression in some breeds, but this does not dictate behavior at the level of the individual due to variability and environmental influences. It can guide us in understanding the type of aggression we may be seeing, and help us treat it, or know how much effort the treatment may take. Many who have dogs in their homes choose a breed based upon numerous factors, including potential levels of aggression. And while breed can indicate a propensity for a dog to be more or less aggressive than another breed, it's important to remember that individuals vary, whether it's due to genetics, epigenetics, or prior experiences, behavioral contexts, or a malady that's actually at the root of the aggressive behavior.

References

AKC (2022) What is Canine Good Citizen (CGC)? Available at: www.akc.org/products-services/training-programs/canine-good-citizen/what-is-canine-good-citizen (accessed 30 October 2023).

Akhtar, N., Smith, M.J., McKirdy, S. and Page, R.E. (2006) Surgical delay in the management of dog bite injuries in children, does it increase the risk of infection? *Journal of Plastic, Reconstructive & Aesthetic Surgery* 59(1), 80–85.

Amat, M., Manteca, X., Mariotti, V.M., Ruiz de la Torre, J.L. and Fatjó, J. (2009) Aggressive behavior in the English Cocker Spaniel. *Journal of Veterinary Behavior – Clinical Applications and Research* 4(3), 111–117.

AMVA (2023a) Dog bite prevention. Available at: www.avma.org/resources-tools/pet-owners/dog-bite-prevention (accessed 30 October 2023).

AMVA (2023b) Why breed-specific legislation is not the answer. Available at: www.avma.org/public/Pages/Why-Breed-Specific-Legislation-is-not-the-Answer.aspx (accessed 30 October 2023).

Arain, M., Haque, M., Johal, L., Mathur, P., Nel, W. *et al.* (2013) Maturation of the adolescent brain. *Neuropsychiatric Disease and Treatment* 9, 449–461.

ASPCA (2023) Position Statement on Pit Bulls. Available at: www.aspca.org/about-us/aspca-policy-and-position-statements/position-statement-pit-bulls (accessed 23 December 2023).

Baack, B.R., Kucan, J.O., Demarest, G. and Smoot, E.C. (1989) Mauling by pit bull terriers: case report. *Journal of Trauma: Injury, Infection, and Critical Care* 29(4), 517–520.

Beck, A.M. and Jones, B.A. (1985) Unreported dog bites in children. *Public Health Reports* 100(3), 315–321.

Blackshaw, J.K. (1991) An overview of types of aggressive behaviour in dogs and methods of treatment. *Applied Animal Behaviour Science* 30(3–4), 351–361.

Boenning, D.A., Fleisher, G.R. and Campos, J.M. (1983) Dog bites in children: epidemiology, microbiology, and penicillin prophylactic therapy. *The American Journal of Emergency Medicine* 1(1), 17–21.

Borchelt, P.L. (1983) Aggressive behavior of dogs kept as companion animals: classification and influence of sex, reproductive status and breed. *Applied Animal Ethology* 10(1–2), 45–61.

Buss, D.M. and Shackelford, T.K. (1997) Human aggression in evolutionary psychological perspective. *Clinical Psychology Review* 17(6), 605–619.

Casey, R.A., Loftus, B., Bolster, C., Richards, G.J. and Blackwell, E.J. (2014) Human directed aggression in domestic dogs (*Canis familiaris*): occurrence in different contexts and risk factors. *Applied Animal Behaviour Science* 152, 52–63.

Centers for Disease Control and Prevention (2003) Nonfatal dog bite-related injuries treated in hospital emergency departments – United States, 2001. *Morbidity and Mortality Weekly Report* 52(26), 605–610.

Collins, K., Reid, P. and Martinez, L. (2012) Bred to fight: evaluating play in pit bull puppies from fighting lines. *Proceedings of the AVSAB/ACVB Meeting*, San Diego, CA, pp. 31–36. Available at: www.academia.edu/download/46980690/THE_PREVALENCE_OF_BEHAVIOR_PROBLEMS_AND_20160703-1452-10rpqbd.pdf (accessed 23 December 2023).

Duffy, D., Serpell, J. and Hsu, Y. (2008) Breed differences in canine aggression. *Applied Animal Behaviour Science* 114(3–4), 441–460.

Dutrow, E.V., Serpell, J.A. and Ostrander, E.A. (2022) Domestic dog lineages reveal genetic drivers of behavioral diversification. *Cell* 185(25), 4737–4755.

Fadel, F.R., Driscoll, P., Malgorzata, P., Wright, H., Zulch, H. *et al.* (2016) Differences in trait impulsivity indicate diversification of dog breeds into working and show lines. *Scientific Reports* 6, 22162.

Flint, H. (2017) Understanding fear and stranger-directed aggression in companion dogs. PhD thesis, University of Guelph, Guelph, Canada. Available at: https://atrium.lib.uoguelph.ca/items/99317b26-3fad-4cb5-9456-c54330c0811a (accessed 11 December 2023).

González-Martínez, Á., Pernas, G., Diéguez, F. and Suárez Rey, M. (2011) Risk factors associated with behavioral problems in dogs. *Journal of Veterinary Behavior: Clinical Applications and Research* 6(4), 225–231.

Hampshire, A., Chamberlain, S.R., Monti, M.M., Duncan, J. and Owen, A.M. (2010) The role of the right inferior frontal gyrus: inhibition and attentional control. *NeuroImage* 50(3), 1313–1319.

Hart, B.L. and Hart, L.A. (1985) *Canine and Feline Behavioral Therapy*. Lea & Febiger, Philadelphia, PA.

Hsu, Y. and Sun, L. (2010) Factors associated with aggressive responses in pet dogs. *Applied Animal Behaviour Science* 123(3–4), 108–123.

Kirkham, N.Z., Cruess, L. and Diamond, A. (2003) Helping children apply their knowledge to their behavior on a dimension-switching task. *Developmental Science* 6(5), 449–476.

Leighton, R. (1911) *The New Book of the Dog: A Comprehensive Natural History of British Dogs and their Foreign Relatives, with Chapters on law, Breeding, Kennel Management, and Veterinary Treatment*. Cassell, London.

Lindner, D.L., Marretta, S.M., Pijanowski, G.J., Johnson, A.L. and Smith, C.W. (1995) Measurement of bite force in dogs: a pilot study. *Journal of Veterinary Dentistry* 12(2), 49–52.

Maddie's Fund (2012) Incorrect breed identification. Available at: www.maddiesfund.org/incorrect-breed-identification.htm (accessed 30 October 2023).

Mehrkam, L.R. and Wynne, C. (2014) Behavioral differences among breeds of domestic dogs (*Canis lupus familiaris*): current status of the science. *Applied Animal Behaviour Science* 155, 12–27.

Mischel, W. and Ebbesen, E.B. (1970) Attention in delay of gratification. *Journal of Personality and Social Psychology* 16(2), 329–337.

Patronek, G., Sacks, J., Delise, K., Cleary, D. and Marder, A. (2013) Co-occurrence of potentially preventable factors in 256 dog bite-related fatalities in the United States (2000–2009). *Journal of the American Veterinary Medical Association* 243(12), 1726–1736.

Presutti, R.J. (1997) Bite wounds. early treatment and prophylaxis against infectious complications. *Postgraduate Medicine* 101, 243–254.

Raaijmakers, M.A.J., Smidts, D.P., Sergeant, J.A., Maassen, G.H., Posthumus, J.A. *et al.* (2008) Executive functions in preschool children with aggressive behavior: impairments in inhibitory control. *Journal of Abnormal Child Psychology* 36(7), 1097–1107.

Reid, P. and Collins, K. (2012) Assessing conspecific aggression in fighting dogs. ASPCA anti-cruelty behavior team, Urbana, Illinois. *Proceedings of the AVSAB/ACVB Meeting*, San Diego, CA, pp. 37–39. Available at: www.academia.edu/download/46980690/THE_PREVALENCE_OF_BEHAVIOR_PROBLEMS_AND_20160703-1452-10rpqbd.pdf (accessed 23 December 2023).

Salonen, M., Sulkama, S., Mikkola, S., Puurunen, J., Hakanen, E. *et al.* (2020) Prevalence, comorbidity, and breed differences in canine anxiety in 13,700 Finnish pet dogs. *Scientific Reports* 10(1), 1–11.

Spanner, H. (2023) Top 10: which animals have the strongest bite. *BBC Science Focus Magazine*. Available at: www.sciencefocus.com/nature/top-10-which-animals-have-the-strongest-bite (accessed 30 October 2023).

Swick, D., Ashley, V. and Turken, A.U. (2008) Left inferior frontal gyrus is critical for response inhibition. *BMC Neuroscience* 9, 102.

Takeuchi, Y., Kaneko, F., Hashizume, C., Masuda, K., Ogata, N. *et al.* (2009) Association analysis between canine behavioural traits and genetic polymorphisms in the Shiba Inu breed. *Animal Genetics* 40(5), 616–622.

Tamm, L., Menon, V. and Reiss, A.L. (2002) Maturation of brain function associated with response inhibition. *Journal of the American Academy of Child and Adolescent Psychiatry* 41(10), 1231–1238.

Thorpe, I.J.N. (2003) Anthropology, archaeology, and the origin of warfare. *World Archaeology* 35(1), 145–165.

Trinkaus, E. and Zimmerman, M.R. (1982) Trauma among the Shanidar Neandertals. *American Journal of Physical Anthropology* 57(1), 61–76.

Våge, J., Wade, C., Biagi, T., Fatjó, J., Amat, M. *et al.* (2010) Association of dopamine- and serotonin-related genes with canine aggression. *Genes, Brain, and Behavior* 9(4), 372–378.

Vaidyanath, V. (2019) Man suffocates neighbor's pit bull to death in front of owner. *International Business Times*. Available at: www.ibtimes.com/man-suffocates-neighbors-pit-bull-death-front-owner-2807321 (accessed 30 October 2023).

Wang, X., Pipes, L., Trut, L.N., Herbeck, Y., Vladimirova, A.V. *et al.* (2018) Genomic responses to selection for tame/aggressive behaviors in the silver fox (*Vulpes vulpes*). *Proceedings of the National Academy of Sciences of the United States of America* 115(41), 10398–10403.

Wright, H.F., Mills, D.S. and Pollux, P.M.J. (2011) Development and validation of a psychometric tool for assessing impulsivity in the domestic dog (*Canis familiaris*). *International Journal of Comparative Psychology* 24(2), 210–255.

Zapata, I., Serpell, J.A. and Alvarez, C.E. (2016) Genetic mapping of canine fear and aggression. *BMC Genomics* 17, 572.

10 Learning, Problem Solving, Training, and Breed Differences

Abstract

Chapter 10 is a scientific review on training and breed differences: how do different dog breeds and breed types react differently to training? How do the motivations between breeds differ? For example, which breeds are considered to be more food-driven than others, and what breeds will likely exhibit more prey drive? Why is it important to keep breed in mind? Case study examples are provided where training is comparatively "easy" and short in duration for some dogs and some breeds, and where training is longer in duration and takes more repetition for others.

The Plight of Potato Chip

What do we mean, scientifically speaking, when we use terms like "learning," "problem solving," and "training?" Well, let's start with the first term, and work our way from there. In psychological terms, "learning" is a relatively permanent change in one's behavior as a result of experience. It's the acquisition of knowledge or skills through experience, study, or by being taught. Think back upon your favorite human examples: a young child learning their ABCs and singing the alphabet song, or excitedly recognizing and identifying a now-familiar letter. A baby learning how to communicate with basic American Sign Language, before he or she can verbalize their wants and needs. These are the building blocks of learning; these children and babies are realizing that there are certain items that correspond to certain terms, denoted by corresponding sounds, signals, or shapes.

Many babies can learn to communicate with American Sign Language by the age of 6 months (Barnes, 2010) and typically developing children learn to verbally identify letters (Adams, 2003) and numbers (van Marle *et al.*, 2014) by the age of 3 or 4 years. Not all of us learn at the same rate, or in the same way, though. Dogs, like human children, sometimes reach developmental milestones at different points, or learn a concept at a later time than their peers. Dr Carly Loyer (discussed in Chapter 7) examined individual differences in learning, with the number of trials that most dogs took to learn a new skill averaging 15 to 30 trials. One dog in her study, a pit bull mix named Potato Chip, earned the unenviable label of "learning outlier": he took 88 trials to master a new task. Potato Chip was the canine poster child for "slow learners," but he got the concepts in the study … eventually!

Along a different evolutionary axis, dogs represent a dramatic behavioral divergence from their wolf ancestors. Dogs, in comparison to wolves, show decreased aggression, increased tameness, and the capacity to learn tasks through their interactions with humans (Hare and Tomasello, 2005; Miklosi, 2009). If you happen to be interested in the topics of canine cognition, emotion, and learning, you'll notice that many of the studies include the work of Dr Brian Hare. Dr Hare's body of canine work focuses on canine emotion and cognition. Hare founded Duke University's Canine Cognition Center in 2009 and, along with his wife, Vanessa Woods, wrote the book, *The Genius of Dogs*, in 2013 (Hare and Woods, 2013). So, what *is* the genius of dogs? Hare points out that the paradoxical dog is not becoming less "important" as technologies advance, but more so. Cases in point: dogs often do a job that other technology, or humans, cannot do, or cannot do as well. This includes therapy dogs that provide reassurance, assistance and service dogs with very specific tasks, working dogs like the Rogue Detection Dogs that help track and protect endangered species, and COVID-detecting canines that can tell if you have the novel coronavirus far sooner than any laboratory could. Hare points out that there are many "types" of intelligence. You've probably heard the adage, "Everyone is a genius. But if you judge a fish

 Breed Differences in Dog Behavior: Why Tails Wag Differently. (R.L. Ha, T.L. Brad and J.C. Ha)
DOI: 10.1079/9781800624566.0010

by its ability to climb a tree, it will live its whole life believing that it is stupid." This refers to finding relevant ways to test one's intelligence, including perception and motivation. For Hare, there are multiple dimensions to measure canine cognition, including empathy, communication, cunning, memory, and reasoning. Some dogs will excel in one dimension, but not in another, but is one necessarily "more intelligent" than another because of this?

Hare has ten games, or tests, of cognition, through his website, www.dognition.com. In eight of the ten tests, there weren't many significant cognitive differences between breed groups or breeds. But in two of them, "arm pointing" and "memory versus smell," there were some surprising results. Purebred dogs were more reliant upon human gestures than mixed breed dogs were, but mixed breed dogs were more reliant on their memory than purebred dogs were. So, who was the smarter dog? Hare notes that these are different ways to measure intelligence, and while there are breed group and breed differences, we still don't have the data set to begin to understand all of these differences. Hare, like many other researchers, found within-breed differences that were almost as striking as between-breed ones. For example, he compared Labrador Retrievers who had been bred to be assistance dogs with those who had been bred to be explosive detection dogs, finding huge cognitive differences between these lines within the same breed. Thus, there's so much variability within a breed that breed itself, and alone, can't reliably communicate a dog's cognitive abilities.

Problem Solving

Agent J

James Edwards was facing the wrong way when the elevator opened from behind. "You're late, sit down," a voice said from behind him. He turned around and sauntered into the testing room, which was filled with 1960s-style egg-shaped chairs and adorned with circular overhead lights that gave the entire room a circle motif. Edwards took a seat beside five other men and looked up at the man who'd admonished him for his tardiness. The man introduced himself as Mr Zen and then informed the sextet of test-takers that they represented the "best of the best" of their respective military branches, with Edwards representing the NYPD. Edwards was the only man among the six who asked why they were there, and then the test commenced. The test packet was flimsy, Edwards' pencil broke when he tried to break the packet seal open, and the egg-shaped chairs were devoid of a surface to write upon. The test-takers struggled as they put holes in their paper and tried to adjust their bodies to provide a writing surface. And then Edwards looked at a large metallic table in the center of the room, which would be perfect for writing on, and pulled it over to his chair. The table elicited a cringeworthy screech as he dragged it along the floor. The other five test-takers watched Edwards, and then, as he sat down again and put his test onto the table, he asked, "You wanna get down on this?" The other men simply stared back.

If you've seen the movie *Men in Black*, you're familiar with Will Smith's famous scene. The Men in Black are putting potential candidates through a litany of tests, and this, the first, assesses their ability to solve a problem (taking the test on the flimsy paper packet, with sharp pencils and no hard writing surface) irrespective of whether it disrupts the room or not. That table was loud, heavy, and far from Edwards' seat, but it solved his problem to take the written test. The answers on the written test were likely looked at, as well, but the Men in Black needed to be able to assess a problem and solve it in the best way possible, just as Edwards did. Agent K, played by Tommy Lee Jones, looked on at the test through two-way glass. If you haven't seen the movie, I won't ruin it for you, but we'll just leave it at this: Edwards (later known as Agent J), like many humans, excelled at problem solving.

Problem solving is such an enjoyable activity for some people that we've even created an industry out of it: puzzles, games, and "escape rooms" that call upon our strongest problem skills. Yes, these escape rooms do provide outside assistance if we get stuck, but the concept behind them is to identify the steps one needs to take to "escape" from a room. But what about our canine companions?

Help!

Dogs are a lot of things, but they aren't going to be problem solvers like Agent J; they look to their humans to help them out of a tight spot. And that's okay—because it's our fault. For generation upon generation, we bred dogs to be "more" of every trait that we desired: more sensitive to our needs and feelings, more driven to herd or pull a sled, more communicative, with their brachycephalic faces; "more" so that they would adapt well to their myriad lifestyles with their humans. But in so doing, we also made them more helpless in certain situations. We've enabled them to rely upon us, and, without realizing it, we have bred our best friend to not perseverate. In free-living situations, perseveration can mean the difference between finding prey or going hungry, or finding shelter or going cold. But perseveration was essentially "bred out of" our dogs as they came to rely upon humans more and more.

Think of this as a species-wide "failure to launch." Our dogs are the middle-aged child who's still at home, financially dependent, happy to have his parents cook his food, do his laundry, and make the big decisions. They can't chart their own course. Sound harsh? Well, maybe that's because it hits too close to home. Our dogs excel at reading our signals, intuiting our emotions, even smelling our low blood sugar and cancer, but if a problem is too difficult to solve, they'd rather phone their best friend rather than continue on. So, while they aren't good at problem solving alone, they're very good at finding ways to elicit human help to solve their problems.

The same isn't the case for their sister taxa, the wolves, who excel at problem solving. Recall the work of Dr Monique Udell from Oregon State University (Chapters 2 and 7). Dr Udell found that wolves will persevere during cognitive tests, while dogs either quit trying or look to their humans for assistance. In a 2015 study, Dr Udell's test had initial conditions: "alone," where the human experimenter left the testing area, and "human-in," where the experimenter remained in the area, stood neutrally, and looked at the puzzle (Udell, 2015). For those canines that failed to solve the puzzle in both the alone and human-in conditions, they were provided with a condition where the human encouraged them to engage with the puzzle. Udell found that wolves could solve this puzzle box task with an 80% success rate, while dogs only had a 5% success rate. While the dogs did increase their contact with the puzzle box when humans encouraged them, there was only a small increase in their success to solve the puzzle (Udell, 2015). Udell concluded that dogs depend on and defer to their humans—and we would agree with this conclusion.

So, while we've established that wolves are better problem solvers than dogs, let's take this a step farther and ask: are there dog breeds who are more like Agent J than the rest of the "best of the best" taking the test? And if so, which breeds are these?

Let's define what we mean by "problem solving." We can refer to it as the thinking and thought processes that are aimed at finding solutions to specific problems. We identify a problem, conceive an idea to solve that problem, and go through the steps, including the mental operations, to execute that solution.

Training

Mixed messages

What does it mean to "train" an animal to learn what these things mean? Let's operationally define what we mean by "training." Dog training can refer to many different things. It typically refers to the application of behavior analysis that uses the environmental events of antecedents and consequences to modify the behavior of a dog, to either assist in specific activities or to undertake particular tasks, or for him or her to participate effectively in contemporary domestic life.

Once more, in layman's terms, please?

Obedience training is a subset of dog training and usually ranges from very basic, specific training, such as teaching the dog to reliably respond to basic verbal commands such as "sit," "down," "come," and "stay," to higher level competition where additional commands, accuracy, and performance are scored and judged.

According to dog trainer Cheryl Frantz of the Zoom Room in Seattle, dog training is "having fun with and games with your dog," and that's a good way to look at it. Think back upon your preschool and kindergarten years. You were learning important concepts, like colors, shapes, letters, days, and months, but you were also immersed in play. Incentivizing learning for young children—and for dogs—with enjoyable activities made learning more accessible and even, dare we say, fun. Learning that a "rectangle" could be the base of a block castle was a lot more fun than being shown a flash card of a rectangle alone, just as being taught the word "come" with an abundance of enthusiasm (and perhaps a toy or two, as well) would also be more pleasant.

The classic Scott and Fuller study

Scott and Fuller (1965) were careful to separate tasks in which the trainer asked the subject animal to perform a specific behavior, such as quieting, leash training, retrieving, motor skills and obedience, from those in which they asked the animal to problem-solve. They clarified that while both "training" and "problem solving" involved "selective reinforcement of successful responses," they believed that the tasks listed above required a more stereotyped response from the animal than problem solving.

They further separated their "training" tasks into those that used reward-based reinforcement and those that used "forced training" where subjects were either prevented from performing the incorrect response, or were mildly punished for doing so. They were particularly interested in what portion of variability, or differences, between subjects was attributable to variation in genetic breed differences, but they also wished to explore correlations between trainability, learning and emotions.

Scott and Fuller's forced training approach and findings

One early training task was to get the puppies to stay still for one minute on a scale for accurate weight measurements. This was done by placing the puppies on a scale and holding the experimenter's hands near the dog to discourage movement. Dogs were not touched unless they moved. As you might imagine, this was a difficult task for very young dogs. By 5 weeks of age, some individuals of some breeds were beginning to learn to stay still on the scale. The biggest differences between breeds were seen at 16 weeks of age, where 70% of the Cocker Spaniels stayed still for 1 minute, but only 10% of the Wirehaired Fox Terriers (Fig. 10.1) could do this. Shelties were the second best at this task, and the Basenjis and Beagles were similar to the Terriers.

Starting at 19 weeks of age, the puppies started learning to walk on a lead (choke collar with chain leash) between their outdoor pens and the laboratory. The goal was for the dogs to walk to the left of the experimenter (without body contact or vocalizations) and with a slack leash. Once a day for 5 days, they walked the course between the two locations and starting on the third day they were also led up a flight of stairs. Then there was a break until 22 weeks of age, when they started another 5 days of leash training (once per day to the laboratory). To evaluate performance of the dogs, they recorded errors in the following categories (balks, fighting or biting leash, being ahead or behind, body contact with handler, and vocalizations), with

Fig. 10.1. Basil, the Wirehaired Fox Terrier. Image from Wikipedia and released by the contributor to the public domain, PD-Self. Available at: https://commons.wikimedia.org/wiki/File:P1030547_(Medium).jpg.

a maximum of three errors in each category per day. Breed differences during leash training were greatest for fighting the leash (Basenjis), body contact with handler (leaping on handler and winding between their legs; Shetland Sheepdogs) and vocalization (howling and wailing; Beagles).

The last "forced" assessment was to get dogs to remain on a stand for 30 seconds and then jump down on command. This was done over 3 days, with the experimenter gradually being more and more distant from the dogs (to increase difficulty). The dogs initially wore a choke collar and lead, but this was removed as they improved, and the experimenter moved away from 1.5, 3, 6, 12, and 14 feet away (the last distance was also behind a screen). Subjects that did not leap when commanded were "gently pushed" off the stand. The Cocker Spaniels were the easiest to train in this task, followed closely by the Wirehaired Terriers, and Basenjis were the most difficult. Beagles and Shelties were intermediate between the other breeds.

Overall, Scott and Fuller concluded that the breeds varied widely in their ability to be trained to be "quiet" or "inactive" on command, but all of them are capable of this training. Cocker Spaniels (Fig. 10.2) were the easiest to train on these behaviors using these methods, and Basenjis and Beagles were consistently more difficult, while the Shelties and Wirehaired Fox Terriers did better or worse depending on the task. They attributed the ease of training the Cocker Spaniels using these methods to the artificial selection of this breed for "training to crouch." Hand signals elicit crouching fairly easily in this breed, and hand signals were used to some extent in all three of the assessments.

Scott and Fuller's reward-based training approach and findings

For reward-based training, they used food and praise, and they trained three different tasks: (i) goal orientation (finding a piece of food in a box; four trials), and then finding again when the box was relocated in the room; (ii) training to retrieve an object; and (iii) motor-skill test (climbing up a stack of boxes, then crossing a bridge to another stack of boxes to get a food reward). They found that all breeds performed fairly similarly to the goal

Fig. 10.2. Cocker Spaniel. This file is licensed under the Creative Commons Attribution-Share Alike 3.0 Unported license. Available at: https://commons.wikimedia.org/wiki/File:EnglishCockerSpaniel_wb.jpg#file.

orientation, but Basenjis were best, perhaps due to their speed. None of the breeds did well on the retrieval tests, and there were no significant differences between breeds. They concluded that perhaps they should have conducted the test earlier in development. They originally started it at 9 weeks of age, but the schedule of other assessments made it logistically difficult, so they stopped and re-did this at 32 weeks of age. Anecdotally they noted that the few pups they trained at 9 weeks of age did the task more easily than all of the dogs at 32 weeks of age. The Basenjis showed an innate climbing ability in their pens prior to the motor-skill assessment. This prompted the motor-skill assessment, and the Basenjis did do well in the climbing test, particularly in the highest elevation tests, but the Cocker Spaniels would run and jump and use momentum to carry them up, so they also did well (second place overall). Beagles and Fox Terriers were intermediate in this assessment and the Shetland Sheepdogs did not perform as well as the other breeds, with many of them showing fear of heights.

Across all of the training assessments, they concluded that it was difficult to target one skill, and that most of their "tests" involved several components of behavior and emotion. Breed differences varied depending on the assessment, and across training trials, and the effect of heredity was difficult to measure.

Scott and Fuller's development and differentiation of problem-solving behavior: findings on breed differences

In order to assess problem-solving behavior, they administered more complex tests, including testing spatial ability (mazes and spatial orientation), barrier or detour tests, cue response, discrimination, delayed-response tests, and tracking. Despite the large number of assessments, the results were mixed. When all of the breeds were compared on all of the tests, no breed came out as "more intelligent" or a better problem solver than another. Even within a test, a breed group might perform poorly early on, but end up with the best results on the last trial. However, there were some patterns suggesting that the four hunting breeds (Beagles, Basenjis, Terriers, and Cocker Spaniels) performed best, and they suggested that this is likely due to food motivation and the use of food rewards in the assessments. The Beagle had the most top ranks on the assessments, which they attribute to the fact that they were bred to hunt without direction. Subjectively, they note that the Shelties gave the impression that they were looking for direction from a human handler, and indeed they are bred to work closely with a handler.

Overall, they suggest that the next step might be to evaluate breeds commonly believed to be highly intelligent, such as Poodles and Border Collies. The tests they conducted with these limited breeds suggest an average intelligence across breeds.

Unfortunately, training has received less scientific evaluation for breed differences since Scott and Fuller's work. Most of the work that has been done is survey-based and asks humans if dog breeds differ in obedience training and other forms of performance training.

Think for yourself

In terms of breed differences in training, let's review the Hart and Hart study of 56 breeds (Hart and Hart, 1985a). They found that these breeds were reported by veterinarians and obedience judges as having "low trainability" (by this, we mean the ease of housebreaking and obedience training. For example, a dog that takes a long time to housetrain and go through obedience training would have low trainability, while a dog that quickly learned these concepts would have "high trainability"): Afghan Hound, Basset Hound, Beagle, Bloodhound, Boston Terrier, Boxer, Chow Chow, Cocker Spaniel, Dalmatian, Elkhound, English Bulldog, English Sheepdog, Great Dane, Husky, Irish Setter, Lhasa Apso, Malamute, Maltese, Pekinese, Pomeranian, Pug, Saint Bernard, Samoyed, Weimaraner and Yorkshire Terrier. In contrast, they found that these breeds were very high in trainability: Akita, Bichon Frise, Doberman, German Shepherd, Poodle, Rottweiler, Shetland Sheepdog, Shih Tzu, Springer Spaniel and Welsh Corgi.

Keep in mind that the study only included 56 breeds, and Border Collies weren't included. The study also relied on expert opinions from veterinarians and dog show judges, so they didn't directly measure these characteristics in the dogs, but rather the opinion of experts about the breeds. Nevertheless, it does provide some information about breed differences that may influence trainability.

How important is breed group, including historical job duties, to a dog's ability to be trained? There's some evidence that "working" breeds are easier to train than non-working breeds, in that they acquire knowledge at a faster rate. Asp *et al.* (2015) surveyed the owners of 3591 purebred dogs registered with the Swedish Kennel Club. Eleven of the breeds were working breed dogs and nine were non-working. Overall, the working breeds were 10% more trainable than the non-working breeds. There was some overlap, however, between the two groups. For example, non-working breeds that showed a close relationship with a handler had high trainability scores: Toller, Golden, Sheltie and Lagotto. The lowest scores for training came from owners with these breeds that do not have a history of working closely with handlers: Chihuahua, Rhodesian, American Staffordshire Terrier, Jack Russell, and Bernese Mountain Dog. The breed differences they found for these owner-reported characteristics were relatively small, but statistically significant.

Similarly, another research group (Serpell and Hsu, *et al.*, 2005) surveyed the owners of 11 common breeds, including the Labrador Retriever, Golden Retriever, Shetland Sheepdog, Rottweiler, English Springer Spaniel, Poodle, Yorkshire Terrier, West Highland White Terrier, Dachshund, Siberian Husky, and Basset Hound. They found significant differences in reported trainability between breeds, with Labrador Retrievers reported as the "most trainable" and Basset Hounds as the "least trainable."

A recent study sought to determine the genetic drivers of breed differences in behavior. To analyze this, they used over 4000 DNA samples from domestic, semi-feral and wild canids, as well as over 46,000 behavioral surveys about individual dogs from their owners (Dutrow *et al.*, 2022). They identified 10 lineages including the African-Middle Eastern, Asian Spitz, Dingo, Herder, Pointer-Spaniel, Retriever, Scenthound, Sighthound, Sled Dog and Terrier. They found multiple lineages were associated with higher trainability, specifically among herder, pointer-spaniel, and retriever breeds. In addition, their findings on broader categories of working and sporting breeds versus non-working or unrecognized breeds showed that working and sporting breeds received higher trainability scores. Interestingly, they found that the Scenthound lineage and was less trainable than other lineages. They attribute this to artificial selection for traits focused on following instincts rather than seeking out human cues/input during training.

Given that these are human reports of trainability in dogs, it's worth questioning whether these statistical differences related to actual differences in the dog's trainability, or are associated with human stereotypes or perceptions. William Helton (2010) designed a series of studies to ask whether differences in physical ability due to breed might affect human perception of training acuity in dogs. To determine the answer, he used national agility statistics, training records for agility, and physical characteristics of these dogs. He found that the representation of breeds in agility competition that were considered to be highly trainable versus low trainable was heavily skewed to breeds considered more trainable, but that the bias in "champions" was not skewed given the ratio of these breeds in competition. He also found that the level of precision in performance competition was similar across breeds, and only speed differed between high and low trainability breeds. This difference was also true in training. Lastly, he found that breeds considered to be highly trainable were more similar in height than breeds that were thought to be lower in trainability. Given these collective findings, he proposed that human perception of differences in trainability is driven more by physical differences between breeds than cognitive differences between them.

Given this concern, studies that do direct observations on dogs to evaluate behavior are warranted. Recall the 2022 study by a group of scientists in Finland that we discussed in the Temperament chapter (Chapter 7). The scientists performed a standardized cognitive assessment of 1002 dogs from 13 different breeds (Junttila *et al.*, 2022): Australian Kelpie, Australian Shepherd, Belgian Shepherd Malinois, Border Collie, English Cocker Spaniel, Finnish Lapphund, German Shepherd, Golden Retriever, Hovawart, Labrador Retriever, Mixed Breed, Shetland Sheepdog and Spanish Water Dog (see Table 7.2 for sample sizes).

They found significant differences between breeds in the cylinder test, which is a measure of inhibitory control. Specifically, they found that Border Collies and mixed breeds showed the highest levels of inhibitory controls compared to the Belgian Shepherd Malinois (the Finnish term for the Finnish lineage of Belgian Shepherd or Belgian Malinois) and the Labrador Retriever. When presented with an unsolvable task, two breeds (the Australian Kelpie and Golden Retriever) spent significantly more time in human-directed behavior (approaching the experimenter or the owner) while attempting to solve the problem. The German Shepherd and the Belgian Shepherd Malinois were more likely to show independence when attempting to solve the problem. They found no significant differences between these 13 breeds for short-term memory or logical reasoning. Limitations of the study include the limited number of breeds evaluated, no controls for differences in training experience in each individual dog, and the potential for geographic differences in breed variability.

What the trainers think

Because this is still a relatively new field of study, the scientific literature on training differences between dog breeds is still pretty scant. This led us to consult some expert trainers for their first-hand experiences working with different breeds. Since these experts do a lot of training of different breeds, we wondered if they had differing levels of success, or if they had different approaches that they used for different breeds. Alternately, we wondered if they found that the individual was the only factor in learning, trainability or problem solving.

Seattle resident Julie Forbes has been a dog trainer for more than two decades. She earned a Bachelor's degree in Animal Science from the University of Vermont and completed the Academy of Canine Behavior's apprentice program in dog behavior and training. For Julie, dog training is very specific to the breed and to the individual. When she's training dogs, there are particular breeds that she has worked with that appear to learn quicker than others. But that may be appearance alone.

"When people ask me if one breed learns quicker than another, my answer is: 'Yes and no.' The thing with breeds now ... it's hard to say if things were different 20 years ago and if my change is perspective-driven, or if dogs are actually different from two decades ago, so I would say yes. There are accurate generalizations that we can make about breeds and breed groups and as valid is that there's so much individual variation within breed groups."

"If you want a Husky or a Husky mix, be prepared for an active, smart dog with a high energy level. Now, that's not the case for every dog of that breed or mix. Is the individual going to be mellow and prove me wrong? Perhaps. Such an important word to have is 'motivation.' It has less to do with 'intelligence' and more to do with what motivates the individual. This is equally applied to differences in breed groups. We have to ask: what was this breed bred to do? Herding breeds were bred to work with humans. They'll listen to direction for 'Left, right, hold them, bring them to me,' all of the whistles and commands. Once the dog is 'clicked in,' they are bred to have an ear out for human direction. For that reason, those dogs are looking for that direction. They want to know: what are we doing? We really see that with our Cattle Dogs, more so than we do with our Labrador Retrievers."

Julie and her wife, Darcie, share their home with both.

"Levi, our male Cattle Dog, is like, 'I'm your guy, I'm the one to do it.' There's so much joy for him in being given direction. He's thirsty to work, and he wants to know what you are doing, and how he can help. On the opposite end of the spectrum are the Scenthounds. They're bred to track the scent; you're just there to keep up with them. They aren't looking for direction from you, but they tend to be food-motivated and you can motivate them in that way. If you bring food into the mix, now this task will make sense to them."

"Speaking especially to obedience, there's a dynamic from human to dog. With a Scenthound, they're asking: why would I do this? It's not in their job description to look to their human. They need that outside motivator, which is usually food. Scenthounds can be tough. Dogs from the Terrier Group are another good example: I've met more Terriers that are actually socially motivated ... go into the hole, and kill that animal. Not a lot of input goes in while getting job done. And with dogs from the Sporting Group, there's a human–dog component, but they can be more compromised, like Retrievers versus Setters."

"With Huskies, when you're first learning to work with them, you learn very quickly that they want to know that you know what you're doing and that you respect them, especially. As we are respecting that they have respect for us, we're working to establish respect ... it's important to be respectful of them. This applies to Malamutes, too, and probably the entire breed group."

Anyone who has ever known, or worked with, Northern breed dogs like Huskies and Malamutes can understand what Julie is referring to. It's akin to Theory of Mind (ToM) (Chapter 8): "... they want to know that you know what you're doing ..." Northern breed dogs are a breed apart.

"It makes sense to think that it's due to them being an ancient breed," Julie continued. "They're like, 'I'm going to save your ass out in the Arctic, but I need to make sure that there's some mutual respect here.' Historically, the amount of responsibility that they have is immense ... I attribute it more to something like, 'You're going to trust me with this job description, and then over here try to train me stuff that I'll pick up, and then also be disrespectful to me?' They know when you don't know what you're doing."

"There are definite differences between breed groups," Julie explained. "If you're having a problem with one particular dog, and they're leash reactive to other dogs … knowing if it's a Border Collie versus a Bull Terrier or Pit Bull, will give me a different feel for the situation, and also knowing that there's a range of typical behaviors."

"With obedience training, it's important to know what the dog's job description was historically, or what it is currently, as that depends on what you're trying to teach them. Is the dog being taught to find someone missing in the woods? If so, a Scenthound or a sporting breed would excel at this, and the Scenthounds are the ones that are especially used in this line of work. Bloodhounds used to be the only dogs whose scent detection could be used in court. If we are trying to train a dog to track, then I want to use a scent dog."

Horse Training

So, do breed differences pertain to dogs and dogs alone? Not at all. Horses have been domesticated for at least 6000 years and have been trained for a wide range of uses, though, so they're an excellent species to compare dogs to.

Alaskan horse trainer Sarah Weideman has a decade of experience as a professional American Riding Instructors Association (ARIA) Level 2 Hunter under saddle, general horsemanship, and Level 1 dressage trainer. She also has an additional 40 hours of mounted patrol training. Sarah has worked with horses ranging from miniatures to the largest draft breeds, but breed doesn't always play the most significant role in a horse's trainability. "Their training backgrounds play a bigger part in the trainability of each horse than their breed does," she said. "With that said, I've found that some breeds are more sensitive to more harsh or softer training methods than others, and some others seem to 'remember' or associate things for a longer time than others. This can change the difficulty of the training of the horse."

"Horses are quick to pick up the action-reward training as well as action-negative training. I prefer action-reward forms of training, but a negative correction (punishment) is sometimes needed, as well, when done fairly (e.g. if you kick me, you will have to move your feet in a manner that you don't want to for a while; I repeat the action that triggered the kick until the kicking stops). Horses tend to respond positively and correct the behaviors quickly. When corrections are done in excess, the Arabian is one breed that I have found who will hold a grudge and at the first tipping of the fair scale, they will react in a survival manner. They will first try to flee, and if that doesn't work, they fight back. Not every resistance or fight is physical, but they do fight back."

"For ease of training, breed-wise, it's much like any other animal that has been selectively bred to do a specific task. It all comes down to what the breed has been bred to do. If you took a Friesian who is bred for dressage and tried to cut cattle with it, the task would be much harder than when done with an American Quarter Horse who has a pedigree going back many generations with proven cattle horses. It's completely possible to train a Friesian or a Belgian draft horse to cut cattle; however, the natural tendency to 'lock on and go' to a cow wouldn't be there and would make the job of the trainer much more difficult."

Learn Their Currency

We've been talking off and on about motivational currency, but just how important is this concept? Anyone who has seen the 1980 movie *The Gods Must Be Crazy* (if you haven't, we highly recommend it) probably recalls the scene with Xi, the hunter-gatherer protagonist from the Kalahari, walking off into the distance with paper money blowing all around him. (Don't worry, this doesn't give much away if you haven't seen the film). For Xi, this money was of no value; it literally wasn't his currency … And this is the case throughout human cultures. What's "valuable" to some is relatively worthless to others. The same is true for other species, as well, including our dogs. While one breed (Labradors) would try to figure out quantum physics for a food treat, another breed might put their nose in the air for that, but would work on finding world peace for a quick game of ball.

"With the dogs I've trained, it depends more upon the connection between the human and their dog and how much time they're playing and having a good time than it does the breed," dog trainer Cheryl Frantz said. "Certain breeds are toy, or food, or praise motivated, but all of them do have a motivation; you just have to find it."

Horseman Gary Pegg of Rainier, Oregon, has trained horses—and Tennessee Walking Horses in particular—for decades.

"I used to think there was a significant difference in the intelligence level of horses from breed to breed; however, I've come to the conclusion that horses, like people, are a collection of individuals, each with their own distinct personality, likes and dislikes. Because horses and people are individuals, the interaction between a human and a horse can vary widely. A person loses a trusted old mount and cannot form the same bond with the new horse or a willful, stubborn horse becomes a willing, gentle partner with a new human. The infinite combinations of people and horses would make any definitive statement that one breed possesses more intelligence than another pretty hard to substantiate."

"Perhaps the truest test of intelligence is what a horse learns without the direct interference of mankind. At our house, the water tank is in the corral. To access the feed from the water trough, a horse has to travel north, out the gate, then around the perimeter of the corral to the feed bunks on the south. I've had horses that have figured out the problem with no assistance and others that will stand and stare longingly at freshly filled feeders from the wrong side of the fence until I take mercy and lead them to the gate. There doesn't seem to be much difference from breed to breed on which figures out the problem easier; some individuals get it, some don't. Oddly enough, the horse that would turn around and run back into the corral, because it was a direct line towards the feeders was also my worst 'Houdini horse,' the one that figured out how to open gates, untie knots and turn on the water; so was he smarter, or just busier?"

"I believe that there are tendencies and characteristics, that run in the different breeds, that strongly influence trainability and the appearance of intelligence. The 'hot' breeds, including Arabians, are more reactive and faster to flight from mild to moderate motivation. The 'cold' breeds, including Belgians, are more stoic and faster to forgive. I think that within each breed different characteristics have been developed for lines of unrelated uses. The laid-back show Quarter Horse is nearly unidentifiable compared to the more amped up cutting horse. The easy going, hardworking Tennessee Walker of last century's small farms is much different from today's high strung, big lick show horse. You can look at many other breeds and find similar disparities."

If what Gary is speaking about feels familiar, perhaps it's because it's very similar to our discussion of different lines within one dog breed. This is true for both horses and dogs, where artificial selection continues to differentiate subgroups (lines) within one group (a breed). And both Sarah and Gary discussed "cold" and "hot" horses—and we see this same concept in dogs.

"Possibly even more important to trainability than how reactive or stoic a horse is, is how strong of a sense of self-preservation that horse has. A horse with a really strong sense of self-preservation doesn't think you have its best interest at heart. Add that to an over reactive personality and you've got a long job ahead of you. Good examples of this are mustangs, mules and zebras."

"As a general rule, the hotter the blood, the slower they initially learn. In the round pen when you shout 'hup' and wave your arms, a hot-blooded horse will take off at a dead run and continue to run with minimal encouragement, to the point that they endanger their well-being. Do the same with a cold-blooded horse, and they will do a lap or two before they pull up, look around and look like they are saying, 'Tell me again why we're running?' If the goal is to teach the horse that it's good to stand still, the cold blood learns much faster."

"We're quick to impose human thought processes onto our pets. That's why for so long I thought the horse with the strong sense of self-preservation was willful, stubborn, or even mean. I was imposing my values onto a horse's reactions without thoroughly considering the true nature of the actions I considered 'poor.' Is it possible that some action that appears to be intelligent could be attributed to another instinct?"

"Horses don't think like people. I'm not sure how they think, but I'm convinced it isn't like you or I. How do you define 'intelligence' in a horse, let alone measure it? A willing, people-oriented horse will learn a new task much faster than a suspicious, nervous horse, even though the nervous horse immediately ran out the gate and around the corral to the food while the willing horse stood rooted, staring mournfully. Does that make the nervous horse smarter for figuring out the maze or the willing horse smarter for recognizing the shortest distance is a straight line? Who is smarter; the Lipizzaners of the Spanish Riding Academy with all their maneuvers or the Mustang who can survive in the wild?"

"Bottom line: for what I do, I've never noticed that much difference in intelligence, but I've seen a large difference in trainability."

It's Not Just the Domesticated Species

Variation in learning and motivational currency are demonstrated by non-domesticated species as well. Jessica Meaghan (Meg) Mas has worked with large-bodied apes, including chimpanzees and orangutans, since 2009. As a chimpanzee and orangutan caregiver at the Center for Great Apes in Wauchula, Florida, she has found distinct differences in how the species approach the same environment.

"Chimpanzees and orangutans are both brilliant, but they often think about things differently," Meg explained. And there are evolutionary reasons for this. "In free-living situations, chimpanzees live in large social groups, whereas orangutans tend to be solitary. Their natural environments are very different."

In their natural environments in Asia, Orangutans primarily eat fruit, foraging alone or as a mother-offspring pair, while in their natural environments in Africa, chimpanzees are omnivores who sometimes hunt meat, with different members of the social group taking on different roles.

"Lots of factors, including these social and environmental differences, contribute to how their minds work," Meg explained, "and I've been fortunate to observe some of these differences."

"As an intern years ago, one of the orangutans had found a small nail and was using it (but only when he was alone) to cause some damage around his enclosure. The caregivers knew that he must have some kind of tool, but couldn't find anything. A few days later, the orangutan presented the nail and received a treat for trading it back to one of his caregivers. It seemed as if he hid the nail until he realized that it wasn't particularly useful, and then he traded it for a high-value treat."

"In the same week, one of the chimpanzees somehow got ahold of a metal bolt, but he didn't hide it; instead, he immediately showed his caregiver, knowing that he would get a treat for returning it."

Anyone who has ever worked with large-bodied apes in captivity can attest to their ability to find things they aren't supposed to have or break things in their enclosures!

"Reflecting back on that after years of working with the apes, I know it can be explained a number of different ways, such as individual differences between that particular orangutan and chimpanzee, or prior experience with those items," Meg said.

"Any time we have moved orangutans into buildings that were previously occupied by chimpanzees, we've had to modify the light structures or put in barriers to protect them. The orangutans will find all those 'weak' areas quickly ... honing in on things that chimpanzees didn't seem to notice or bother with for years. Orangutan mischief and curiosity never ceases to amaze me!"

"Speaking broadly, I think orangutans have a bit more patience and tend to think about things longer. Chimpanzees appear to make their decisions more quickly and sometimes impulsively. They use their intelligence in different ways. The biggest differences, though, are individual. There are orangutans who don't bother with light fixtures, and there are chimpanzees who will hide bolts. They're all different, and they're all brilliant."

Feeling Anxious

Grover barked. And barked. And barked. The pitch of his cries increased as he heard the footfalls coming closer to the door. How could such a loud bark come from such a small dog? His lips peeled back, revealing his snarling teeth, and the whites of his eyes showed as he barked with increasing intensity.

"It's okay, buddy," his pet sitter, Brooke, said to him gently as she approached the door. Even when he saw Brooke, the little dog didn't stop barking. Grover knew and liked Brooke, and she'd watched him many times, but any time any person came to the door, Grover barked incessantly like this. This regular crescendo of cries made him very unpopular with Brooke's landlord, and other people weren't exactly knocking down the door to be Grover's pet sitter, either. Complicating the matter was the fact that Grover's owner, Hilary, didn't discourage or redirect the behavior; rather than realizing that the barking was anxiety-based, and working on a way to help him cope with this anxiety, she played up the anxiety and then praised Grover for barking. So, what can be done for a dog like Grover?

When working with an anxious dog, the basic treatment is always the same, but a couple of factors influence the details of the treatment for canine anxiety. The first has to do with the trainability of the dog, and Hart and Hart (1985b) showed that it can be tied to the breed. Of course, there is always individual variability, so this isn't set in stone. The other factor—and this can be the tough one—is the owner compliance in the treatment plan.

The recommended first step is behavior modification and training to eliminate anxiety, specifically, by using counter-conditioning. Recall from Chapter 9 that counter-conditioning involves altering how one, in this case, one's dog, reacts to or feels about a given stimulus. For example, a trainer could pair a positive stimulus, such as praise and calmly petting the dog, with a stimulus that makes a dog react anxiously, such as a delivery person coming to the front door or the sound of fireworks.

If it's an anxiety case in a Beagle, or other "less trainable" dog, then the training ability that is required to effectively modify this behavior has to be at the top level, and might take a while. Clients may need to hire a professional trainer. If they can't afford a professional trainer, and they struggle with the training, then they may need to talk to their vet about anti-anxiety medications. If you have really compliant owners, you may not need the medication. But with average levels of compliance, a Beagle, and high severity, it is more likely that medication is part of the solution.

On the flip side, in cases of Australian Shepherds with the same level of anxiety and same level of owner compliance, it's much less likely that we recommend talking to their veterinarian about medication. This is due to high trainability. If you have an Australian Shepherd with an issue, you can show the owners counter-conditioning, and it works ... because it does not take much training to correct.

We do find individual variability between breeds. Some Australian Shepherds are slow learners, but it's also hard to separate this out from whether that individual dog is having a hard time learning, or if the owner is doing it wrong. Usually, in these cases, it's that the owner is doing it wrong, and not the dog having a hard time learning.

When a dog is learning slower than would be anticipated, it's also a huge indication to go back and look at their medical condition. We have to find out: what's interfering with their learning? What's making it harder to train them? Often, the answer is: pain. If we can correct these issues, we can help them return to normal. How quickly you go to a medication depends upon the specific situation. If the dog is a "slow learner," and the owner is doing everything right, training-wise, it will more than likely be time to use medication.

Conclusion

Of all the topics related to breed differences, training is where breed differences in behavior are the hardest to predict, and it's where we make our weakest argument in this book. There's just so much variability, and there was likely some selection for obedient and potty trainable dogs across breeds. That's not to say that there isn't evidence, it's just that there's not *quite enough* yet, which is one reason why Dr Hare is advocating for more citizen scientists to take the helm with this line of research. While there is some evidence for differences between extreme breeds, such as the Beagle and the Australian Shepherd, we are also finding significant within-breed differences (recall the Labrador Retrievers who were bred to be assistance or detection dogs). So while we see a lot of between-breed differences in temperament and motivation, the same is not the case (at least, not yet!) for learning and training. This can partially be explained because learning and cognition are very complicated and encompass a lot of different parts of the brain. We are dealing with complex gene effects, complex interactions with the environment, and there are also a lot of complications—think, for example, of the variability between owners and the variability between the relationships that dogs have with their people. Consider, too, the difficulty in even defining "trainability" and "intelligence." You might have one owner who recognizes anxiety and helps a dog deal with it, while you might have another owner mistakenly reinforcing the anxiety. Also, different learning/intelligence traits are still being selected for in different breeds, and within a breed.

For dogs, like horses, we have to agree with horseman Gary Pegg that "the infinite combinations of people" and *dogs*, "make any definitive statement that one breed possesses more intelligence than another pretty hard to substantiate."

References

Adams, M.J. (2003) Alphabetic anxiety and explicit, systematic phonics instruction: a cognitive science perspective. In: Neuman, S.B. and Dickinson, D.K. (eds) *Handbook of Early Literacy Research*. Guilford Press, New York, pp. 66–80.

Asp, H.E., Fikse, W.F., Nilsson, K. and Strandberg, E. (2015) Breed differences in everyday behaviour of dogs. *Applied Animal Behaviour Science* 169, 69–77.

Barnes, S.K. (2010) Sign language with babies: what difference does it make. *Dimensions of Early Childhood* 38(1), 21–30.

Dutrow, E.V., Serpell, J.A. and Ostrander, E.A. (2022) Domestic dog lineages reveal genetic drivers of behavioral diversification. *Cell* 185(25), 4737–4755.

Hare, B. and Tomasello, M. (2005) Human-like social skills in dogs? *Trends in Cognitive Sciences* 9(9), 439–444.

Hare, B. and Woods, V. (2013) *The Genius of Dogs: How Dogs Are Smarter Than You Think*. Plume, New York.

Hart, B.L. and Hart, L.A. (1985a) Selecting pet dogs on the basis of cluster analysis of breed behavior profiles and gender. *Journal of the American Veterinary Association* 186, 1181–1185.

Hart, B.L. and Hart, L.A. (1985b) *Canine and Feline Behavioral Therapy*. Lea & Febiger, Philadelphia, PA.

Helton, W. (2010) Does perceived trainability of dog (*Canis lupus familiaris*) breeds reflect differences in learning or differences in physical ability? *Behavioural Processes* 83(3), 315–323.

Junttila, S., Valros, A., Maki, K., Vaataja, H., Reunanen, E. *et al.* (2022) Breed differences in social cognition, inhibitory control, and spatial problem-solving ability in the domestic dog (*Canis familiaris*). *Scientific Reports* 12(1), 22529.

Miklosi, A. (2009) *Dog Behaviour, Evolution, and Cognition*. Oxford University Press, New York.

Serpell, J.A. and Hsu, Y.A. (2005) Effects of breed, sex, and neuter status on trainability in dogs. *Anthrozoös* 18(3), 196–207.

Scott, J.P. and Fuller, J.L. (1965) *Genetics and the Social Behavior of the Dog*. The University of Chicago Press, Chicago, IL.

Udell, M.A.R. (2015) When dogs look back: inhibition of independent problem-solving behaviour in domestic dogs (*Canis lupus familiaris*) compared with wolves (*Canis lupus*). *Biology Letters* 11(9), 20150489.

van Marle, K., Chu, F.W., Li, Y. and Geary, D.C. (2014) Acuity of the approximate number system and pre-schoolers' quantitative development. *Developmental Science* 17(4), 492–505.

11 What We Know, What We Don't, and Where We're Going

Abstract

Chapter 11 summarizes the scientific evidence that breeds are under selection and that breed genetics can affect behavior. While we have some evidence about breed differences, we still don't have enough information to make sweeping statements about all members of a particular breed. This lack of evidence, paired with a propensity to classify dogs based upon physical appearance and assumed recent genetic heritage (breed), rather than on their behavior, is why breed-specific legislation (BSL) is still so dangerous. BSL is breed discrimination; it's canine eugenics. Breed bans and eugenics likely fuel the fire against belief in the possibility of breed-specific differences in behavior. This chapter includes a discussion of the future of this species, including the lasting repercussions of continued inbreeding and welfare implications.

Flux, Flow, and the Future

Romeo. Annie and Charlie. Banjo and Pepper. Keroua and Haza. Bella, Buddy and Elke. Their stories tell us so much about the rich history and potential future of *Canis familiaris*, including the lasting repercussions of continued inbreeding and its welfare implications. Throughout this book, we have asserted that breed and pedigree are just as important as the environment, including epigenetics and early life experiences, in shaping a dog's behavior. While this may have appeared at the onset to be a bold claim, we have provided ample evidence of the interrelationship of these factors to answer the question: is a dog … just a dog? The answer to this is "no," but with an asterisk.

Many, if not all, of the topics that we discussed in this book come back to one important point: we can't define a species, much less a breed. And depending upon their geographic location, breeds continue to be in flux. While we have attempted to quantify breeds as accurately as possible, different organizations recognize different numbers of breeds and breed groups. A breed might be recognized by one organization, but not by another; it's a dynamic process.

This is an exciting time to be studying humans' first friend. Though we've made some exciting discoveries thus far, there's still so much to learn. The study of dogs, and cats, and horses is no longer in the shadow of studying free-living animals alone; we can see the intrinsic value in studying those animals that evolved via natural selection, with minimal human influence, and those animals that we have shaped the most. Those species that are most familiar to us are becoming more attractive for scientific study and so, too, are the replicated studies that will clarify our research questions. It's our hope that this book contributes to a growing body of research that helps us better support and understand our dogs.

While it might seem duplicitous, there *are* breed differences, but breed can only be used as a point of departure to make generalizations and identify patterns. Since the first time that humans and the ancestors of modern dogs met, we have been selecting dogs for different roles. Two hundred years ago, we didn't have the variety in this species that we do today; as we continue to breed for increasingly exaggerated traits, we will see increasing differences between dog breeds. This sparks a conversation about the moral and ethical issues pertaining to breeding dogs that have health issues for the sake of aesthetics and status.

Much of the scientific literature about the differences between dog breeds contains sound but often non-replicated methodologies, making it difficult to compare results. As research continues to come out, it's important to pay close attention to their methods. If two studies appear to provide conflicting answers, we must determine: are the methodologies comparable, or do they differ so much in their definitions and methods that you can't compare the

 Breed Differences in Dog Behavior: Why Tails Wag Differently. (R.L. Ha, T.L. Brad and J.C. Ha)
DOI: 10.1079/9781800624566.0011

two? We'd like to see more studies that use direct observation of dogs rather than surveys of owners to determine behavioral differences, as well. Otherwise, we may only be measuring breed bias instead of real differences.

What We Know

Ancient breeds are genetically and behaviorally distinct from modern breeds

In this book, we answered the not-so-simple question "What is a dog?" in the best way that we know how. No other known species has as much morphological diversity as *Canis familiaris*, and that's almost completely due to human influence. Dogs are our first friends, and we have molded them to be our perfect companions, while they have influenced our development, as well. The genus *Canis* is the most abundant of the terrestrial carnivores and comprises species and subspecies including coyotes, the Dingo, jackals, wolves, and dogs. Modern domesticated dogs did not descend from what we know as modern wolves; they define a monophyletic group (a group of organisms consisting of all descendants of a common ancestor) that is sister to Old World wolves (Freedman *et al.*, 2016). Thus, the two modern species are sister taxa and share ancient ancestors in common. Even within the diverse family of Canidae, the domestic dog is unique in the animal kingdom in purpose and in the influence of artificial selection on creating breeds (Wayne and Ostrander, 2007). Wild canids range from the 8-inch-tall fennec fox to the 32-inch-tall gray wolf, but domesticated dogs vary even more, ranging from the 5.9-inch-tall Chihuahua to the massive 44-inch tall Great Dane. Their extreme phenotypic diversification has yielded 400—and by some accounts, 1000 or more!—distinct dog breeds (Mehrkam and Wynne, 2014). Dogs also have a wider variety of coat lengths and colors, leg lengths, face shapes, tail lengths and curvatures, ear positions, and body types than their wild canid cousins do. In fact, dogs have more phenotypic diversity than all of the other species of carnivore (Freedman *et al.*, 2016).

Ancient dog breeds predate their documentation, but the first historical record of distinctive dog breeds appeared around 3000–4000 years ago (Brewer *et al.*, 2001). Many of the breed groups we know today had been identified by the Roman period, and Europe saw a sizable propagation of breeds during the Middle Ages (Clutton-Brock, 2016). Estimates on the origin of dog breeds and breed types range from ancient times (breeds that are thousands of years old) to more modern ones, such as those created during the Victorian era (1830–1900). Dogs have a unique demographic history and have experienced multiple breed-specific bottlenecks, which in turn has left genetic signatures in their genomes: many modern breeds have genomic signatures from early ancestors (Ostrander *et al.*, 2017). A 2010 study examined 48,000 single nucleotide polymorphisms in 85 breeds. The researchers found that modern dogs, including the Herding Dogs, Mastiff-type breeds, Retrievers, Scenthounds, Sighthounds, small Terriers, Spaniels, Spitz breeds, and Toy Dogs, had distinct genetic clusters that corresponded to their function or phenotype. The study also found 13 breeds that genetically diverged from modern ones: the Afghan Hound, Akita, Alaskan Malamute, American Eskimo Dog, Basenji, Canaan Dog, Chinese Shar-Pei, Chow Chow, Dingo, New Guinea Singing Dog, Saluki, Samoyed, and Siberian Husky (vonHoldt *et al.*, 2010). The researchers found evidence of three dog groups that were distinct and divergent from modern ones: an Asian group, which included admixture with Chinese wolves, and included the Akita, Chow Chow (Fig. 11.1), Dingo, New Guinea Singing Dog, and Shar-Pei; a Northern group, which included the Alaskan Malamute and Siberian Husky; and a Middle Eastern group, which included the Afghan Hound and Saluki (vonHoldt *et al.*, 2010).

Genes affect behavior, and these can vary between breeds

Breeds were created by selecting individuals with the desired characteristics and breeding them to other individuals with those characteristics, and repeating that for multiple generations. Unfortunately, this sometimes leads to inbreeding, and some breeds, such as Bull Terriers and Dalmatians, have developed issues related to a lack of genetic diversity. Fortunately, some breeders of inbred lines have attempted to reduce the likelihood that offspring will have genetic abnormalities by "outcrossing" or "outbreeding" dogs that do not share a common ancestor within the last four generations or more. These efforts are critical for both physical and behavioral health of dog breeds. Unfortunately, the focus has only recently been on the importance of breeding for behavioral health, but it is clear from

Fig. 11.1. Chow Chow, Asian Group. Photograph courtesy of Remigiusz Józefowicz. The image is licensed under CC-BY-SA license and is provided by Wikipedia. Available at: https://commons.wikimedia.org/wiki/File:ChowChow2Szczecin.jpg.

the dog genome work that genetics influences behavior in complex, but undeniable ways, and that selection acts on the genetic underpinnings of behavior.

Despite the difficulty of predicting the behavior of one individual due to recessive traits, mutations, and the influences of the environment on genetic expression, most individuals demonstrate behavioral traits that are highly similar to the rest of the population (at the species level), as well as the breed level (e.g. the friendly Golden Retriever), as well as one or both parents. Overall, it is better to think about the likelihood of genetic predispositions or tendencies given a species, breed, and an individuals' parents. This information, while imperfect due to the complexities and exceptions discussed here, is generally useful and accurate, and can be a guide in selecting a puppy or helping with a behavioral issue.

The environment and genes interact in complex ways to produce unique individuals

Genetics, environmental influences, including epigenetics, temperament, and early life experiences, work together to create the variation that we see within individuals. These disparate sources influence our ability to interpret species and breed differences, as well. Individual variation is particularly important to the story of the domesticated dog—how and why they entered our lives, and what perfect storm of factors likely led to that. Traits such as curiosity and courage likely characterized the ancestors of the canines who first came to share their lives with us. Individuals that had those traits were likely attracted to human camps and were more likely to mate with one another. Their pups were also more likely to share those traits with their parents. Those that didn't likely moved off and away from humans. Individuals with these similar phenotypes mated with one another. Even after humans began directly manipulating which dogs bred, and creating breed standards, there were also often subgroups, or "lines," within one breed that differed in coat color, temperament, and behavior. Poodles are a great example of this. In addition to being grouped by size, Poodles can only be registered if they have certain coat colors, and those colors are required to be uniform throughout (Fig. 11.2). Coat color, paired with size restrictions, has fragmented the genetic diversity of Poodles into five distinct groups. The genetic patterns observed in the eight comparison breeds indicated that fragmentation, driven by breed standards, is likely common among many dog breeds.

Fig. 11.2. A silver Miniature Poodle. This file is licensed under the Creative Commons Attribution 2.0 Generic license. Available at: https://commons.wikimedia.org/wiki/File:Silver_Miniature_Poodle_stacked.jpg.

You can contrast the example of fragmentation in the "Poodle" with lumping of dogs considered "Pit Bulls." "Pit Bull" is actually an umbrella category for several different "bully" breeds, many of which have been bred for different purposes. The data on purpose-bred Pit Bulls is difficult to find, however, as breeding Pit Bull-type dogs to fight other dogs is both unethical and illegal. Thus, while we know that some Pit Bulls are bred for show or companionship and others for fighting, we're only able to comment on these trait-level differences anecdotally. What we do know, given data from other lines of dogs, is that dogs who have been bred for a certain purpose will generally act differently from dogs who were bred for another purpose.

Breeds vary in temperament

Indeed, a recent paper used the Canine Behavioral Assessment & Research Questionnaire (C-BARQ) survey and the dog genome to examine breed *average* phenotypes (MacLean *et al.*, 2019). The researchers used data from more than 14,000 dogs representing 101 breeds, finding a high level of among-breed heritability for 14 behavioral traits. "Breed average phenotypes were associated with certain regions of the genome," Dr Serpell explained. "There were 131 different regions of the genome that were strongly associated with the differences between the different breeds." These 131 different regions account for approximately 15% of a dog breed's personality, with the most heritable traits being trainability, aggression toward strangers, and chasing. "Most of these genome regions are known to be expressed strongly. This is exciting because it suggests that there are likely to be genes that are strongly influential, in terms of behavior differences. This is the kind of study that you can't really do in other species, because many dogs have been bred for behavior—for a particular kind of activity: barking, hunting, or fighting, all the uses that a dog has been put to. We have this really unique animal that you can do these kinds of studies in and reveal genetic associations that would be hard to track down." Given our genetic manipulation of dogs, it makes sense that there will be differences in temperament between breeds and breed groups.

Artificial selection has produced significant genetic differences between breeds

Visually, a German Shepherd Dog and a Miniature Pinscher, or a Collie and a Bloodhound, are strikingly dissimilar, but just how different *are* dog breeds from one another? While variation can arise within populations, it usually takes thousands of years—or longer—for speciation to occur. Most dog breeds originated during the Victorian era, falling far short of that typical timeline, although humans have accelerated this differentiation process. During speciation, a slow accumulation of mutations causes heritable changes to the phenotype. Those mutations that confer an advantage for individuals will be selected for, while those who don't have advantageous mutations would have lower rates of survival, and subsequently, their lineages would have had decreased rates of reproductive success. For the domesticated dog, breeders often chose individuals with the most exaggerated traits and bred them with other dogs who were similarly unusual. When a mutation arose that owners found novel or aesthetically pleasing, they ensured the proliferation of this trait through selective breeding. Many of these traits, such as brachycephalic faces and truncated limb length, would have been deleterious for these dogs in a free-living situation, but given human assistance (including Cesarean sections for many breeds), they're allowed to propagate. While they aren't different species—yet—their current trajectory is toward speciation if we continue to make breeds more dissimilar from one another.

Most differences in dogs, even among the most dissimilar within the species, are driven by relatively few loci, or regions, in the genome. These loci have a large phenotypic effect, yielding strong variation among breeds. Researchers have found 14 single nucleotides that are breed-specific in 60 breeds of dogs, and molecular variance indicated that the among-breed variation accounted for more than 27% of the total genetic variation, and genotypes could be used to correctly assign 99% of the dogs to specific breeds (Parker *et al.*, 2004). Even with the relatively short (evolutionarily speaking, of course) timespan between breeds emerging from ancestral dogs, the study found that dog breeds are distinct and that individual dogs can be assigned to breeds based upon their genotypes.

Breeding for a positive trait sometimes leads to another unrelated neutral or negative trait

Genes that are close together on the same chromosome are more likely to stay linked together when the recombination of male and female DNA result in a fertilized egg. Genes that are next to one another may be coding for very different traits, such as a physical trait and a behavioral trait. This can result in some interesting inheritance patterns where particular traits tend to co-vary with one another. This linkage between behavior and morphology is exemplified in domesticated animals, where the domestication of mammals appears to produce similar differences in appearance and behavior. Scientists have shown that breeding for "tameness" can result in morphological differences as well. In fact, the changes in morphology are similar *across* species, and include traits such as floppy ears and wavy or curly coats. This has certainly been the case in mammalian studies of domestication and suggests that something about "tameness" is genetically linked to some physical traits. We also see the same kind of genetic linkage when selecting for long bodies and inadvertently producing dogs with hip dysplasia.

We've adapted the social behavior of some breeds to better suit our needs

Zimen's work in the 1960s and 1970s made it clear how different domesticated dogs are from wolves, and more recent work suggests this even between breeds of domesticated dogs. We can see the effects of selective breeding on social behavior when a Pug affectionately warms their person's lap, a Border Collie gets down low and herds our livestock, a Rottweiler vigilantly patrols their people's home, a Dalmatian provides companionship to the firefighters of a firehouse, a Cairn Terrier ferrets out a fox, a Greyhound flattens out and glides across the ground, or a Labrador Retriever gently brings back a duck. All of these behaviors better suited the needs of humans, but the partnership between human and dog suited dogs, as well. None of these behaviors would have been adaptive in the absence of humans, and they are all distinctly different from the behavior of wolves, and, logic would follow, from the behavior of the last common ancestor that wolves and dogs shared.

The relationship between breed and aggression is complicated because "aggression" is a complex topic—there are many types of aggression and aggressive behaviors are often context-specific

Early in the domestication process, canines that demonstrated friendliness and lower levels of reactivity were selected to breed with one another and share our homes, but aggression wasn't "bred out" of dogs; many dogs had early roles, such as hunting, herding, and guarding, in which aggression gave them an advantage. While aggression is often considered to be one thing, there isn't just one type of it; there are numerous context-specific instances of aggression. For dogs, most people are thinking about either human-directed aggression or dog–dog aggression, and there is some evidence that this varies among breeds.

Breed differences in training, if any, are subtle, and may relate more to motivation

Unlike aggression, the data on learning, training and problem solving does not support breed differences. The best work in science is typically well-controlled studies, like those of Scott and Fuller (1965), yet they found relatively few differences between the breeds they studied, despite great effort in controlling the environment and upbringing. Overall, they suggested that the next step to uncovering breed differences in these areas, might be to evaluate breeds commonly believed to be highly intelligent, such as Poodles and Border Collies. The tests they conducted with these limited breeds suggest an average intelligence across breeds.

Most of the work that has been done subsequently, has been based on surveys of owners, etc., about what breeds they think are the easiest to train, most intelligent, etc. Differences have been found in these areas, but direct observational work by scientists of whether breed differences occur is harder to come by. One reason for this is the expensive nature of the work done by Scott and Fuller (1965), and another is that motivation factors can be intertwined with "intelligence" measures, thus complicating the interpretation of any differences that are discovered.

Going Rogue

You can make breed generalizations, but breed is only one point of departure from "general" canine behavior. We'd be remiss if we failed to mention the example of the Rogue Detection Team dogs, canines of varied genetic backgrounds that all share one important characteristic in common: high drive. The dogs of the Rogue Detection Teams exemplify both breed group differences *and* individual differences and the importance of individual variability. The Rogue Detection Teams, based out of Rice, Washington, have a mission to bring the conservation detection dog method to the forefront of scientific research. Their high-drive dogs and human partners, called "bounders," have circumnavigated the globe in the name of conservation. From Nepal to Malaysia, from Southern California to Southern Africa, and from British Columbia to France, the Rogues have reached some of the world's most remote ranges and as of this writing, have more than 75 years of collective experience. These aren't purpose-bred, pedigreed pups, though: the Rogue dogs were all rescued from shelters (Table 11.1 and Fig. 11.3). Some of them, like Pips, had been returned to the shelter up to six times before their particular set of skills made them the perfect partners for their human bounders (think of them as the Liam Neesons of the canine world, with his famous line "But what I do have are a very particular set of skills" in the movie *Taken*).

"The Rogue dogs tend to be a lot of the working breeds, because of their obsessive, high-energy behavior, but they're also dogs that don't look like they'd be working dogs," explained bounder Jennifer

Table 11.1. The Rogue Detection Team dogs.

Dog	Breed	Age (years)	Weight (pounds)
Beckett[a]	Ewok/no idea	10	23
Cricket	Border Collie mix	3	43
Dio[a]	Australian Cattle Dog	11	35
Filson	Australian Cattle Dog	9	36
Hugo	German Shepherd	4	Large
Indy	Australian Cattle Dog	3	30
Jack[a]	Australian Cattle Dog	11	45
Jeckyll	Australian Cattle Dog	3	41
Maple	German Shepherd Dog mix	3	56
Pips[b]	Australian Cattle Dog	14	37
Ranger	Herding Dog mix	6	Large
Violet	Pit Bull mix	3	45
Whisper	Australian Shepherd	5	44
Winnie[b]	Black Lab/mix	12	54

[a]Semi-retired. [b]Retired.

Fig. 11.3. A Rogue Detection Team dog. Photograph courtesy of Holly Cook Photography, LLC.

Hartman. Take Beckett, for example: he's a small, black, fluffy dog, who looks like he's equal parts Ewok, Teddy Bear, and Papillon. Not exactly the hardy, working type. "When we're looking for a Rogue Detection Dog, we're looking at personality traits. Our selection process isn't breed-based," Jennifer said. "In the shelter world, our dogs are typically 'unadoptable' due to their high-energy, obsessive desire to play fetch—but this obsessive energy is perfect for us, because we pair this energy with detecting an odor, and reward our dogs with their ball for locating the odor. They get to do the one thing that makes them most happy, and in the process, we're able to survey across vast and unique landscapes to locate cryptic and elusive odors for conservation initiatives. We adopt the misfits. These are 'bad' dogs for a good cause: 'bad' because they didn't thrive in a home environment, but they do thrive in the great outdoors, doing what they love with their people."

The Rogue Detection Teams adopt any breed as long as they are obsessed with playing fetch. Dogs tend to be between 3 and 4 years old, with long lives ahead of them. "Some people are surprised by that because they assume that we'd want to work with puppies to imprint them while they are young. What we've found, though, with the obsession to play fetch, is that while a puppy can be playful, there are only a few dogs who grow up to be obsessed with fetch," explained Jennifer. Obsessive fetch behavior typically manifests when the dogs are older and out of puppyhood. "Because we look in shelters, most of our dogs tend to be mixes and mutts of all shapes and sizes. It's really the individual that matters, not the breed, because there are so many dogs in shelters who would love to hike, explore, sniff, and play all day. They are the diamonds in the rough, and while they may take a little extra to search for, once we bring them into our pack, we never look back."

"We're always outsiders, even within the biology world," said Jennifer. "The traditional method for tracking animals is using a radio collar or trap, but we use scent detection work to find them."

And that non-traditional approach—using high-drive dogs rescued from area shelters—has yielded extraordinary results; the dogs can detect a diverse range of species on land, sea, and air, including the wolverine and the gray wolf, the Southern resident killer whale and the transient killer whale, the storm petrel and the silverspot butterfly, the rubber boa and the cascade red fox, the Western toad and the wild boar, the African lion and the giant armadillo, the clouded leopard and the Crau plain grasshopper, and more than three dozen other animal species, so far. The dogs are additionally able to detect Western X disease (or little cherry disease), and polychlorinated biphenyl toxins, as well as ivory, shark fin, wildlife carcasses, and snares, which is proving to be invaluable in the fight against animal trafficking.

We love the Rogue Dogs and their work, and we love that they demonstrate an important point for this book. Note that the *majority* of these dogs come from working breeds, and particularly the high-energy ones. That fits with the breed differences we've been talking about. You can also see that there are exceptions to the rule, however, like Beckett, the "Ewok". Some dogs have high drive personalities that have nothing to do with breed. This is what makes some people think that breed differences aren't real. Of course, we would argue that the large number of working breeds points to the relationships we're talking about. Breed is a factor.

Threads and Inklings: Where Is the Direction of Dog Research Going?

From the earliest days of dog research, we've taken a lot of wrong turns, met with evolutionary dead ends, and found ourselves on unexpected detours when we didn't know which questions to ask—or how to ask them. But as researching our closest companions has gained traction, resulting in an increasingly impressive body of canine literature, we have righted the course of canine research. History and science alike are filled with fateful missteps. Sometimes, one wrong turn can change the shape of history, but with each miscalculation and dead end, we can recalibrate and determine a new route.

The driver realized too late that he'd made a wrong turn. The motorcade had been traveling at a fast clip, trying to avoid any potential threats after a bomb had missed its mark earlier that day. In their haste to pay a visit to an officer who had been wounded in the bombing, they had mistakenly turned down tiny Appel Quay instead of taking a main road. It was there that 19-year-old rebel Gavrilo Princip recognized the Archduke and his wife. Princip raised his pistol and fired two shots at point-blank range. Within minutes, Archduke Franz Ferdinand and his wife, Sophie, died from their injuries. Ferdinand was the heir presumptive to the Austro-Hungarian throne and the Habsburg Empire. His death during an official visit to Sarajevo on June 28, 1914 would light a spark that eventually flamed into the outbreak of World War I that August. That same year, the American Kennel Club (AKC) recognized its 73rd breed into its registry: the Welsh Springer Spaniel.

There have been two constants in our lives: our love of dogs and the imminent threat of warfare. Over the course of our evolutionary histories, war has shaped both of our species, and we have shaped one another. The Barbet and the Dogo Argentino had been recognized as unofficial breeds for a number of years, but their official recognition added validity to their status. Over the course of the last century, we have added 122 breeds to the AKC's registry, with each breed having its own official standards. But how much have we really learned about the different breeds and breed groups?

Lines within breeds (future work)

While we have some evidence about breed differences, we still don't have enough information to make sweeping statements about all members of a particular breed. This lack of evidence, paired with a propensity to classify dogs based upon physical appearance and assumed recent genetic heritage (breed) rather than on their behavior, is why breed-specific legislation (BSL) is still so dangerous. BSL is breed discrimination; it's canine eugenics. Many breeds, such as Golden Retrievers, are bred for certain vocations (e.g. show or field), and in recent years, dogs falling under the "Pit Bull" umbrella have been bred for show and, unfortunately, for fighting. This has been used to bolster BSL arguments, but again, we lack the scientific evidence for differential temperament based upon different breeding; this remains anecdotal.

Throughout this book, we have been discussing differences between dog breeds, but we would be remiss to not mention the differences between lines of dog breeds, as well. We have been selecting particular lines of one breed for so long that we are now witnessing distinct differences between lines, for example, show dogs versus field dogs. A survey of dog owners in the UK revealed that a Labrador Retriever's working status, as pets, gun, or show dogs, was the most significant predictor of their behavior (Lofgren *et al.*, 2014). Specifically, in comparison to both of the other lines, the pet lines were more likely to show the following characteristics: agitation when ignored, barking and excitability, human-, object-, and noise-fear, non-owner aggression, and unusual behaviors (also referred to as stereotypies), as reported by their owners. In comparison to one or both of the other lines, gundogs were reported as more likely to exhibit: agitation when ignored, attention seeking, fetching, and human and object fear, as well as trainability. Show dogs were reported as less trainable than either pets or gundogs, and more likely to show unusual behaviors than gundogs (Lofgren *et al.*, 2014). Another UK survey of dog owners reported that working Labrador Retrievers and Border Collies were more interested in the environment and easier to train than show Labrador Retrievers and Collies (Fadel *et al.*, 2016).

The differences between American and UK Labrador Retrievers are not recognized by any of the major national breed clubs (e.g. the AKC or the UK Kennel Club); there is only one standard for the Labrador Retriever. In practice, the American Labradors have been bred for hunting and competing in field trials, where high energy, drive, and stamina are desired for champions, whereas English Labradors have been bred for conformation (show). Unfortunately, we could find no scientific research looking at any behavioral differences between these lines. This is likely due to the logistics of getting a large sample of each line in one geographical region.

It's difficult to interpret the owner-reported differences between these lines of dogs because behavior is influenced by both genetics and the environment, and the expectations of the owners for gun or show dogs may be different than those of pet dog owners. So these results could be driven by owner expectation or by relaxed selection (breeding requirements) for pet dogs, compared to dogs bred for task-specific work. However, a genomic study of UK Labrador Retrievers found a genetic split between working (field and gun) and show (conformation) lines, and pet dogs were a mixture between the two lines (Wiener *et al.*, 2017). These specific differences included some areas of the genome that have previously been associated with craniofacial morphology and some areas associated with neuronal or neurological function. Given the current limitations of behavioral genetics work, the authors in this study were not able to confirm that the differences they found that were associated with the brain had a direct impact on behavior. Other studies based on surveys of owners have found differences in impulsivity between working and show lines (Fadel *et al.*, 2016). While the link between genetic differences and behavioral differences in breed lines has not been directly made in a rigorous scientific study, the cumulative evidence strongly suggests that at least part of the differences seen between lines are the result of breeding and genetics.

There may be two separate, distinct populations of Pit Bull-type dogs. The first type is the normally inhibited, wonderful family dog that many people know, and the second type is the potentially dangerous dog that lacks, or has markedly reduced, inhibition of aggression and bite. This second type most often shows up very clearly in dogs that were shipped in from areas with a high incidence of dog fighting activity who were later adopted from a shelter. These dogs are far more likely to exhibit reduced inhibition than dogs surrendered locally from, say, a family breeding situation, or of course, from a show breeder. Family dogs and show dogs have the "pittie" temperament that most of us are familiar with. Dogs bred for fighting other dogs, however, were selected over generations to breed with one another because they had comparatively lower rates of inhibition and higher rates of reactivity. On top of that, they haven't been obedience trained, and then they are often abandoned. These are the dogs that can have the potential to become dangerous—it's the combination of breeding, lack of training, and lack of socialization.

The most valuable tool to distinguish between these two lines of Pit Bulls is an assessment for inhibition. For instance, you can assess their ability to learn a new task, since learning requires well-developed inhibition ("Don't do what I want, do what I am supposed to do to get the reward"). If there has been an aggressive incident, bite severity reflects inhibition: did the dog "pull its punch," or did they go all out? The idea of trait-level characteristics, genetic selection, and the development of "lines" within a breed is an important one for understanding these sharp differences in what superficially may appear to be similar animals.

Timing of teens

There is anecdotal evidence that breeds differ in the timing of adolescence. There appears to be a size relationship where larger breeds are slower to mature. We think it has been shifted earlier in smaller and highly derived breeds. This warrants more study as the timing of adolescence has clinical implications for behavioral issues in dogs. Dogs that lack inhibition and/or are unruly when they are young, are often improved by time. At about 18–24 months of age, brain development and maturation become complete and these dogs show significantly improved control over their behavior. Your training is more likely to "stick" after that point, and J.C.H. has had many dog clients that were relieved to discover that their dog was indeed trainable and a better companion for their family after they "settled" down. It is also notable that all professional dog training organizations (e.g. service dogs) do their specialized training and assignments after 18 months, and most of them wait until 2 years of age. This makes sense if trainability and impulse control is not mature until up to 24 months in dogs.

Littermate syndrome

This concept refers to the tendency for aggression between littermates that are adopted together. This phenomenon is fairly well known in the dog world, but we need to study this to determine if the aggression is just chance between littermates or just aggression that would happen anyway. It's possible that being littermates is "unnatural" in that it wouldn't occur in the wild where sexes disperse for breeding. So, it could be a frustration of a drive to disperse from littermates. It may be particularly bad between males who may be naturally competitive for access to females. There is some suggestion that this syndrome may vary between breeds. For example, it happens with Huskies, but not with Poodles.

Behavioral genetics

In 2008, Tyrone Spady and Elaine Ostrander, a leader of the Canine Genome Project, wrote a paper summarizing what we knew at the time about breed-specific behavior, and suggested that the future would likely yield more examples of breed-specific behavior, given the success in mapping and sequencing the dog genome (Spady and Ostrander, 2008). This has certainly come to pass in the scientific literature, but has not been widely accepted in the public.

MacLean *et al.* (2019) used the C-BARQ survey to suggest not only a genetic basis for traits like personality, but significant breed differences in behavior, as well. There were 14 significant breed differences in behavior, as shown in Table 11.2. These differences were found at the breed level, and do not evaluate individual level differences. Additionally, they found that genetic polymorphic loci (single nucleotide polymorphisms) were "disproportionately expressed in the brain and involved in pathways related to the development and expression of behavior and cognition" (MacLean *et al.*, 2019).

Table 11.2. Behavioral differences between breeds.

Behavioral measure	Heritability estimate
Trainability (frequency)	0.73
Stranger aggression (intensity)	0.68
Chasing (severity and frequency)	0.62
Attachment/attention seeking (frequency)	0.56
Dog aggression (intensity)	0.52
Stranger fear (intensity)	0.50
Owner aggression (intensity)	0.49
Touch sensitivity (intensity)	0.48
Dog rivalry (intensity)	0.47
Energy (frequency)	0.46
Separation problems (frequency)	0.46
Excitability (frequency)	0.42
Dog fear (intensity)	0.40
Non-social fear (intensity)	0.34

Created from data of MacLean *et al.* (2019).

Heritability refers to the percentage of the variation in the trait caused by genetics. Of the behaviors listed in Table 11.2, trainability, stranger aggression, chasing, and attention seeking were the most heritable, with heritability estimates of 56% or higher. These behaviors are notable in that they are likely the ones targeted in the creation of modern breeds, and they line up nicely with behavioral patterns found in functional breed groups (MacLean *et al.*, 2019).

Which breeds are most likely to have behavioral effects? This depends upon how long they have been isolated, how long the breed has been around, and how extreme the artificial selection was.

Natural selection of domesticated species

Domesticated dogs are also finding their way into the wild as free-ranging dogs that are not dependent on humans for their survival. Studies are ongoing in regard to their impact on other species (as predators) as well as whether they behave adaptively, as their canine cousins would (MacDonald and Carr, 1995). Early evidence suggests that ecology shapes their social group sizes and interactions, but whether these influences are affected by breed is not known (MacDonald and Carr, 1995). The study of feral dog populations is complicated by human interference, including poisoning and shooting of feral dogs, as well as the influx of stray dogs into feral dog populations, and the potential mixing of dogs with the Dingo, wolves, etc. Most populations are not truly independent of humans, and rely on scavenging. There is evidence, however, of a truly feral population of dogs in the Galapagos, so there is the potential for natural selection to occur (Reponen *et al.*, 2014). Future study may determine whether behavioral differences between breeds affect the success of such populations.

Where do we go from here?

One obvious solution to filling in the speculation with facts, and the partial evidence with more data, is more rigorous science. If we could repeat the work of the observational studies of Scott and Fuller (1965) today, with controlled study of individual dogs, across many breeds, and raised in a humane indoor and outdoor facility, then we could measure individual and breed differences across many aspects of behavior and behavioral development. This type of study requires a great deal of planning, oversight, and funding to do well, and the resources are typically not available. Hence, there are more studies being done with surveys rather than direct observations of dog behavior by trained scientists. Unfortunately, one inkling we have right now is that most breeds will become more inbred. This is likely because it takes a lot of effort to outbreed, particularly given that lines are already limited and somewhat inbred now. We can also hypothesize, however, that population sizes of mutts will decrease due to shelter work and spay neuter efforts.

Conclusion

Clearly, behavior varies by individual and by breed. You probably wouldn't use a Bloodhound as guard dog, or a Teacup Poodle for search and rescue or agility. But training and cognition does not seem to vary by breed! Maybe because, unlike "guarding" and "herding", we didn't artificially select for "smart"!

We *do* have evidence for breed differences between working lines and show/pet lines, We also have some studies showing breed differences in aggression (human directed and dog directed) in some common breeds, but the methods vary. It's also clear that there are some temperament or reactivity breed differences. Again, the caveat is that there is a limited number of breeds studied and their methods vary. And, while we indicated that we don't have any evidence of training/learning differences between breeds, canine researchers could take Scott and Fuller's advice and compare Border Collies to some other breeds before we are sure of that finding.

And, while we have some evidence about breed differences, we still don't have enough information to make sweeping statements about all members of a particular breed, and given the important influences of early environment and individual variation, that should not be surprising.

In this, the final chapter, we tied everything together. We'll be the first to admit that we didn't tie it with the tidiest—or the tightest—knot, but that's because we still have so much to learn. As long as we struggle to define a species, and as long as we disagree on discrete categories like breed groups, we'll still struggle with defining breeds and breed differences.

Throughout this book, we have attempted to quantify breeds as accurately as possible. The processes of artificial selection we're using on dog breeds is like natural selection on steroids. We already have dogs that are so dissimilar in behavior, appearance, and recent selection that they appear to be different species. If you follow that logic through, and some breeds become isolated from other breeds or intermix with wolves or coyotes, it's highly likely that some breeds will diverge enough from *Canis familiaris* to become new species. So, if we could time travel, which of the domesticated dog breeds might become a new species?

Evolution involves descent with modification, but it's a one-way street: it isn't a process where you can go backwards. Once a species or breed goes extinct, you can't "back-breed" to resurrect them. Just look at the attempts to bring back the woolly mammoth, using modern elephants and woolly mammoth DNA. Since 2015, Harvard geneticist George Church has been working on this project, attempting to create a "mammophant." His laboratory has isolated 44 woolly mammoth genes and they're attempting to use gene editing and an artificial womb made from vascularized decidua stem cells (Dormehl, 2018).

"We have recently shown that we can create over 13,000 edits per genome," Dr Church stated. "But this isn't directly relevant to the type and number of mutations that we seek in elephants. We have also found a way to get more accurate reading of ancient DNA."

Woolly mammoths thrived during the Pleistocene epoch, disappearing from the planet approximately 4000 years ago, some of which were later discovered in the Siberian permafrost. Let's just not refer to the experiment as Pleistocene Park. While the impending return of the mammoth has been heralded over the past few years, it hasn't happened yet, and Dr Church isn't the first scientist to attempt this feat. To be clear, however, this wouldn't be a mammoth; it would be a hybrid, our closest approximation to the mammoths of ages past. Endangered species often have that classification because they have limited opportunities for genetic variation: as their populations continue to dwindle, the likelihood of breeding closely related individuals increases. We're looking at a similar problem with inbred lines of dogs; as we continue to breed closely related individuals, we diminish the strength of that gene pool, increase the risk of deleterious genetic mutations, and thus risk losing that breed. The more we learn about dogs, the more we must strive to be responsible stewards of this species. This begins with responsible breeding, including cessation of breeding animals with known health issues, actively discouraging backyard breeding, increasing owner education, and having policies that reflect the latest scientific findings.

In *On the Origin of Species*, Darwin (1859) stated, "I think it inevitably follows, that as new species in the course of time are formed through natural selection, others will become rarer and rarer, and finally extinct. The forms which stand in closest competition with those undergoing modification and improvement will naturally suffer most." Many dog breeds have already become extinct, and others, such as the Bull Terrier, have changed so much that they are unrecognizable in comparison to the founding members of their breeds. It logically follows that some breeds will continue to become rarer, and eventually go extinct, while others will likely speciate. We have learned so much from the fossil finds of canids from sites such as Koster and Stilwell II. In 20,000, 30,000 or 50,000 years from today, what breed or species of canid will be accompanying the unearthed skull found at the site of "Washington I?"

References

Brewer, D., Clark, T. and Phillips, A. (2001) *Dogs in Antiquity: Anubis to Cerberus, the Origins of the Domestic Dog*. Aris and Phillips, Warminster, UK.

Clutton-Brock, J. (2016) Origins of the dog: the archaeological evidence. In: Serpell, J. (ed.) *The Domestic Dog: It's Evolution Behaviour, and Interactions with People*. Cambridge University Press, New York, pp. 7–21.

Darwin, C. (1859) *On the Origin of Species by Means of Natural Selection, or, The Preservation of Favoured Races in the Struggle for Life*. J. Murray, London.

Dormehl, L. (2018) Harvard is getting ready to resurrect the woolly mammoth. *Digital Trends*. Available at: www.digitaltrends.com/cool-tech/harvard-woolly-mammoth-2018-update (accessed 11 December 2023).

Fadel, F.R., Driscoll, P., Pilot, M., Wright, H., Zulch, H. *et al.* (2016) Differences in trait impulsivity indicate diversification of dog breeds into working and show lines. *Scientific Reports* 6, 22162.

Freedman, A.H., Lohmueller, K.E. and Wayne, R.K. (2016) Evolutionary history, selective sweeps, and deleterious variation in the dog. *Annual Review of Ecology, Evolution, and Systematics* 47(1), 73–96.

Lofgren, S.E., Wiener, P., Blott, S.C., Sanchez-Molano, E., Woolliams, J.A. *et al.* (2014) Management and personality in labrador retriever dogs. *Applied Animal Behaviour Science* 156, 44–53.

MacDonald, D.W. and Carr, G.M. (1995) Variation in dog society: between resource dispersion and social flux. In: Serpell, J. (ed.) *The Domestic Dog: It's Evolution Behaviour, and Interactions with People*. Cambridge University Press, New York, pp. 199–216.

MacLean, E.L., Snyder-Mackler, N., vonHoldt, B.M. and Serpell, J.A. (2019) Highly heritable and functionally relevant breed differences in dog behaviour. *Proceedings of the Royal Society B: Biological Sciences* 286(1912), 20190716.

Mehrkam, L.R. and Wynne, C.D.L. (2014) Behavioral differences among breeds of domestic dogs (*Canis lupus familiaris*): current status of the science. *Applied Animal Behaviour Science* 155, 12–27.

Ostrander, E.A., Wayne, R.K., Freedman, A.H. and Davis, B.W. (2017) Demographic history, selection and functional diversity of the canine genome. *Nature Reviews Genetics* 18(12), 705–720.

Parker, H.G., Kim, L.V., Sutter, N.B., Carlson, S., Lorentzen, T.D. *et al.* (2004) Genetic structure of the purebred domestic dog. *Science* 304(5674), 1160–1164.

Reponen, S.E., Brown, S.K., Barnett, B.D. and Sacks, B.N. (2014) Genetic and morphometric evidence on a Galápagos Island exposes founder effects and diversification in the first-known (truly) feral western dog population. *Molecular Ecology* 23(2), 269–283.

Scott, J.P. and Fuller, J.L. (1965) *Genetics and the Social Behavior of the Dog*. The University of Chicago Press, Chicago, IL.

Spady, T.C. and Ostrander, E.A. (2008) Canine behavioral genetics: pointing out the phenotypes and herding up the genes. *American Journal of Human Genetics* 82(1), 10–18.

vonHoldt, B., Lohmueller, K.E., Han, E., Parker, H.G., Quignon. P. *et al.* (2010) Genome-wide SNP and

haplotype analyses reveal a rich history underlying dog domestication. *Nature* 464(7290), 898–902.
Wayne, R.K. and Ostrander, E.A. (2007) Lessons learned from the dog genome. *Trends in Genetics* 23(11), 557–567.
Wiener, P., Sánchez-Molano, E., Clements, D., Woolliams, J.A., Haskell, M.J. *et al.* (2017) Genomic data illuminates demography, genetic structure and selection of a popular dog breed. *BMC Genomics* 18(1), 609.

Index

Note: All index entries refer to dogs, unless specified otherwise. Specific breeds have *only* been indexed where there is significant information, and not just lists of breeds.
Page numbers in **bold** refer to figures; those in *italics* refer to tables.

Printed and bound by CPI Group (UK) Ltd, Croydon, CR0 4YY
13/09/2026
14961115-0002